# 3rd International Conference on Mechanical Instrumentation and Automation (ICMIA 2022)



Journal of Physics: Conference Series Volume 2264

Sanya, China
25-27 March 2022

ISBN: 978-1-7138-5517-0
ISSN: 1742-6588

**Printed from e-media with permission by:**

Curran Associates, Inc.
57 Morehouse Lane
Red Hook, NY 12571

CURRAN ASSOCIATES INC.
proceedings
.com

**Some format issues inherent in the e-media version may also appear in this print version.**



Institute of Physics
Dirac House, Temple Back
Bristol BS1 6BE UK

Phone: 44 1 17 929 7481
Fax: 44 1 17 920 0979

techtracking@iop.org

**Additional copies of this publication are available from:**

Curran Associates, Inc.
57 Morehouse Lane
Red Hook, NY 12571 USA
Phone: 845-758-0400
Fax: 845-758-2633
Email: curran@proceedings.com
Web: www.proceedings.com

# TABLE OF CONTENTS

# Preface

Due to recent pandemic, 2022 3rd International Conference on Mechanical Instrumentation and Automation (ICMIA 2022) was held virtually online during March 25-27, 2022. The decision to hold the virtual conference was made in compliance with many restrictions and regulations that were imposed by countries around the globe. Such restrictions were made to minimize the risk of people contracting or spreading the COVID-19 through physical contact. There were 80 individuals who attended this on-line conference, represented many countries including Malaysia, Singapore, Indonesia, Brazil and China.

ICMIA 2022 is to bring together International Symposium on Mechanical Instrumentation and Automation to a common forum. And another goal is to promote scientific information interchange between researchers, developers, engineers, students, and practitioners working all around the world. The conference will be held every year to make it an ideal platform for people to share views and experiences in International Symposium on Mechanical Instrumentation and Automation and related areas.

During the conference, the conference model was divided into three sessions, including oral presentations, keynote speeches, and online Q&A discussion. In the first part, some scholars, whose submissions were selected as the excellent papers, were given about 5-10 minutes to perform their oral presentations one by one. Then in the second part, keynote speakers were each allocated 30-45 minutes to hold their speeches. In the second part, we invited four professors as our keynote speakers. Prof. Hamid Reza Karimi, Politecnico di Milano, Italy. His original research and development achievements span a broad spectrum within the topic of control systems, mechatronics, intelligent systems, and fault diagnosis with applications such as vehicles, robotics and manufacturing systems. And then we had Prof. Na Li, Xidian University, China. Her research focused on the error theory of electronic equipment, multi-scale electromechanical coupling problems, and the design and optimization of new nanoelectronic equipment structures. Assoc. Prof. Mohamed Arezki Mellal, M'Hamed Bougara University, Algeria. His research interests include developing new nature-inspired optimization methods for solving engineering problems of system dependability, design, manufacturing, and energy efficiency. Finally, we had Assoc. Prof. Lisu Yu, Nanchang University, China as our finale keynote speaker. Their insightful speeches had triggered heated discussion in the third session of the conference. Every participant praised this conference for disseminating useful and insightful knowledge.

The proceedings are a compilation of the accepted papers and represent an interesting outcome of the conference. Topics include but are not limited to the following areas: Instrumentation, Precision Instruments and Machinery, Automation and Electronic Information Engineering and other related topics. All the papers have been through rigorous review and process to meet the requirements of international publication standard.

We would like to acknowledge all of those who supported ICMIA 2022. The help and contribution of each individual and institution was instrumental in the success of the conference. In particular, we would like to thank the organizing committee for its valuable inputs in shaping the conference program and reviewing the submitted papers.

We sincerely hope that the ICMIA 2022 turned out to be a forum for excellent discussions that enabled new ideas to come about, promoting collaborative research. We are sure that the proceedings will serve as an important research source of references and knowledge, which will lead to not only scientific and engineering findings but also new products and technologies.

The Committee of ICMIA 2022

# Committee member

**Committee Chair**
Prof. Na Li, Xidian University, China
Prof. Yuhao Wang, Nanchang University, China

**Publication Chair**
Prof. N. Rajathi, Kumaraguru College of Technology, India

**Technical Program Committee Chair**
Prof. Zhenghai Wang, Nanchang University, China
Assoc. Prof. Lisu Yu, Nanchang University, China

**Technical Program Committee**
Prof. Shuming Yang, Xi'an Jiaotong University, China Fellow of the International Society for Nanomanufacturing (ISNM), China
Prof. Go-Long Tsai, National Taipei University of Technology, Taiwan, China
Prof. Yansong Wang, Shanghai University of Engineering Sciencc, China
Prof. Kuzmin Alexander, St Petersburg State University, Russia
Prof. Liu Qiuxin, Wuhan University of Science and Technology, China
Assoc. Prof. Maíra Martins da Silva, University of São Paulo, Brazil
Assoc. Prof. Tian Soon Lee, Multimedia University, Malaysia
Assoc. Prof. Chua Kian Jon Ernest, National University of Singapore, Singapore
Assoc. Prof. Shu-San Gan, Petra Christian University, Indonesia

Journal of Physics: Conference Series **2264** (2022) 011002 doi:10.1088/1742-6596/2264/1/011002

# Peer Review Statement

All papers published in this volume have been reviewed through processes administered by the Editors. Reviews were conducted by expert referees to the professional and scientific standards expected of a proceedings journal published by IOP Publishing.

- **Type of peer review:** Single Anonymous
- **Conference submission management system:** Morressier
- **Number of submissions received:** 61
- **Number of submissions sent for review:** 65
- **Number of submissions accepted:** 33
- **Acceptance Rate (Submissions Accepted / Submissions Received × 100):** 54.1
- **Average number of reviews per paper:** 2
- **Total number of reviewers involved:** 6
- **Contact person for queries:**
  **Name:** Xuexia Ye
  **Email:** xx.ye@keoaeic.org
  **Affiliation:** AEIC Academic Exchange Information Centre

# Design and Implementation of Adjustable Voltage Source Module for Ion Chromatography

**Biao Wang** [1, 2], **Zhongping Ming** [1, 3], **Jing Chen** [4*], **Bin Zhou** [4]

[1]School of Instrument Science and Opto-electronics Engineering, Hefei University of Technology,Hefei, 230009, Anhui, China.

[2]Anhui Province Key Laboratory of Measuring Theory and Precision Instrument, School of Instrument Science and Optoelectronics Engineering, Hefei University of Technology, Hefei 230009, Anhui, China.

[3]Anhui Wanyi Science and Technology Co., Ltd. Hefei, 230065, Anhui, China.

[4]Anhui Electrical Engineering Professional Technique College, Hefei, Anhui, 230051, Anhui, China.

[*]Corresponding author's e-mail: chenjingdz024@163.com, 13721073546.

**Abstract.** Undeniably, adjustable power supply has many applications in ion chromatography. One example would be that a safe and stable adjustable power supply is indispensable for the driving part of the suppressor in ion chromatography. However, the traditional design of adjustable power supply suffers from high complexity, leading to its incompetence for practical engineering under the background of the continuous shortage of chips. Therefore, a method for an adjustable power supply module for ion chromatography is proposed and studied. Firstly, the microprocessor is used to output the MCPWM wave to control the output voltage of the flyback transformer. Secondly, the OCADC is used to collect the feedback voltage. Thirdly, the duty cycle of the MCPWM is adjusted using the PID algorithm to achieve constant voltage output. Further, hardware overcurrent protection is realized through MCPWM fast interrupt function. By this method, an adjustable power supply with a wide voltage output range, high current, and low voltage ripple can be output with less chip resources. Both the theoretical design scheme and the specific hardware circuit of this design is presented. Moreover, a simulation verification analysis is made. Finally comes the verification of the feasibility of this method.

## 1. Introduction

It is by comparison that ion chromatography (IC) performs the identification of compounds, entailing extremely strict requirements on the consistency of experimental conditions. As one of the common detection methods of ion chromatography, conductance detection must take usage of a suppressor to reduce the background conductivity. Therefore, the power supply driving the suppressor, especially the adjustable power supply, occupies an extremely important position in many fields such as electrochemical analysis [1-2].

Ion chromatography is a technique for separating ions from polar compounds. It can dissolve the analyte in a suitable solvent, and then bring it into the detector through the mobile phase to detect the content of the analyte. The mobile phase usually consists of an aqueous solution, and the conductivity detector is a detection mode based on the conductivity of the electrolyte in ion chromatography. The difficulty of this method is to identify the signal generated by the ionic species in the sample from the total signal. In order to achieve direct measurement, the ionic charge in the mobile phase must be as low as possible, because the ions in the mobile phase will produce background conductance rate, making it difficult to measure the conductivity of the analyte ions. When using a conductance detector, in order to improve the signal ratio, a suppressor is usually installed before the detector, aiming to selectively remove the interference of the ion signal in the mobile phase [3-5]. At present, the commonly used suppressor is a membrane suppressor that generates regenerated ions by on-line water electrolysis. The suppressor needs to provide a constant current for electrolyzing water to generate hydrogen ions and hydroxide ions [6].

The constant current source drive of the traditional suppressor usually uses a fixed voltage source output, and the MOS tube is used as the current output switch. In order to achieve sufficient current output, it is generally according to the maximum output current that the constant source drive voltage is selected. Under this condition, it follows that when the suppressor needs to provide a large current, the MOS tube heats up badly because it is bearing the most of the voltage. Conductivity detectors are extremely sensitive to temperature, and usually require a control accuracy of 0.01°C. The heat source in the same module will inevitably affect the detection results. At the same time, when the constant current source drives the fixed voltage output without current monitoring or the software fails, the suppressor will be damaged, and the accuracy of the output voltage also directly affects the repeatability of the detection [7-9].

Looking at the design of the suppressor constant current source drive power supply, it is not difficult to find that in the selection of the circuit, most designers only focus on the realization of the suppressor constant current source drive power supply function. However, in the actual use of the suppressor, traditional fixed voltage source output will greatly reduce the detection efficiency, accuracy and repeatability of the conductance detector. Aiming at this problem, this paper presents a method for a self-protected adjustable voltage source driven by a constant current source of the suppressor. Firstly, the single-chip microcomputer is used to output MCPWM waves with different duty ratios to control the flyback converter, and the target voltage is output through the output coil. Secondly, the fast interrupt pin of MCPWM wave is utilized to form hardware protection; Thirdly, the voltage of the feedback coil is collected as feedback for PID to adjust the output. Combined with the device data, the output analysis is given, and then the method is compared with the measured data. Further, the influence on the output under different loads was studied, and the technology of the adjustable voltage source for ion chromatography with self-protection function was finally optimized.

## 2. Theoretical and Numerical Methods

### *2.1. Theory of Flyback Converter*

The output of the adjustable power supply is usually completed by the switching power supply. The flyback switching power supply possesses the characteristics of few components, simple circuit and multiple outputs. Taking consideration of that, it meets the design requirements. As shown in Figure 1, when Q1 is turned on, all rectifier diodes are reversely cut off, and the output capacitor supplies power to the load. T1 is equivalent to a pure inductance, the current flowing through Np rises linearly and reaches the peak value Ip. When Q1 is turned off, all winding voltages are reversed. This flyback voltage turns the output diode into a conducting state, and $1/2LI^2$ of the energy stored in the primary is transferred to the secondary, providing the load current while charging the output capacitor [10].

The formula of the ampere-turn ratio of the primary winding and the secondary winding:

$$I_{S1} \times N_P = I_{S2} \times N_S \tag{1}$$

In the formula, $I_{S1}$ is the peak current when the primary winding is on, $N_P$ is the number of turns of the primary winding, $I_{S2}$ is the peak current when the secondary winding is on, and $N_S$ is the number of turns of the secondary winding. When several secondary windings are turned on at the same time, the sum of the ampere-turns of all secondary windings is equal to the primary ampere-turns.

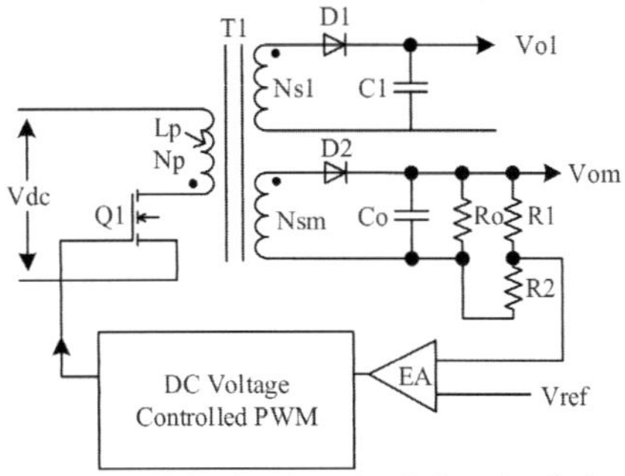

Figure 1. Illustration of the basic principle of a flyback converter

According to the conservation of energy, the power supplied by the power supply $V_{dc}$ in one cycle is:

$$P = (V_{dc} \times T_{on})^2/(2 \times T \times L_p) \quad (2)$$

In the formula, P is the power provided by the power supply $V_{dc}$ in one cycle, $T_{on}$ is the conduction time, T is the cycle time, and $L_p$ is the primary winding inductance[11]. In the converter, the two windings are wound on the same magnetic core, the magnetic flux Ø is the same, and the magnetic flux change rate d Ø /dt is also the same[12].therefore,

$$V_P/N_P = V_S/N_S \quad (3)$$

In the formula, $V_P$ is the voltage of the primary winding, $V_S$ is the voltage of the secondary winding, $N_P$ is the number of turns of the primary winding, and $N_S$ is the number of turns of the secondary winding

### 2.2. Output Capacitance vs. Voltage Ripple

When the frequency is lower than 500KHz, the ripple flowing into the output capacitor can be regarded as determined by the capacitor equivalent series resistance (ESR) and the ideal pure capacitor [13], then

$$V_{rr} = dI \times R_0 \quad (4)$$

$$V_{cr} = I_t/C_0 \quad (5)$$

In the formula, $V_{rr}$ is the resistive voltage ripple value generated by ESR, $dI$ is the peak-to-peak value of the output current ramp, $R_0$ is the ESR value, $V_{cr}$ is the inductive voltage ripple value generated by the ideal pure capacitor, and $I_t$ is the triangular wave current during the off-time of the switch. The average value of $C_0$ is the pure capacitance value.

### 2.3. The cross section is a series-parallel formula of a single-layer helical coil

$$L_{total} = L_{base} \times S^2 \quad (6)$$

$$I_{max} = I_{base} \times P \quad (7)$$

In the formula, $L_{total}$ is the total inductance of the windings in series and parallel, $I_{max}$ is the total capacitance of the windings in series and parallel, $L_{base}$ is the inductance of one winding, $S$ is the number of windings in series, $I_{base}$ is the maximum rated current of a single winding, and $P$ is the number of windings in parallel.

## 3. Scheme design and circuit construction

### 3.1. Design

According to the above theory, the adjustable voltage source scheme as shown in Figure 2 is determined.

The scheme is composed of one MCU(LPC1768), power supply, working indicator light, crystal oscillator, programming interface, drive circuit, overcurrent protection circuit, voltage acquisition circuit and transformer (749196321). First, the single-chip microcomputer is used to output the MCPWM wave with adjustable duty cycle, and the output voltage of the transformer is controlled after the drive circuit; the feedback circuit collects the voltage and connects it to the on-chip ADC pin OCADC of the single-chip microcomputer. After PID adjustment, it controls the duty of the next round of MCPWM waves. The overcurrent protection circuit can output a trigger signal to the motor PWM fast interrupt pin MCABORT of the microcontroller when the current is too large, so as to realize the fast shutdown of the MCPWM wave. Depending on the actual deployment, the maximum output voltage of this design is 30V, the maximum output current 0.3A.

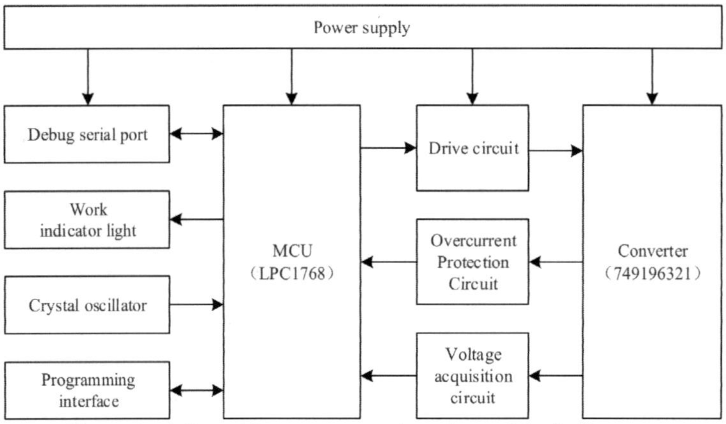


Figure 2. Adjustable power supply scheme block diagram

### *3.2. Circuit construction*

In order to verify the correctness of the theory and scheme design, we designed the following hardware circuit. Figure 3(a) shows the main circuit diagram while PCBA of the adjustable power supply module is showed in figure 3(b).

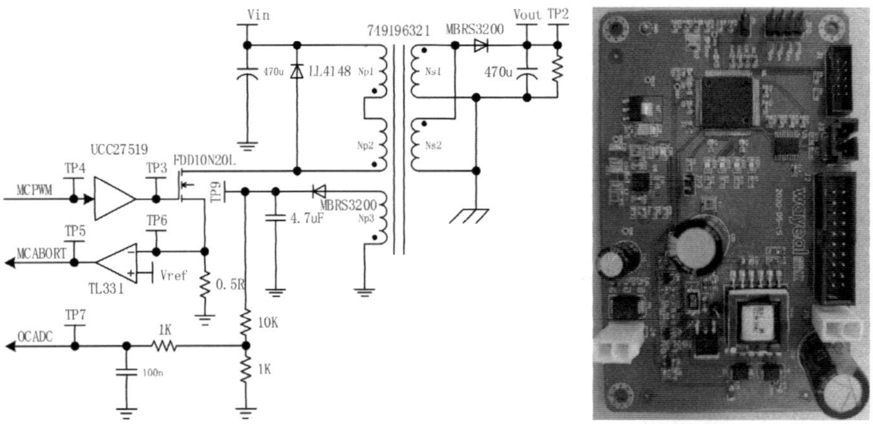


(a) (b)

Figure 3. The main circuit diagram of the adjustable voltage source (a), PCBA (b)

In this circuit, after the MCPWM wave is boosted by the voltage, the MOS tube works in the constant current region. According to formula (1) (2) (3), when the input voltage is constant, as long as the MCPWM duty cycle of the feedback loop is kept constant change, the output voltage can be kept constant. The input current is fed back to the MCABORT pin of the microcontroller by the comparator.

After the MCPWM wave with a frequency of 100KHz passes through the gate driver UCC27519, the amplitude is increased from 3.3V to 5V, which is used to drive the MOS tube to turn on. The input current limit uses a 0.5Ω sampling resistor. When the design input current exceeds 1A, the output of the comparator TL331 is low-leveling, triggering the fast interrupt function of MCPWM as hardware feedback. The maximum output voltage is 36V, which is divided by 1/11 of the voltage. After being adjusted by the internal PID of the microcontroller, the duty cycle of the next cycle of MCPWM is controlled.

A transformer with a single-layer spiral coil in cross section is used, and the inductance value of each turn is 14.7uH. The primary side coils Np1 and Np2 are connected in series, as the power input coil, and the Np3 coil is used as the feedback coil. The ADC collects the voltage; the secondary side coils Ns1 and Ns2 are connected in parallel as the power output coil. According to the above formulas (1) (3) (4), the inductance ratio of the power input coil, power output coil and power feedback coil is 4 :1:1, the turns ratio is 2:1:1, the voltage ratio of the power supply output coil and the power supply feedback coil is 1:1, and the output current can be doubled by using the parallel connection.

The output capacitor uses a low-resistance electrolytic capacitor, its ESR value is 27mR, the MCPWM frequency is 100KHz, and the target output current of the adjustable power supply is 0.5A. According to formula (4), the resistive ripple voltage is 2.7mV, according to formula (5). ) capacitive ripple voltage is 5mV.

## 4. Experimental verification

### *4.1. ADC voltage acquisition*

The on-chip ADC is 12 bits, and the voltage divider ratio of TP9 test point voltage and TP7 test point voltage is 11:1. After the ADC collection value Vadc is determined, the calculation formula of the feedback voltage value at TP9 is: Vol=(Vadc/4096)*3.3*11. First, configure the software to output a fixed voltage, measure the feedback voltage value of the TP9 test point with a multimeter, and compare it with the voltage value calculated by the ADC. If the two values are equal, the AD voltage acquisition is normal. Set the working clock frequency of ADC to 1.0MHZ. AD conversion can be completed in 65uS after each trigger sampling. According to experience, the sampling rate of each channel of about 1KSPS suffices to meet the requirements.

### *4.2. Voltage test*

#### *4.2.1. Switching test between load and light load*

The circuit may switch between load and light load during operation. Now do the following test:test conditions: input voltage 36V, output voltage set to 30V, remove dummy load, use electronic load LIST function, set step current change. Electronic load current switches between notches of 0~0.1A, 0~0.2A and 0~0.3A periodically, each lasting 1s. The oscilloscope is used to record the voltage waveform of the TP2 test point as Fig.4.

From the Fig.4, it can be deduced that: When the load current changes in steps of 0~0.1A and 0~0.3A, it is equivalent to the change of load and light load. According to the above data analysis, the larger the load, the greater the deviation between the output voltage of the TP2 test point and the set voltage. When there is a load, the voltage can quickly follow; the average value of the output voltage waveform of the TP2 test point is close to the set value. When the load is light, since the load terminal is equivalent to an open circuit, the voltage is unstable. During the charging process of the large-capacity capacitor at the output terminal, the voltage waveform is a slowly rising state.

#### *4.2.2. Different resistance value dummy load test*

Generally, it is not encouraged to leave the output of the flyback converter unloaded, otherwise the semiconductor device will be subject to damage. In order to prevent the voltage of the TP2 test point from rising too high and exceeding the withstand voltage value of the output capacitor of the TP2 test point, a dummy load is added to the output end of the TP2 test point to form a closed loop. If the dummy load resistance value is too small, the power consumption will be too large. To verify the appropriate dummy load resistance value, do the following tests: The input voltage is 36V, the output voltage is set to switch from 0V to 30V, and the dummy load resistance values are 1K, 2K, 3K, 5K, and 7K, respectively. Use an oscilloscope to record the voltage waveform of the TP2 test point as Fig. 5 and Fig.6.

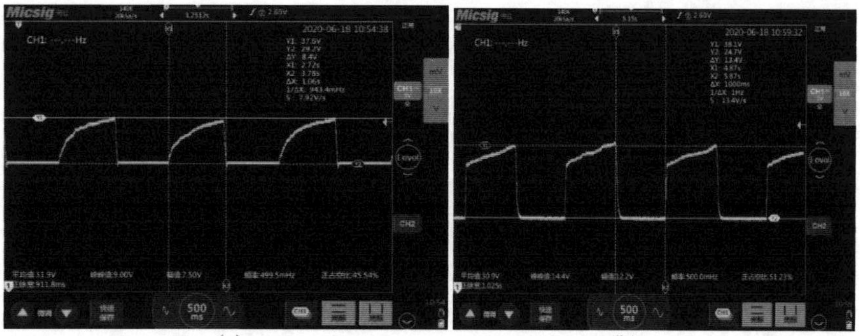

(a) (b)

Figure 4. Electronic load 0A~0.1A step change (a), electronic load 0A~0.3A step change (b), TP2 test point voltage waveform

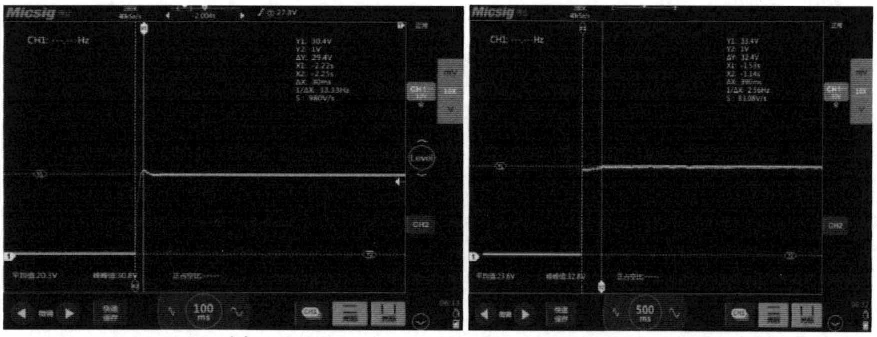

(a) (b)

Figure 5. The dummy load is 1K (a), 3K (b), the voltage waveform of TP2 test point

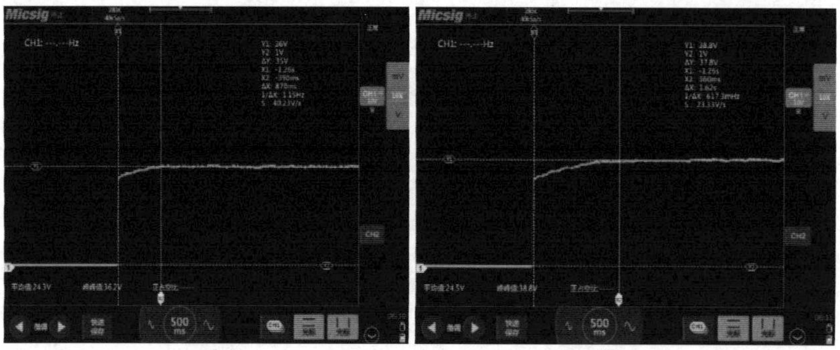

(a) (b)

Figure 6. The dummy load is 5K (a), 7K (b), the voltage waveform of TP2 test point

According to the analysis of the Fig. 5 and Fig.6, when the dummy load is 3K, the effect of suppressing the voltage rise is obvious, so the resistance value of the dummy load is 3K; when the output voltage of the TP2 test point is 30V, the current consumed by the dummy load is 10mA.

### *4.2.3. Different load switching test*

There may be different load switching situations in actual work. In order to verify the specific usage scenarios, the following tests are now performed: Test conditions: the input voltage is 36V; the output voltage is set to 30V; the dummy load is 3K; the electronic load LIST function is used; the step current change is set; and the electronic load current is respectively 0A~0.1A, 0A~0.3A, 0.1A~0.2A, 0.1A~0.3A, switched periodically with each time interval being 1s. Use the oscilloscope to record the voltage waveform of the TP2 test point as Fig.7 and Fig.8.

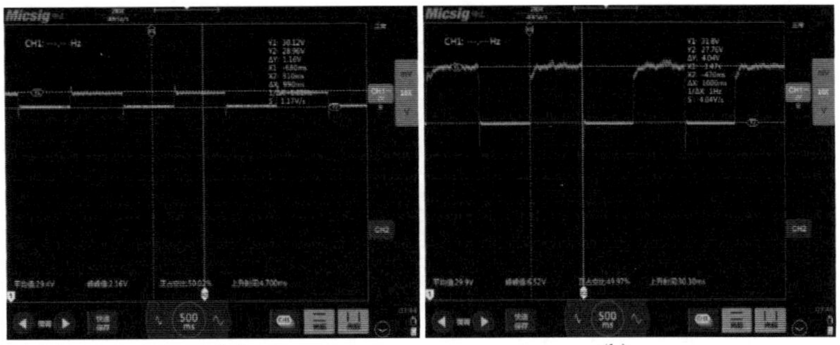

(a) (b)

Figure 7. Electronic load 0A~0.1A step change (a), electronic load 0A~0.3A step change (b), TP2 test point voltage waveform

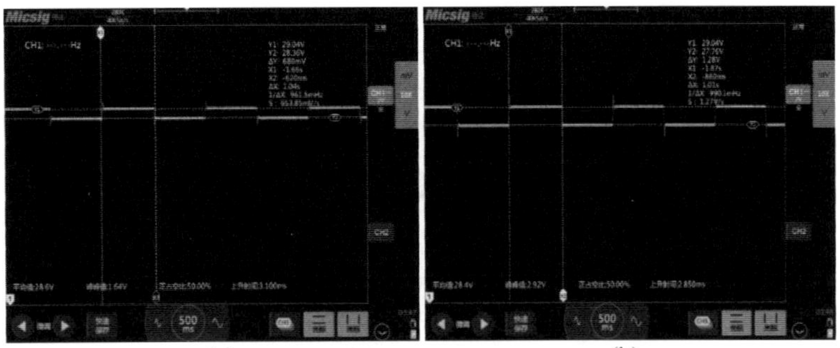

(a) (b)

Figure 8. Electronic load 0.1A~0.2A step change (a), electronic load 0.1A~0.3A step change (b), TP2 test point voltage waveform

Scrutinizing the Fig.7 and Fig.8, it can be extrapolated that: When the electronic load current changes at 0A~0.1A, 0A~0.3A, 0.1~0.2A, 0.1~0.3A, the voltage can quickly follow; due to the influence of the transformer coil and Schottky diode impedance, when the load is large, the output voltage of TP2 test point is relatively low, and when the load changes, there will be a voltage overshoot due to the overshoot of the duty cycle.

## 5. conclusion

In this paper, based on the flyback converter, combined with the use of power supply in ion chromatography, a design method of ion chromatography adjustable voltage source module considering cost and performance is proposed. This method takes advantage of the MCPWM output built in the microcontroller to control the on-off of the MOS tube. Then the flyback transformer is utilized to attain the voltage output. Followingly, the feedback coil voltage acquired by the on-chip ADC of the microcontroller is managed to help the PID adjust the MCPWM, achieving constant voltage as desired. With the help of the MCPWM fast interrupt pin, the hardware protection function is realized. Further, this paper analyzes the input and output waveforms of the voltage, and the performance of the circuit in different application environments. It can be concluded that it has advantages of simple circuit, small output voltage ripple, and low cost.

**Acknowledgments**
This study was supported by the Natural Science Foundation of Anhui Province for Higher Education Institutions (Grant No. KJ2021A1574), Young Teachers Teaching Research Project of Hefei University of Technology (Grant No. JYQN2015) and National College Students' innovation and entrepreneurship training program (Grant No. 202010359054).

**References**

[1] Zhang Yuping,Li Ying,Cui Chengxing, (2018) Chromatographic Analysis. Chemical Industry Press, Beijing, 169-173.
[2] Mou Shifen, Zhu Yan, Liu Kena, (2019) Ion Chromatography Method and Application. Chemical Industry Press, Beijing, 155-163.
[3] A.J.Bard and L.R.Faulkner, (2005) Electrochemical Methods Fundamentals and Application. Chemical Industry Press, Beijing, 96-100.
[4] Liu Yongqiang, Yu Hong, (2015) Application of ion chromatography in the analysis of anions and cations in ionic liquids. Journal of Analysis and Testing, 34: 734-743.
[5] Zhao Xinying, Qu Feng, Mu Shifen, (2017) Important progress of ion chromatography and the development of my country in recent years. Chromatography, 5:223-228.
[6] Huang Yuan, Mou Shifen, Hou Xiaoping, (2000) The working principle, development and application of ion chromatography membrane suppressor. Analytical Instruments, 43:36-42.
[7] Abraham I. Pressman and Keith Billings and Taylor Morey, (2009) Switching Power Supply Design. Electronic Industry Press, Beijing, 73-83.
[8] Yin Yue, Wang Dong, Gao Ce, Shu Xin, Fan Jinzhu, Tian Yu, (2018) Design and Realization of On-Line Adjustable Power Modules. Analytical Instruments, 16:5-8.
[9] Sanjaya Maniktala, (2015) Switching Power Supplies.People's Posts and Telecommunications Press, 81-93.
[10] Zeng Zhiqiang, Zheng Xincheng, Chen Wei, (2012) A Design of Flyback Transformer Considering Efficiency and Electromagnetic Compatibility. Electrical Technology, 36:1899-1921.
[11] Tohsaka Toshiaki, (2006) Design of Measuring Resistance Circuits. Science Press, Beijing, 121-125.
[12] Cheng Hui, Ren Yongfeng, Wang Qiang, (2000) Measurement and Suppression of Power Supply Ripple. Power Supply Technology, 43:36-42.
[13] Ji Sanfei, Tan Wangjing, Lei Ruiying, Wang Qing, (2020) Analysis of Influence of Flyback Transformer Leakage Inductance and Distributed Capacitance. Electronic Measurement Technology, 43:66-70.

# Research on assembly clearance of precision spherical coupling parts by point cloud analysis

**Xiaoyu Pan[1, a], Zhibing Liu[1,b*], Lu Zhang[1,c*], Xibin Wang[1,d]**

[1] School of Mechanical Engineering, Beijing Institute of Technology, Haidian, Beijing, China.

[a]mail: panxiaoyu2019@bit.edu.cn, [b]mail: liuzhibing@bit.edu.cn,

[c]mail: zhanglu@bit.edu.cn, [d]mail: wangxibin@bit.edu.cn

**Abstract:** Assembly clearance has a great influence on the performance of precision parts. Improper assembly clearance affects the working state and induces failure of parts. A method to analyze the assembly clearance of the precision spherical coupling parts by analyzing the points cloud was proposed. The least square center was used to find a rough center. Accurate center was searched near the least square center by the particle swarm optimization (PSO) algorithm based on the minimum zone method. Radical clearance and Z clearance of precision spherical coupling parts were calculated. Furthermore, the dynamic change of assembly clearance was also calculated by rotating the parts. The sphericity error evaluation results were improved obviously compared with other results. The accuracy of the evaluation results improved 4.51%~28.97%.

## 1. Introduction

Precision coupling parts are composed of spheres and bowls. The assembly clearance takes an important role on the behavior of precision spherical coupling parts. Improper clearance not only affects the working status, but also can cause the early failure of the parts. Therefore, it is important to measure the assembly clearance. Franciosa et al. [1] predicted the gap between the two connectors during welding with the assumption that the position of the welding torch is known. Mao et al. [2] established an improved quasi dynamic model to study the effect of annular deformation on the assembly clearance between roller bearing and shell. The model was used to study the dynamic performance of the bearing. Yue et al. [3] proposed a novel method for the assemblability evaluation of cylindrical parts assembly considering surface deviations. A search algorithm is established to obtain the boundary points of assemblable posture space, which can realize the quantitative evaluation of assemblability.

Sphericity error has a great influence on the assembly clearance. So, it is necessary to evaluate the sphericity error accurately and rapidly.Sphericity error is the variation of the actual measured sphere to its ideal sphere. And the error value equals to the radius difference between the two concentric spheres containing all measured points. The general methods of sphericity error evaluation include the Least Square Sphere method (LSS), the Minimum Zone Sphere method (MZS), the Minimum Circumscribed Sphere method (MCS) and the Maximum Inscribed Sphere method (MIC). Roque [4] presented a new method for sphericity of minimum zone tolerance. A model based on the regression of the error by each CMM axis was incorporated into the sphericity measurement model. The method can effectively contribute to the improvement of sphericity measurement from point coordinates in precision and accuracy.

Journal of Physics: Conference Series **2264** (2022) 012002 doi:10.1088/1742-6596/2264/1/012002

In the recent years, the intelligent algorithm like genetic algorithms (GA) [5], and particle swarm optimization (PSO) algorithm [6,7] are widely used in the form error evaluation. Gu et al. [8] improved the particle swarm optimization algorithm. A calculation process suitable for the evaluation of journal errors was designed. Then, the improved particle swarm optimization algorithm was used to evaluate the roundness of the corrected data.

Sphericity error has a great influence on the assembly clearance of precision coupling parts, and the assembly clearance has a significant impact on the performance of precision coupling parts. Improper assembly clearance will affect the working state of parts and cause part failure. The traditional calculation method for solving sphericity error require amounts of computation when the data is large and the accuracy is low. In this paper, a method combining particle swarm optimization algorithm with LSS method was proposed to evaluate the sphericity error of precision ball bowl pairs. On this basis, the prediction of assembly clearance was realized by analyzing the point cloud of parts.

## 2. Evaluation of sphericity error

The part measured for sphericity error evaluation was precision spherical coupling. It is composed of sphere and bowl and the structure is shown in Figure 1. The clearance fit adopted between the sphere and bowl is demanding, so the sphericity error requirements for individual parts are very strict.

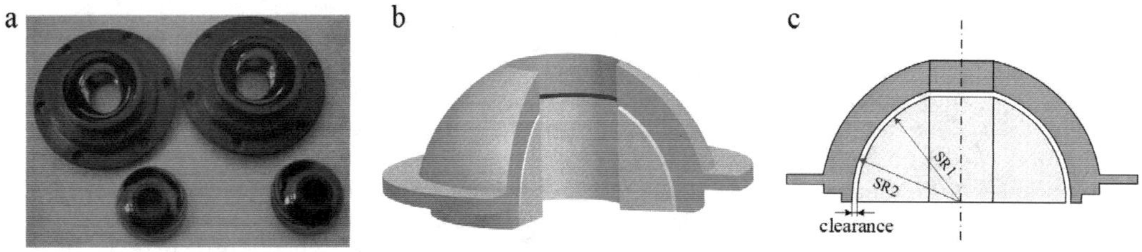


Figure 1. Sphere, ball and the assembly relationship between them. a Sphere and bowl. b 3D assembly relationship. **c** 2D assembly relationship

The most accurate way to evaluate the sphericity error is to find the best center of the sphere and bowl. The least square center was calculated firstly. Then the PSO with changed inertia factor was used to search the best center near the least square center.

### *2.1 Least square sphere method*

Least square method is widely used to evaluate profile errors. The purpose of the least square sphere is to make the square difference of the radius minimum.

Assuming the equation of sphere after fitting is as follows:

$$(x_i + x_0)^2 + (y_i + y_0)^2 + (z_i + z_0)^2 = r^2 \tag{1}$$

Where ($-x_0$,$-y_0$,$-z_0$) is the center of the sphere in assuming, ($-x_i$,$-y_i$,$-z_i$) is the i-th measuring point, i=1,2,3... N is the number of the measured point on the sphere.

Radial deviation of every point is:

$$\Delta r = \sqrt{(x_i + x_0)^2 + (y_i + y_0)^2 + (z_i + z_0)^2} - r \tag{2}$$

The sphericity error is:

$$\mathrm{LS} = \Delta r_{max} - \Delta r_{min} \tag{3}$$

### *2.2 Particle swarm optimization*

Particle swarm optimization was invented by Kennedy and Eberhart in the United States in 1995. It was widely used to solve optimization problems. Compared with other intelligent algorithms, PSO has the advantages that the parameters are less, the implementation is easier and the convergence rate is faster, which is suitable for scientific research and engineering application.

Mathematical description of PSO is Eq. (4).

$$\begin{cases} v_{ij}(t+1) = \omega v_{ij}(t) + c_1 r_1 \left[ (pb_{ij})_j(t) - y_{ij}(t) \right] + c_2 r_2 \left[ gb_j - y_{ij}(t) \right] \\ y_{ij}(t+1) = y_{ij}(t) + v_{ij}(t) \end{cases} \quad (4)$$

Among them, $c_1$, $c_2$ are accelerating factors, $r_1$, $r_2$ are random numbers between [0,1], and $\omega$ is the inertia factors. The position and velocity of j-th particles and i-th particles are $y_{ij}$ and $v_{ij}$ respectively. The basic flow of PSO is Figure 2.

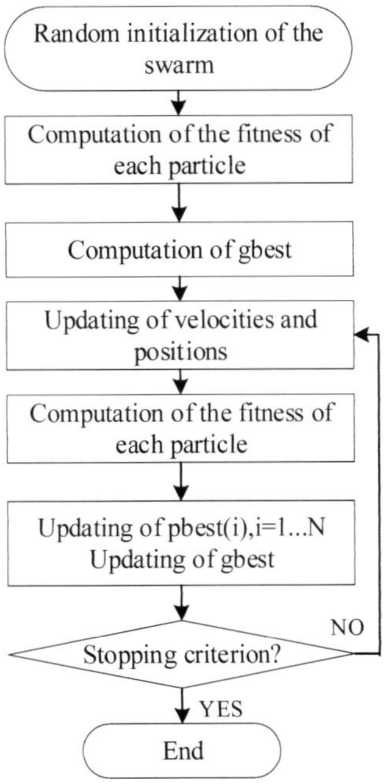


Figure 2. Flow chart of PSO (gbest means global best, pbest means personal best)

Research shows that the inertia factors $\omega$ has a large influence on the optimization performance. It is benefit to find the gbest (globe best) when $\omega$ is relatively high. Meanwhile, it is easy to get the convergent solutions when $\omega$ is relatively low. Compare with the traditional PSO algorithm, the gradual reduction $\omega$ was used, which could avoid getting the pbest (personal best) instead of gbest and accelerate the iteration speed. The formula is Eq. (5).

$$\omega(t+1) = \omega_{max} - (\omega_{max} - \omega_{min})/(t/t_{max})^{1/\pi^2} \quad (5)$$

$\omega_{max} = 0.95, \omega_{min} = 0.05; t_{max}$ is the maximum number of iterations.

To calculate accurate sphericity error, the particle swarm algorithm was applied. The searching area was set near the least square center. The optimization variable is the sphere center coordinates $(x_c, y_c, z_c)$. Figure 3. shows the searching area. Where O is the least square center, S is the least square sphere.

Figure 3. The searching area of the method

According to the MZS, the optimization objective function is Eq. (6).

$$F(x_c, y_c, z_c) = min\{\max\{d_i\} - \min\{d_i\}\} \tag{6}$$

$d_i$ is distance of from i-th measuring point to sphere center.

## 3. Fitting of precision coupling parts

The points cloud of four spheres and two bowls were obtained by the high accuracy CMMs (Leitz PMM-C, Hexagon, China, accuracy: 0.3μm). The 2000 points were distributed on the outer surface of the sphere and the inner surface of the bowl, which were divided into 10 layers. The distribution of measurement points was uniform on each layer. The measurement paths of the sphere and bowl are shown in Figure. 4.

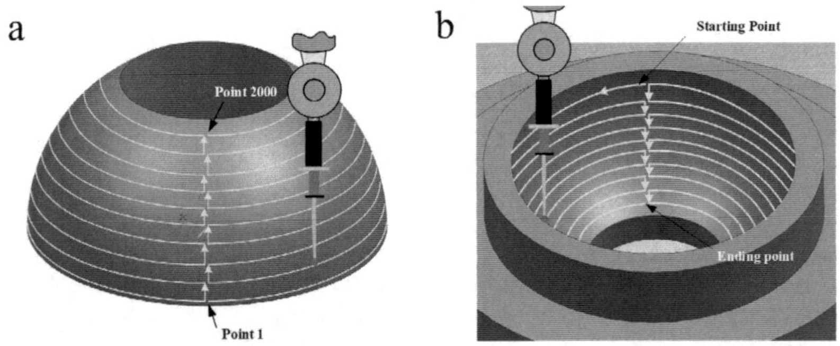


Figure 4. Measurement paths of the sphere and bowl. **a** Measurement paths of sphere. **b** Measurement paths of bowl

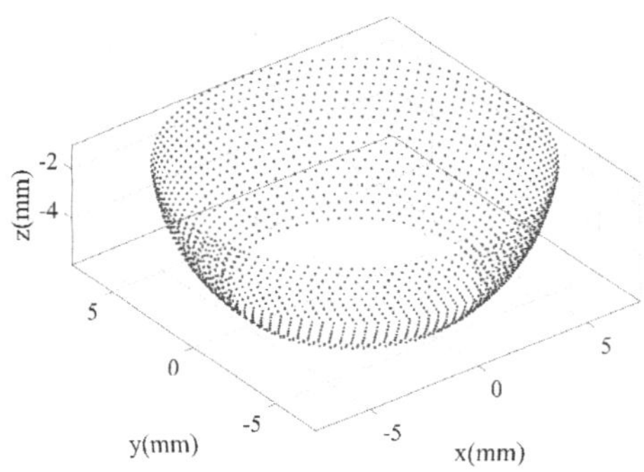


Figure 5. Cloud points distribution

After the measurement was completed, the points cloud file can be fitted by LSS to get the fitted center and radius. The fitting center of least square sphere is shown in Table 1.

Table 1. The center of least square sphere

| | sphere 1 | sphere 2 | sphere 3 | sphere 4 | bowl 1 | bowl 2 |
|---|---|---|---|---|---|---|
| X(μm) | -0.033 | -0.073 | 0.077 | -0.015 | 19.645 | 9.054 |
| Y(μm) | 0.498 | 0.584 | -0.364 | -0.462 | 30.537 | 24.709 |
| Z(μm) | 0.798 | 1.681 | -0.216 | 0.545 | -172.625 | -195.376 |

New center was searched near LSS center by PSO. And the center would change and got more accurate sphericity during the optimization process. The process of optimization is shown in Figure 6.

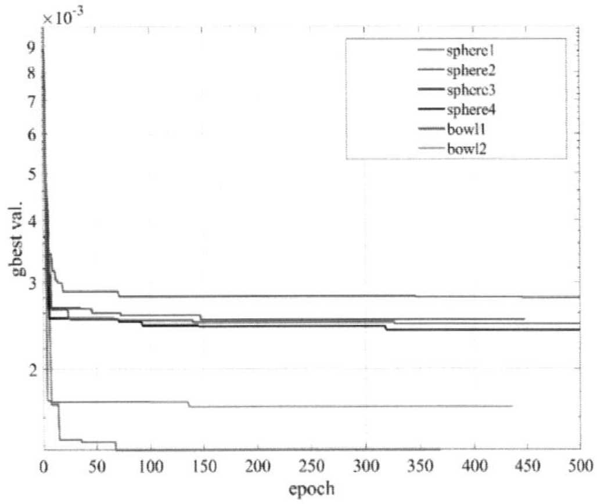


Figure 6. The optimization processes

After the calculation of four spheres and two bowls, Table 2 and 3 were got. The optimization center is shown in Table 2.

Table 2. The center of PSO

| | sphere 1 | sphere 2 | sphere 3 | sphere 4 | bowl 1 | bowl 2 |
|---|---|---|---|---|---|---|
| X(μm) | -0.082 | -0.239 | 0.086 | -0.506 | 19.928 | 9.272 |
| Y(μm) | 0.940 | 1.094 | -0.146 | -0.064 | 30.453 | 24.578 |
| Z(μm) | 0.176 | 1.314 | 0.187 | 0.518 | -172.599 | -195.526 |

The sphericity error of PSO and LSS is shown in Table 3.

Table 3. Sphericity error of PSO and LSS

| | sphere 1 | sphere 2 | sphere 3 | sphere 4 | bowl 1 | bowl 2 |
|---|---|---|---|---|---|---|
| LSM sphericity (μm) | 3.224 | 2.997 | 1.413 | 3.863 | 2.614 | 1.799 |
| PSO sphericity (μm) | 2.498 | 2.352 | 1.341 | 2.744 | 2.496 | 1.649 |

The sphericity error obtained by the improved PSO algorithm is smaller and is closer to the minimum zone compared with LSS and PSO. It is found that improved PSO based on the MZS has a higher accuracy than LSS. The accuracy of six parts improves 22.5%, 21.53%, 5.10%, 28.97%, 4.51% and 8.34%, respectively.

To observe the distribution of the distance from point cloud to the center, all the points were counted. The statistics show that the average radius of the parts is 7.4996mm, 7.4994mm, 7.4998mm, 7.4986mm, 7.5009mm and 7.5001mm, respectively. The average radius of spheres is lower than design radius, which is 7.5mm. The average radius of bowls is higher than design radius. This situation confirms to the production requirement.

The distance from each point to the center of the sphere or bowl is shown in Figure 7.

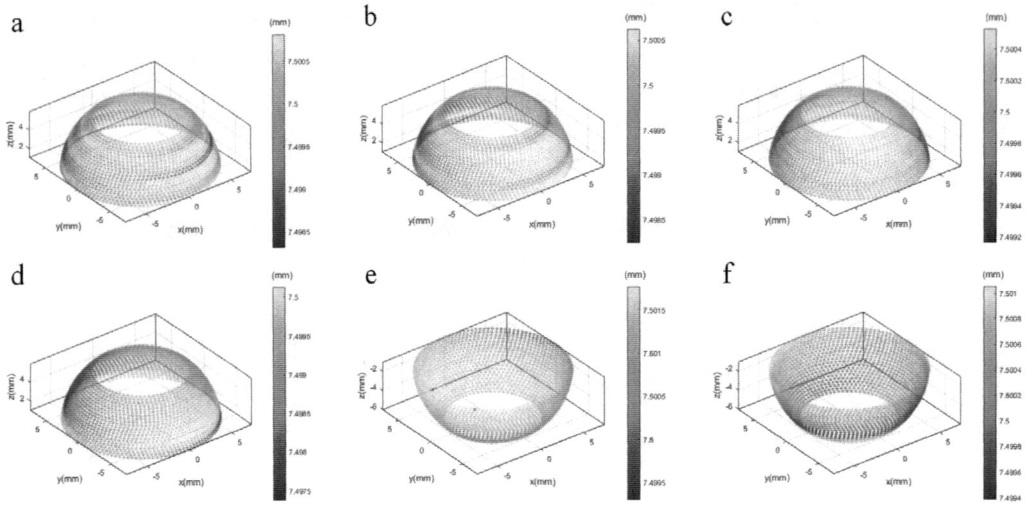


Figure 7. Distance from each point to the center of the spheres and bowls. **a** Sphere 1. **b** Sphere 2. **c** Sphere 3. **d** Sphere 4. **e** Bowl 1. **f** Bowl 2

The distribution of the distance followed the Normal distribution. Eq. (7) is Normal distribution.

$$f(x) = \frac{1}{\sqrt{2\pi}} e^{-\frac{(x-\mu)^2}{2\sigma^2}}, -\infty < x < +\infty \tag{7}$$

Where, μ is the means of the variable, σ2 is the variance of the variable. According to the calculation, each parts follow the Normal distribution: N(7.4996, 0.2912×10-6), N(7.4994, 0.3323×10-6), N(7.4998, 0.0801×10-6), N(7.4986, 0.4085×10-6), N(7.5009, 0.1990×10-6) and N(7.5001,0.1022×10-6). So the center obtained from the PSO can be regarded as the true center.

## 4. Analysis of assembly clearance

The assembly clearance, generally includes radial clearance and Z clearance, could be analysed after the center was found. The points cloud obtained by the measurement needs to be moved to the standard position, where the center of the sphere and bowl is (0,0,0), and the first measurement point is located on the x axis, as shown in Figure 8.

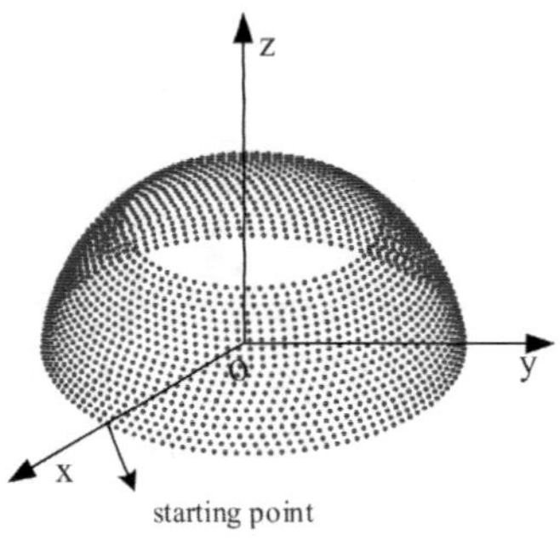


Figure 8. Standard posture

Because the distance from the points on the sphere to the same numbered points on the bowl is much smaller than the distance from the measured points to its neighboring measured points. Therefore, the distance from the measured points on the bowl to the bowl center and the distance from the same number measured point on the sphere to the sphere center were compared. Taking the sphere 1, sphere 2, sphere 3, sphere 4 and bowl 1 assembly clearance as an example, the distributions of radial clearance are shown in Figure 9.

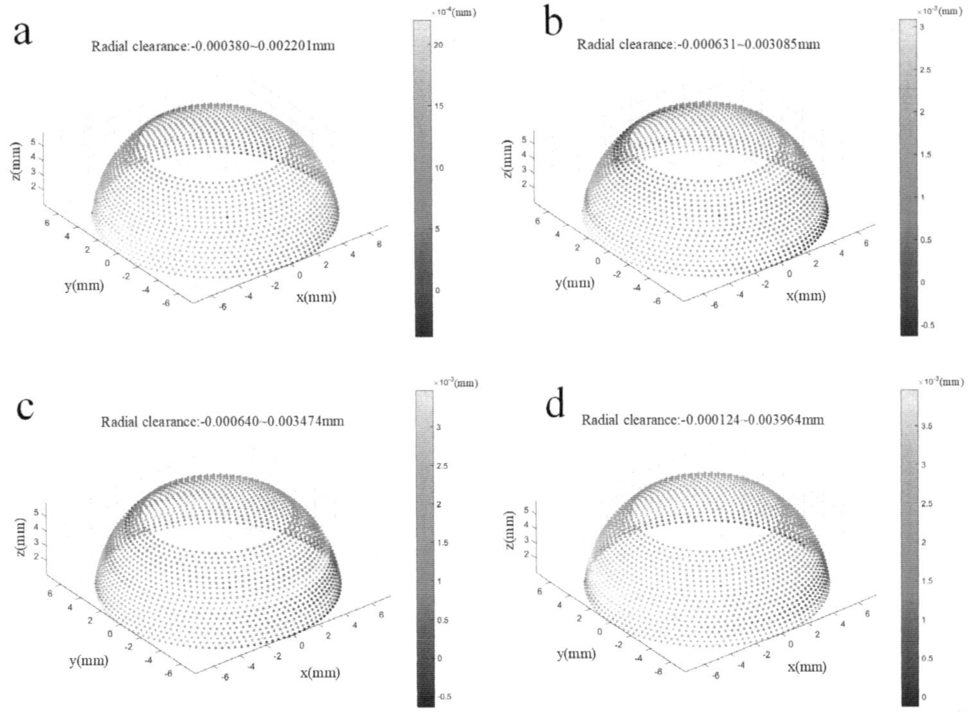


Figure 9. Radial clearance between each sphere and bowl 1. **a** Match of sphere 1and bowl 1. **b** Match of sphere 2 and bowl 1. **c** Match of sphere 3 and bowl 1. **d** Match of sphere 4 and bowl 1

Z clearance was obtained by comparing the z value of the measured points on the bowl and z value of the same number measured points on the sphere. The distribution is shown in Figure 10.

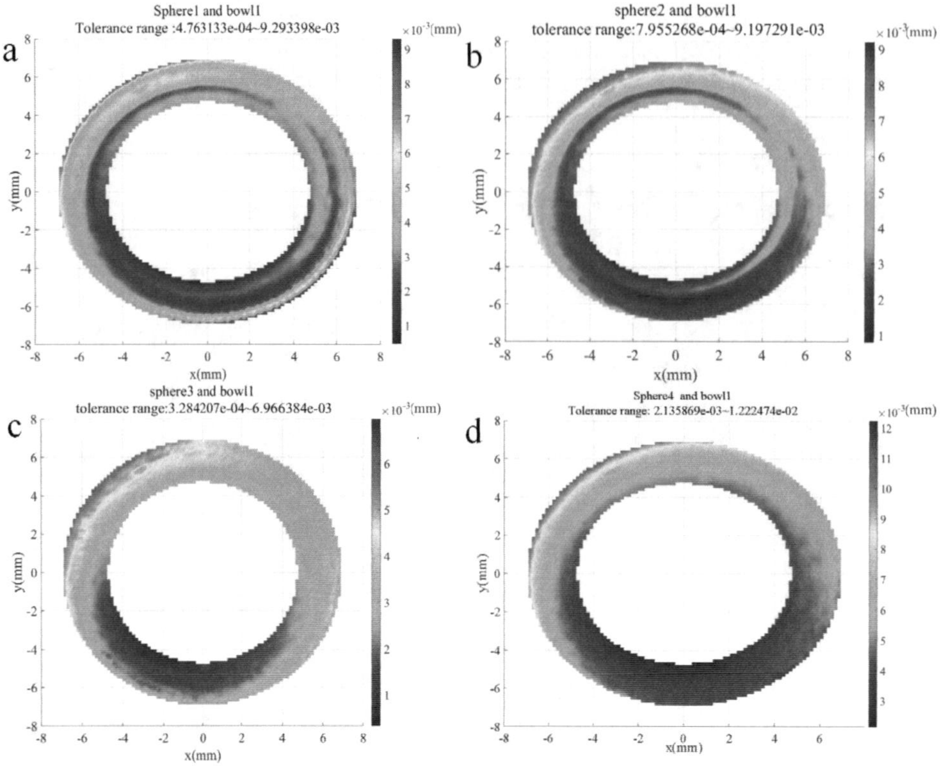


Figure 10. Z clearance between each sphere and bowl 1 (top view). **a** Match of sphere 1and bowl 1. **b** Match of sphere 2 and bowl 1. **c** Match of sphere 3 and bowl 1. **d** Match of sphere 4 and bowl 1

To further analyze the clearance of spheres and bowls, spheres were rotated. And the Z clearance was used in rotating analysis. Figure 11 shows the dynamic change of the assembly clearance and the root mean square of assembly clearance when spheres rotate 360° relatively.

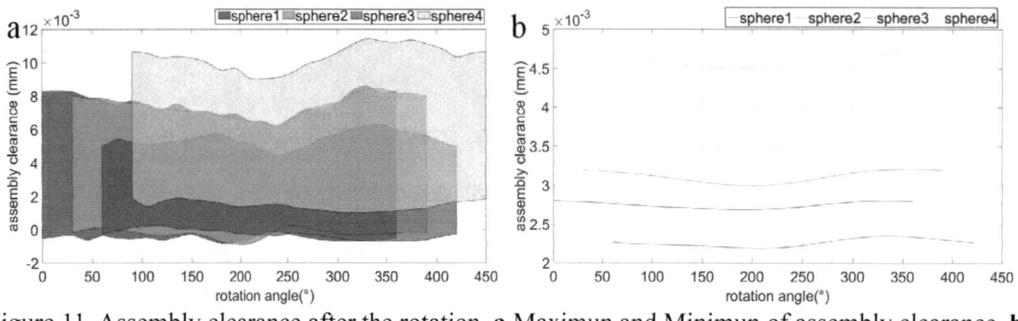


Figure 11. Assembly clearance after the rotation. **a** Maximun and Minimun of assembly clearance. **b** Root mean square error of assembly clearance

Above all, the assembly clearance of the precision spherical coupling parts was obtained. Taking the assembling of four spheres with bowl 1 as an example, It is found that the assembly of sphere 3 and bowl 1 got the smallest clearance. Furthermore, the distribution the assembly clearance is obtained. While the unqualified parts could be repaired in purpose by the analysis results, which can improve production behavior.

## 5. Conclusions

Assembly clearance affects the quality and working life of the precision spherical coupling parts. A method was proposed to get the assembly clearance by analyzing the points cloud of the precision spherical coupling parts. According to the sphericity error evaluation, the accurately center was searched by the particle swarm optimization with changed inertia factor near the least square center. The statistics of distance show that the center obtained from the PSO could be regarded as true center. Besides, the sphericity error evaluation results are improved obviously by comparing with other results. The accuracy of the evaluation results improves 4.51%~28.97%.

Based on the PSO center, the radical clearance and Z clearance were obtained. And the dynamic change of the assembly clearance by rotating the sphere of precision spherical coupling parts was analyzed. Taking the assembling of four spheres with bowl 1 as an example, the clearance of sphere 3 and bowl 1 is the best. With this method, the actual parts do not need to be tested one by one and the clearance could be obtained, which would greatly reduce the workload of the worker and promote the production efficiency.

In this paper, the prediction model of assembly clearance of precision coupling parts was established and discussed. The influence of other parts involved in the assembly on the assembly clearance could be considered based on this paper in a further study.

**Acknowledgements**

This work is supported by the National Natural Science Foundation of China (grant no. 51375055).

**References**

[1] Rameau J.F. Serre P. Moinet M. (2017) Clearance vs. tolerance for rigid overconstrained assemblies [J]. Computer-Aided Design. 97: 27-40.
[2] Mao Y.Z. Wang L.Q. Zhang X. (2018) Influence of ring deformation on the dynamic characteristics of a roller bearing in clearance fit with housing [J]. International Journal of Mechanical Sciences. 138: 122–130.
[3] Yue L. Qiao L.H. Huang Z.C. Nabil A. (2021) A Novel Method for Assemblability Evaluation of Non-Ideal Cylindrical Parts Assembly [J]. Computer-Aided Design. 134:103-120.
[4] Roque C. (2019) Sphericity measurement through a new minimum zone algorithm with error compensation of point coordinates [J]. Measurement. 138: 291-304.
[5] Meo A. Profumo L. Rossi A. Lanzetta M. (2013) Optimum dataset size and search space for minimum zone roundness evaluation by genetic algorithm [J]. Measurement Science Review. 3: 100-107.
[6] Akemi G. Andrés I. (2013) A new iterative mutually coupled hybrid GA–PSO approach for curve fitting in manufacturing [J]. Applied Soft Computing Journal. 13: 1491-1504.
[7] Zhao X.Y. Zhang C.M. Xu L. Yang Bo. Feng Z.Q. (2013) IGA-based point cloud fitting using B-spline surfaces for reverse engineering [J]. Information Sciences. 245: 276-289.
[8] Gu T. Qian X. Lou P. (2022) Research on Roundness Error Evaluation of Connecting Rod Journal in Crankshaft Journal Synchronous Measurement [J]. Applied Sciences. 12:315-341.

# Research on structural health monitoring method based on the inertial inclination sensor array

**Jianpeng Tong[a], Xiaolong Chen[b]* and Haoyu Yu[c]**

Xi'an Key Laboratory of Intelligent Instrument and Packaging Test, Xidian University, Xi'an ,Shaanxi , 710071,China

[a]e-mail: tong39398@gmail.com

* Corresponding author: [b]e-mail:xlchen@mail.xidian.edu.cn,

[c]e-mail: hyyu@xidian.edu.cn

**Abstract.**With the development of human industrial processes, the infrastructure in different fields has been significantly improved, but potential failure will bring huge security risks. Structural health monitoring (SHM) using cloud computing and IoT technology has become a necessary safeguard to ensure the safe and reliable operation of large structures such as aircraft wings, pipelines, bridges, and dams. This paper innovatively proposes a new fitting method based on space deformation, which uses the inclination sensor array to obtain angle information, calculates the node deflection, and performs rational cubic spline interpolation fitting, invert the state of the measured component. At the same time, the theoretical and measured error analyses are carried out. In the length of 1800mm pipe, the error can be controlled below 5% when the number of sensors is greater than or equal to 5, which has an important engineering value.

## 1. Introduction

Structural health monitoring is the latest emerging technology for structural deformation and damage detection and is generally considered to be one of the means to improve the safety and maintain the efficiency of engineering structures. It uses sensors to accurately perceive structural deformation and strain, plays an important role in solving the problems, such as the high maintenance cost of engineering structures and the dislocation of infrastructure security in various countries. At present, it has been highly concerned by government scientists in various countries.

According to incomplete statistics, the total number of highway bridges in the United States has exceeded 600,000, of which about 10% of the bridges have been damaged[1]. The safety operation of the road network is faced with a great potential threat; According to the "2021 Oil and Gas Pipeline Transportation Industry Development Status and Prospect Analysis", by the end of 2020, China's cumulative pipeline mileage for crude oil, refined oil, and natural gas will be 29,000 kilometers, 29,000 kilometers, and 81,000 kilometers, respectively. More than 80% of these oil and gas pipelines have reached or exceeded the design life. On the other hand, the operating environment of the aircraft is harsh, it is greatly affected by external forces, and the operating environment is special. Abnormalities are not allowed in every link. Therefore, structural health monitoring plays an important role in the monitoring of mesoscale structures such as aircraft wings, pipelines, bridges, and dams[2].

In the process of realizing "full coverage of large measured components and facilitating monitoring" by using inclination sensor, it is an urgent problem to solve how to convert the angle information collected into deflection information and reconstruct the deformation of components in space combined with node coordinates. At the same time, deformation monitoring based on multi-node inertial inclination sensor needs to acquire a lot of data as support, so data acquisition and sensor node distribution are very important. It is an important problem that what conditions should be met for the location distribution of sampling nodes and the number of sensors sets to reconstruct component deformation[3]. On the one hand, the distance between sensors should not be too large or too small. If the distance between sensors is too large, the accuracy of the reconstructed model cannot be guaranteed and reconstruction distortion may be caused. If the distance between sensors is too close, the interval between sampling points is small, and as the number of sensors increases, the complexity of the system increases and the stability cannot be guaranteed, and the cost will increase as the number of sensors increases. Therefore, based on the sexual inclination sensor array and reconstruction algorithm, this paper realizes the state inversion of the measured component, and discusses the rationality of the distribution density to ensure fitting accuracy.

## 2. Data collection and fitting method

This section discusses research on inclination data acquisition and spatial curve fitting algorithms. Due to the rapid development of microelectromechanical systems (MEMS), wireless sensor networks and MEMS sensor integration are gradually being applied to structural health monitoring[4-5]. Constructing a suitable higher-order interpolation spline method with adjustable parameters using the inertial inclination solution set can realize structural health monitoring[6].

### *2.1. Data collection*

The sensor array acquisition system designed in this paper is mainly composed of a data acquisition terminal, row and column signal gating, system main controller, and PC display terminal. The main purpose of the data acquisition terminal is to acquire the inclination data of each node. The function of row-column gating is to obtain single row/column inclination node data. Serial communication is mainly used to connect the terminal and the PC and transmit the collected data to the PC through serial communication to analyze and transform the collected data. At the same time, the serial communication also needs to receive some related commands from the PC. As shown in Figure 1. The test element used in the data acquisition terminal is the MEMS acceleration sensor MPU-6050. The MPU6050 chip can simultaneously collect the linear acceleration and angular acceleration data of the three coordinate axes[7]. With stable performance and small size, it is widely used in industrial and control fields. The controller selects the STMicroelectronics STM32F103ZET6 main control chip, which has a built-in serial communication register, which is convenient for transferring data to the PC, and collects the data generated by the MPU6050 chip through the IIC protocol. Cooperate with multiple groups of MPU6050 modules to collect the inclination data of each node.

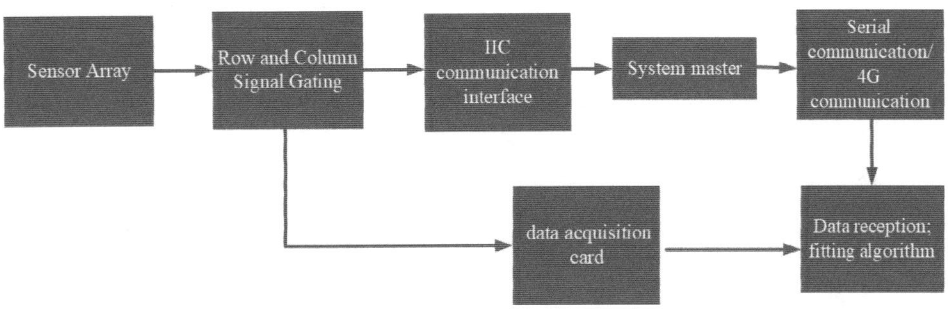


Figure 1. Inclination data collection

### 2.2. Interpolation fitting model

The main used method in this paper is the rational cubic spline interpolation function. Different from ordinary least squares interpolation, linear interpolation, and Hermite interpolation, this method can adjust the local shape according to actual need. Multiple sets of sensors can be selected. The selection of the number of sensors, the local shape adjustment of the fitting curve is flexible[8].

For curve $p(x)$ fitting, in the interval $\left[x_i, x_{i+1}\right]$ the rational cubic spline function $p(x)_{[x_i,x_{i+1}]}$ satisfies the following equation:

$$p(x)_{[x_i,x_{i+1}]}=\frac{5k_i f_i(1\text{-}t_i)^3+V_i(1\text{-}t_i)^2 t_i+W_i(1\text{-}t_i)t_i^2+f_{i+1}t}{5k_i(1\text{-}t_i)+t_i} \tag{1}$$

$$\begin{cases} V_i=(10k_i+1)f_i+5k_ih_im_i \\ W_i=(5k_i+2)f_{i+1}-h_im_{i+1} \\ h_i=x_{i+1}-x_i \\ t_i=\dfrac{x-x_i}{h_i} \\ \mathrm{m}_i=\tan\theta_i \end{cases} \tag{2}$$

Where $k_i$ is an adjustable parameter, and $k_i \neq \frac{1}{5}$, $m_i$ is the first derivative of $p(x)$ at the position of $x=x_i$, which can be measured by the inclination sensor, $f_i$ is the deflection at the position of the $x=x_i$ segment point, and $p(x)$ satisfies The basic formula for interpolation is as follows:

$$\begin{cases} p(x_i)=f_i \\ p'(x_i)=m_i \end{cases} \quad \begin{cases} p(x_{i+1})=f_{i+1} \\ p'(x_{i+1})=m_{i+1} \end{cases} \quad (i=0,1,\ldots,n-1) \tag{3}$$

The condition that $p(x)$ can be determined after both $m_i$ and $f_i$ at all breakpoints are measured or solved. According to the deployment of sensors, only $m_i$ can be measured. Find a way to solve $f_i$ according to $m_i$, let $p(x)$ be second-order derivative, one can get the system of equations for $f_{i-1}, f_i, f_{i+1}$:

$$\alpha_i f_{i-1}+\beta_i f_i+\gamma_i f_{i+1}=\delta_i \tag{4}$$

$$\alpha_i := \frac{20h_ik_{i-1}}{h_{i-1}}+\frac{2h_i}{h_{i-1}} \tag{5}$$

$$\beta_i := \frac{2h_{i-1}}{h_i}+\frac{4}{5}\frac{h_{i-1}}{h_ik_i}-\frac{20h_ik_{i-1}}{h_{i-1}}-\frac{2h_i}{h_{i-1}} \tag{6}$$

$$\gamma_i := -\frac{2h_{i-1}}{h_i}-\frac{4}{5}\frac{h_{i-1}}{h_ik_i} \tag{7}$$

$$\delta_i := -10h_ik_{i-1}m_{i-1}-\left(2h_{i-1}+\frac{2}{5}\frac{h_{i-1}}{k_i}+10h_ik_{i-1}+2h_i\right)m_i-\frac{2}{5}\frac{h_{i-1}m_{i+1}}{k_i} \tag{8}$$

Supplementary boundary conditions:

$$\begin{cases} f_0 = \delta_0 \\ f_n = \delta_n \end{cases} \tag{9}$$

Construct the matrix form of equations and observe the existence of solutions,and let

$$AF = \Delta \tag{10}$$

$$A = \begin{bmatrix} 1 & 0 & \cdots & & & 0 \\ \alpha_1 & \beta_1 & \gamma_1 & \cdots & & 0 \\ 0 & \alpha_2 & \beta_2 & \gamma_2 & \cdots & 0 \\ \cdots & & & & & \cdots \\ 0 & \cdots & & \alpha_{n-1} & \beta_{n-1} & \gamma_{n-1} \\ 0 & \cdots & & & 0 & 1 \end{bmatrix}_{(n+1)\times(n+1)}, F = \begin{bmatrix} f_0 \\ f_1 \\ \cdots \\ f_{n-2} \\ f_{n-1} \\ f_n \end{bmatrix}, \Delta = \begin{bmatrix} \delta_0 \\ \delta_1 \\ \cdots \\ \delta_{n-2} \\ \delta_{n-1} \\ \delta_n \end{bmatrix} \tag{11}$$

According to the deflection value $f_i$, the interpolation spline function $p(x)_{[x_i, x_{i+1}]}$ of each section is calculated to complete the fitting of the whole curve. In order to calculate the deflection by using the node inclination information, it is necessary to install multiple sets of inertial inclination sensors on the line. As shown in Figure 2(a), seven groups of inclination sensors MPU6050 were arranged on the 1800mm pipe equidistant from each other to measure the inclination of each node position. At the same time, a displacement sensor HG-C1200 is installed at the bottom to measure the real deflection change during the experiment. Fixed constraints are given at both ends of the pipe. Figure 2(b) shows the bending deformation after being stressed.

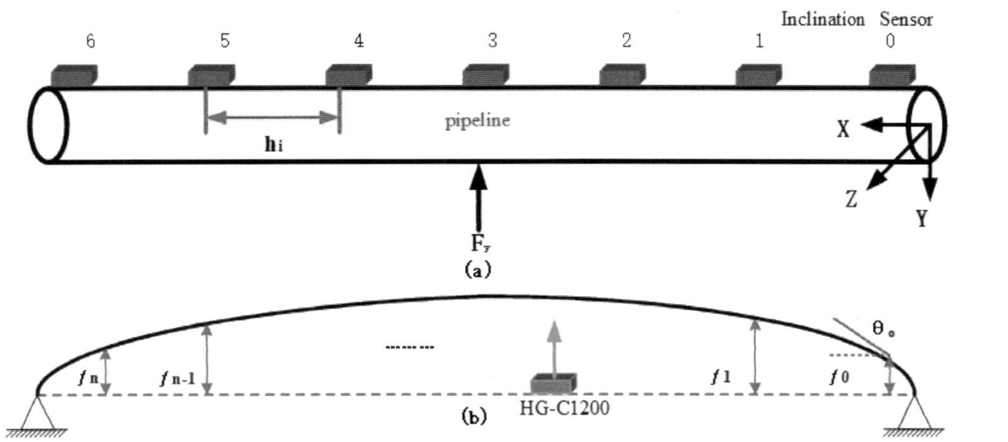


Figure 2. Forced bending of pipeline and placement of sensors

Using the above-mentioned interpolation mathematical model, the pipeline is divided into 7 measurement units, and the coordinates are denoted as $x_i$, the inclination angle corresponding to the sensing unit is denoted as $\theta_i$, the tangent value is denoted as $m_i$, the distance between the defined quantity nodes is $h_i$, and the deflection value at the node is $f_i$ so as to determines the interpolation function of each defined interval, and draw the curve posture of the entire pipeline according to the interpolation spline $p(x)$ on each interval. Figure 3 is the bending curve after fitting the data collected by equidistant distribution of 7 sets of inertial inclination sensors. Table 1 is the angle value of each acquisition.

Table 1. Angles collected by different position sensors

| | $\theta_0$ | $\theta_1$ | $\theta_2$ | $\theta_3$ | $\theta_4$ | $\theta_5$ | $\theta_6$ |
|---|---|---|---|---|---|---|---|
| First collection | 4.21 | 3.68 | 2.65 | 0.27 | -2.52 | -3.48 | -4.43 |
| Second collection | 4.19 | 3.39 | 2.63 | 0.28 | -2.15 | -3.09 | -4.34 |
| Third collection | 4.27 | 3.97 | 2.73 | 0.31 | -2.27 | -3.75 | -4.26 |
| Ave | 4.22 | 3.68 | 2.67 | 0.29 | -2.31 | -3.44 | -4.34 |

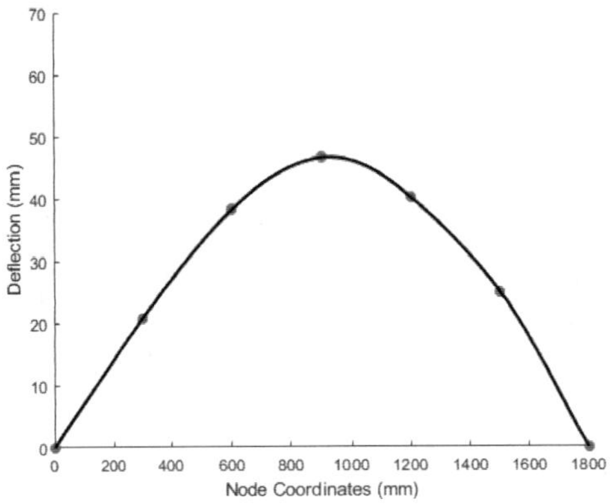


Figure 3. Fitting curve of the solution set of 7 groups of inclination sensor data

## 3. Error analysis

### *3.1. Theoretical error analysis*

If we want to apply it in engineering, it is necessary to analyze the fitting accuracy. According to the deflection equation with the theories of "Elastic Solid Mechanics" and "Material Mechanics", the theoretical value of the bending deflection $w$ at point $x$ equation is as follows. $F$ is the applied force.

$$w(x)=\begin{cases}\dfrac{Fbx}{6EIl}(l^2-x^2-b^2)(0\le x\le a)\\ \dfrac{Fb}{6EIl}\bullet\left[\dfrac{l}{b}(x-a)^3+(l^2-b^2)x-x^3\right](a\le x\le l)\end{cases} \tag{12}$$

The angle in the interpolation function $p(x)$ is calculated by the rotation angle equation in material mechanics:

$$\theta=\begin{cases}-\dfrac{Fb}{6EIl}(l^2-b^2-3x^2)(0\le x\le a)\\ -\dfrac{Fb}{6EIl}(l^2-b^2-3x^2+\dfrac{3l}{b}x^2-\dfrac{3l}{b}\cdot 2\cdot x\cdot a+\dfrac{3l}{b}\cdot a^2)\ (a\le x\le l)\end{cases} \tag{13}$$

$a$ is the location where the force $F$ is applied, the length of the pipe is $l$ , $l-a=b$ . $E$ is the elastic modulus, and $I$ is the inertia moment of the material cross-section to the bending neutral axis. The calculation formula of the theoretical error is as follows. In the error calculation process, $F$ , $E$ , and $I$ can all be simplified.

$$E_t=\left|\frac{100(p(x)-w(x))}{w(x)}\right|\% \tag{14}$$

The constraint condition is that the sensors are equidistantly distributed; the positions of all nodes are known, and the deflection value at the two endpoints is 0 by default; the k value of each point in the interpolation function is equal. The position where the force is applied is $l/2$. The number of sensors is 3-7. At the position of x=1000mm, the influence of the number of sensors and k on the fitting accuracy was explored. As shown in Figure 4. The adjustable parameter k is set to {0.01, 1, 100}, when k=1, the error is relatively small, when the number of inclination sensors is 3, the fitting error is 24.5%, and when the number of sensors is 4-7, the error between the actual deflection and the fitted deflection is {5.2%, 3.0%, 1.6%, 2.8%}, which theoretically proves that the inclination angle can be used to fit the curve deformation when the number of sensors is guaranteed.

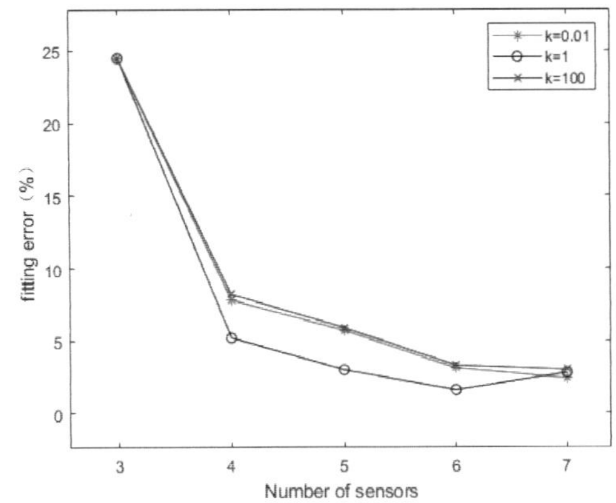


Figure 4. The effect of the number of sensors on the fitting accuracy

*3.2. Experimental verification*

In order to verify the above theoretical error calculation, We carried out five sets of experiments. On a pipe with a length of 1800mm, we arranged 3-7 sensors at equal distances respectively, measured the angles of the corresponding nodes and fitted the curve $p(x)$ by rational cubic spline interpolation. In the model, the adjustable parameter is set to 1. The fitting result is shown in Figure 5(a). Fitting with different numbers of sensors(3-7), the fitting deflection values $p(x)$ at $x$=1000mm are 40.024mm, 47.923mm, 42.836mm, 42.627mm and 46.083mm. Using the displacement sensor HG-C1200 to obtain the actual deflection $s(x)$ at $x$=1000mm, the measured true deflection value $s(x)$ is 44.5mm. The deflection relative error is verified as follows:

$$E_v = \left| \frac{100(p(x) - s(x))}{s(x)} \right| \% \tag{15}$$

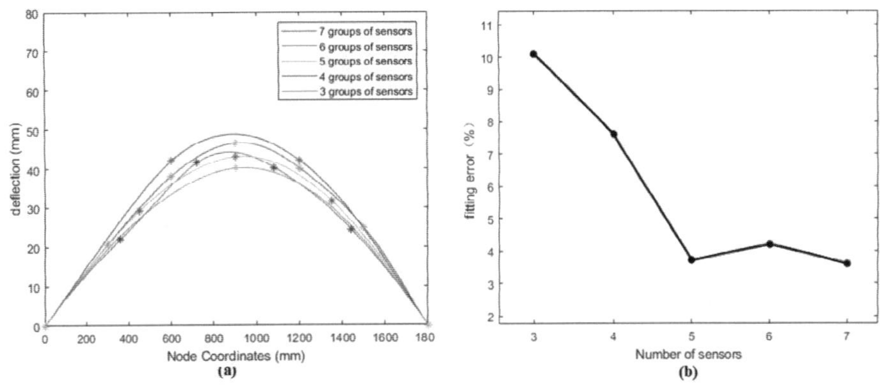


Figure 5. Curve fitting and error validation

The error results are shown in Figure 5(b). When the number of sensors is 3, the error is 10.1%, and when the number of sensors is 4-7, using angle fitting deflection and using displacement sensor to measure deflection, their relative errors are {7.7%, 3.7%, 4.2%, 3.6%}, not much different from theoretical calculations. The feasibility of the method is verified by experiments. In engineering practice, the error is usually limited to 10% [9], and the method used in this paper can limit the error to less than 5% under the condition that the number of sensors is guaranteed.

## 4. Conclusion

In the process of completing the above research work, the inclined sensor and rational cubic spline interpolation method are used to fit the pipe bending. The influence of location distribution of sampling nodes and number of sensors on the fitting accuracy is discussed, and the errors are verified by displacement sensors. When the number of sensors is greater than or equal to 5, the error can be controlled below 5%. In the future research work, the state inversion of surface in space will be explored by using sensor array information, and the reconstruction of surface deformation will be realized by using pitch angle matrix, roll angle matrix and coordinates in space.

## Acknowledgments

The author acknowledges the National Natural Science Foundation of China with Grant No. 61727804, Key R & D plan of Shaanxi Province with Grant No. 2020ZDLGY07-12.

## References

[1] Rizzo, P., Enshaeian, A. (2021) Challenges in bridge health monitoring: A review. Sensors, 21(13):4336.

[2] Annamdas, Madhav, V.G., Bhalla. (2017) Applications of Structural Health Monitoring Technology in Asia. In: International Workshop on Structural Health Monitoring.Stanford. pp. 324-346.

[3] Ghazi, R. M., Chen, J. G., Büyüköztürk, O. (2017) Pairwise graphical models for structural health monitoring with dense sensor arrays. Mechanical Systems and Signal Processing, 93: 578-592.

[4] Abdulkarem, M., Samsudin, K., Rokhani, F.Z. (2020) Wireless sensor network for structural health monitoring: a contemporary review of technologies, challenges, and future direction. Structural Health Monitoring, 19(3): 693-735.

[5] Ha, D.W., Kim, J.M., Kim, Y. (2018) Development and application of a wireless MEMS-based borehole inclinometer for automated measurement of ground movement.Automation in Construction, 87:49-59.

[6] Yu, H., Chen, X., Ren, M. (2021) A coupled bend-twist deformation monitoring method based on inclination measurement and rational cubic spline fitting.Mechanical Systems and Signal Processing, 147: 107084.

[7] Yeh, C.Y. (2019) A Low-Power Monolithic Three-Axis Accelerometer with Automatically Sensor Offset Compensated and Interface Circuit. Microelectronics Journal, 86(APR): 150-160.

[8] Yu, H., Chen, X., Ren, M. (2021) An online pipeline structural health monitoring method based on the spatial deformation fitting. IEEE Transactions on Industrial Electronics, 69: 7383-7393.

[9] Hou, X.M., Yang,X.S., Huang, Q. (2004) Using inclinometers to measure bridge deflection.Bridge Construction, 10(5): 564-569.

# Introduction of an Elevator Guide Shoe with Position and Lateral Force Monitoring Function

**Shuangchang Feng***, **Jie Chen and Yanchun Liang**

Shanghai Institute of Special Equipment Inspection and Technical Research, Shanghai, 200062, China

*Corresponding author's e-mail: fengsc@ssei.cn

**Abstract.** When the elevator guide rail is deformed or the building body is inclined, the car will produce a certain transverse force relative to the guide rail, resulting in serious wear on one side of the guide shoe. Since the existing guide shoe is fixed, when the car generates transverse force relative to the guide rail, only the transverse position of the guide shoe relative to the car can be adjusted. However, in practical application, it is difficult for elevator maintenance personnel to find this phenomenon in time. In most cases, when it is found that one side of the guide shoe is seriously worn, the guide shoe has been worn to an unusable state. This paper introduces an elevator guide shoe with position and lateral force monitoring function. The elevator guide shoe comprises a base fixed on the elevator car, a sliding support horizontally sliding connected with the base, a nylon block fixedly connected with the sliding support and in sliding contact with the elevator guide rail, and a displacement sensor and a force sensor arranged on the base to monitor the transverse sliding distance and transverse force between the sliding support and the base. The device has the advantages of position and transverse force monitoring function, compact overall structure, no need to change the original car layout, safety and reliability.

## 1. Introduction

Elevator guide shoes are used to fix the car on the guide rail so that the car can only move up and down [1]. Each elevator car is equipped with four sets of guide shoes, which are respectively set on both sides of the upper beam and under the safety gear seat at the bottom of the car. Four sets of counterweight guide shoes are installed at the bottom and upper part of the counterweight beam. The guide shoes fixed on the car can move back and forth along the fixed guide rail installed on the wall of the building shaft to prevent the car from deflecting or swinging during operation [2].

When the elevator guide rail is deformed or the building body is inclined, the car will produce a certain transverse force relative to the guide rail, resulting in serious wear on one side of the guide shoe. Since the existing guide shoe is fixed, when the car generates transverse force relative to the guide rail, only the transverse position of the guide shoe relative to the car can be adjusted [3]. However, in practical application, it is difficult for elevator maintenance personnel to find this phenomenon in time. In most cases, when it is found that one side of the guide shoe is seriously worn, the guide shoe has been worn to an unusable state. In order to overcome the defects of the prior art, this paper provides an elevator guide shoe with position and transverse force monitoring function.

## 2. Technical scheme of elevator guide shoe with position and lateral force monitoring function

The device relates to an elevator guide shoe with position and transverse force monitoring functions. The elevator guide shoe comprises a base fixed on the elevator car, a sliding support connected with the base in a transverse sliding manner, and a nylon block fixedly connected with the sliding support and in sliding contact with the elevator guide rail, And a displacement sensor and a force sensor arranged on the base to monitor the transverse sliding distance and transverse force between the sliding support and the base [4].

The base is an L-shaped structure composed of a vertical plate and a bottom plate, and two parallel stiffeners connecting the vertical plate and the bottom plate are arranged on the front of the base to avoid plastic deformation or damage to the base.

The lower part of each stiffener is provided with a rectangular trapezoidal slot hole to place the displacement sensor and force sensor [5]. The lower part of the vertical plate is provided with two horizontal long holes, and the front of the sliding support is located at the two positions directly opposite the long holes, and a screw and a support arm are vertically arranged respectively.

The fixed end of the displacement sensor is installed on the sensor mounting seat arranged at the left stiffener and fixed through a removable guard plate, and the movable end of the displacement sensor is fixedly connected with the support arm passing through the horizontal long hole through a nut.

The fixed end of the force sensor is fixedly installed in the slot hole of the left stiffener, and the movable end is fixedly connected with the screw passing through the horizontal long hole through a nut. The displacement sensor and the force sensor are respectively connected with the control alarm module to send out alarm information when the monitoring data of the displacement sensor and the force sensor are abnormal . The bottom plate of the base is provided with two mounting holes for fixing the elevator car.

A plurality of rows of convex ribs with smooth surface are horizontally arranged on the back of the vertical plate of the base to reduce the friction with the sliding support .The upper and lower edges of the sliding support are respectively provided with hook grooves, which are horizontally sliding matched with the protrusions on the upper and lower edges of the vertical plate of the base.

## 3. Working principle of elevator guide shoe with position and lateral force monitoring function

As shown in Figure 1-4, this paper provides an elevator guide shoe with position and transverse force monitoring function. Description of marks in the figure:

1. Base, 2. Force sensor, 3. Stiffener, 4. Displacement sensor, 5. Sliding support, 6. Nylon block, 11. Mounting hole, 12. Sensor mounting seat, 13. Long hole, 31. Slotted hole, 14. Convex rib, 41. Guard plate, 51. Hook groove, 52. Screw, 53. Support arm. The nylon block 6 of the guide shoe is designed to slide laterally, and the transverse sliding distance and transverse force are detected by corresponding sensors. Through the measured transverse sliding distance and transverse force, the car or elevator guide rail with transverse movement can be found in time.

The elevator guide shoe comprises a base 1, a force sensor 2, a displacement sensor 4 and a sliding support 5. The base 1 is L-shaped, and the bottom plate of the L-shaped base is provided with two mounting holes 11 for fixing the base 1 to the lift car. The bottom plate and the vertical plate of the base 1 are reinforced by two stiffeners 3 to avoid plastic deformation or damage to the base 1.

Each stiffener 3 is provided with a slot hole 31, which is generally rectangular trapezoid, and its connection strength and rigidity are not reduced as much as possible. The force sensor 2 and the displacement sensor 4 can be placed inside the slot hole 31 at the same time, and the lower end of the slot hole 31 is also provided with a sensor mounting seat 12. The sensor mounting base 12 mounts and fixes the displacement sensor 4 through a removable guard 41. The displacement sensor 4 and force sensor 2 are electrically connected to the corresponding control alarm module. When abnormal parameters are detected, the alarm information can be sent out.

The back of the sliding support 5 is buckled with a nylon block 6, which is used to directly contact the elevator guide rail. The upper and lower ends of the sliding support 5 are provided with hook grooves 51. The upper and lower hook grooves 51 hook the upper and lower edges of the vertical plate of the base 1, and the protrusions of the upper and lower edges of the base 1 slide and connect the hook grooves 51 laterally respectively.

A plurality of rows of convex ribs 14 arranged horizontally are arranged on the back of the vertical plate of the base 1, and the surface of the convex ribs 14 is smooth, which is used to reduce the friction with the sliding support 5.

Two horizontally distributed long holes 13 are vertically arranged on the vertical plate of the base 1, and the front face of the sliding support 5 is vertically provided with a screw 52 and a support arm 53 for passing through the long hole 13 and fixedly connected with the movable ends of the force sensor 2 and the displacement sensor 4 respectively. The fixed ends of the force sensor 2 and the displacement sensor 4 are fixed on the opposite stiffener 3 by screws.

During operation, the base 1 is installed in the elevator car through the mounting hole 11 and bolts, and the nylon block 6 is clamped and installed on the elevator guide rail. When the elevator car moves laterally relative to the elevator guide rail, the sliding support 5 slides laterally relative to the base 1. At this time, the force sensor 2 measures the force generated by the lateral sliding, and the displacement sensor 4 measures the lateral sliding distance between the two.

Since the telescopic stroke of the force sensor 2 is small, it may have a certain impact on the measurement results of the displacement sensor 4, while the sliding stroke of the sliding support 5 is limited, so the overall impact is small. Alternatively, the displacement sensor 4 is connected to the sliding support 5 through the long hole 13 to minimize the influence of the force sensor 2 on the measurement result of the displacement sensor 4.

When the force sensor 2 detects that the lateral force between the car and the elevator guide rail is too large, or the lateral sliding distance between the sliding support 5 and the base 1 is large, it will send an alarm in time to inform the motor maintenance personnel to deal with it as soon as possible to avoid excessive wear of the nylon block 6.

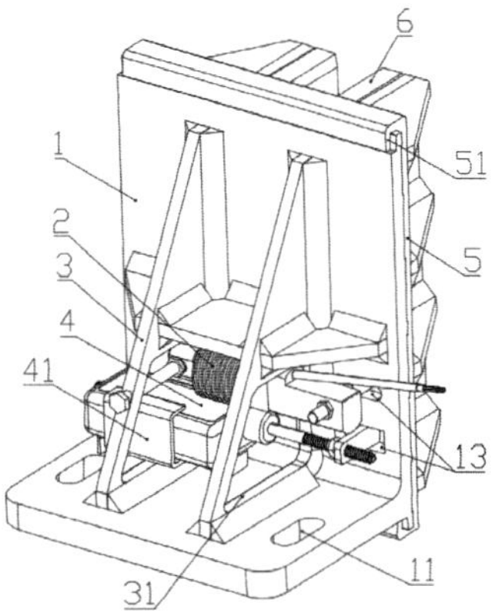


Figure 1 . Schematic diagram of front structure

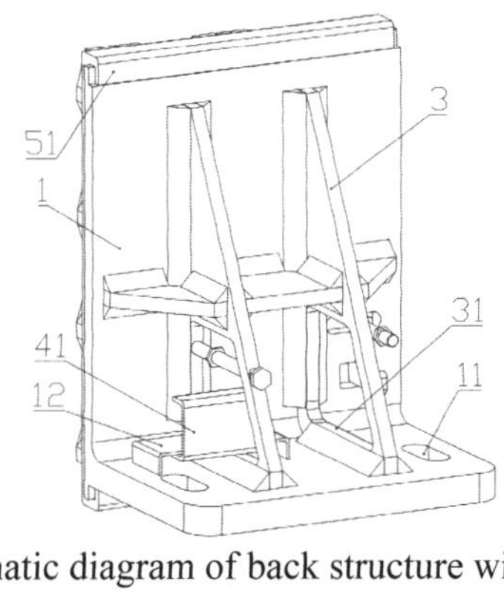


Figure 2 . Schematic diagram of back structure without nylon block

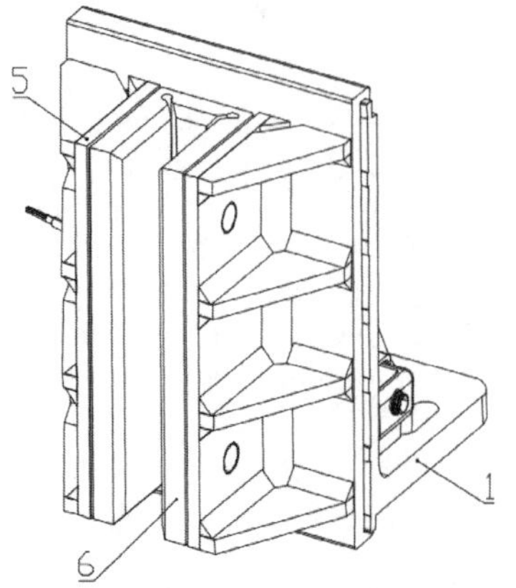


Figure 3. Schematic diagram of back structure

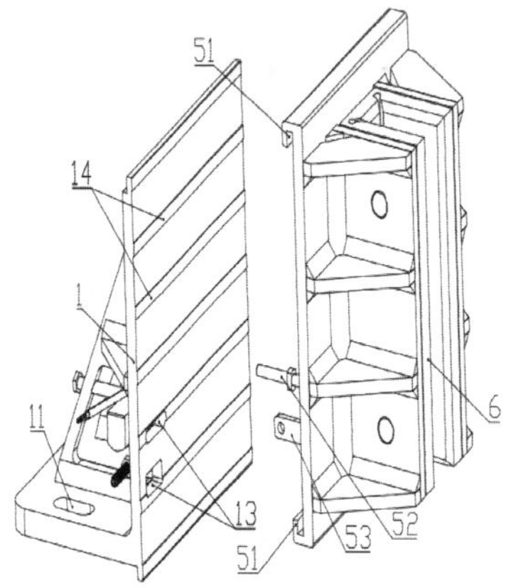


Figure 4 . Installation diagram of base and sliding support

## 4. Conclusion

The elevator guide shoe comprises a base fixed on the elevator car, a sliding support horizontally sliding connected with the base, a nylon block fixedly connected with the sliding support and in sliding contact with the elevator guide rail, and a displacement sensor and a force sensor arranged on the base to monitor the transverse sliding distance and transverse force between the sliding support and the base. Compared with the prior art, the device has the following advantages:

- The device can detect the lateral distance of the elevator car relative to the guide rail at the same time. When the sensor detects that the lateral force or lateral offset of the elevator car is large, it will send an alarm signal to remind the elevator manager in time.
- The overall structure of the device is compact, the force sensor and displacement sensor are hidden and installed inside the stiffener, and there is no need to change the original car layout.

**Acknowledgments**

This research was supported by The Science and Technology Planning Project of Shanghai Municipal Administration of market supervision (2021-23).

## References

[1] Wang, F.X. (1991) Use and maintenance of elevator guide shoes [J] Equipment management and maintenance, 2: 14-16.

[2] Zhang, X.G., Li, H.G., Meng, G. (2011) Experimental study on friction characteristics of elevator guide shoe [J] Mechanical strength, 33: 6-9.

[3] Chen, H., Feng, Y.H., Feng, Y.B.(2013) Design and analysis of elevator rolling guide shoe with braking function [J] Mechanical design and manufacturing, 6: 36-38.
[4] Liu, L., Guo, Y.S., Wei, F.P. (2018) New elevator roller guide shoe [J] Hoisting and transportation machinery, 8: 16-19.
[5] Qian, L.(2012) Shi Xi Measurement of friction and energy consumption of elevator sliding guide shoe system [J] Machinery manufacturing and automation, 5: 4-6.

# Research on a new designed upward turning comb of escalator

**Shuangchang Feng*, Jie Chen and Wenhao Shen**

Shanghai Institute of Special Equipment Inspection and Technical Research, Shanghai, 200062, China

*Corresponding author's e-mail: fengsc@ssei.cn

***Abstract***—Escalator is an important transportation method for people in their daily life. The comb fixed at the entrance and exit is an important component of the escalator. Its function is to facilitate the transportation of passengers and prevent the escalator from being damaged by foreign matters to a certain extent. In order to avoid harm to passengers caused by the comb, the end of the comb should be made of easily broken material traditionally such as plastic and aluminum to ensure that the object can be released when clamped. This greatly increases the loss of comb and the repair time of maintenance personnel and leaves potential safety hazards. The article researches a new designed upward turning comb of escalator. The device comprises comb plate, mounting base, guide rod, spring, comb teeth, rotating shaft, upper baffle, chute, notch, mounting hole, long hole, guide hole, lower baffle, positioning block and arc push block. Compared with the prior technology, this new comb has the advantages of simple structure, convenient operation, repeated use, difficult to damage.

## 1. Introduction

Comb located at both ends of the escalator is an important component of the escalator. The comb is fixed at the entrance and exit and meshed with the steps. Its function is to facilitate the transportation of passengers and prevent the escalator from being damaged by foreign matters to a certain extent [1]. The end of the comb shall be processed into rounded corners, and the shape shall ensure that the risk of foot clamping between the comb and the step is minimized. Nevertheless, there are still many injuries caused by comb on escalators. Some softer or smaller objects, such as slippers and coins, will still enter the interior of the comb and cause accidents [2].

In the prior technology, in order to avoid harm to passengers caused by the comb, the end of the comb should be made of easily broken material such as plastic and aluminum to ensure that the object can be released when clamped, as shown in Figure 1. But the disadvantage is that the comb is easy to be damaged. Whenever an escalator accident occurs, the comb will be damaged with a high probability. This greatly increases the loss of comb and the repair time of maintenance personnel [3]. At the same time, in the daily use of the escalator, the replacement of the broken comb is easy to be ignored by the maintenance personnel, which leaves potential safety hazards [4].

Some escalators also have sensors installed behind the comb. When the sensor detects abnormality, it will apply emergency braking to the escalator. However, the escalator will not stop immediately. There is a certain buffer distance. The buffer distance will further harm passengers and the escalator. Moreover, the emergency braking may also cause the gravity center of passengers standing on the steps to be unstable, resulting in continuous falls [5-6].

The purpose of the new comb researched in this article is to provide a special structure of the comb that the teeth can turn up in order to overcome the defects of the above existing technology.

This new comb can turn upward to avoid possible harm to the passenger. The device comprises a comb plate with comb teeth at the front end and a mounting seat pivoted with both sides of the comb to turn the comb upward. The two sides of the mounting seat are respectively provided with a long hole, and the two sides of the rear of the comb are respectively provided with a rotating shaft. Each rotating shaft is respectively installed in the corresponding long hole and slides under the guidance of the long hole, as shown in the Figure 2. Compared with the prior technology, this new comb has the advantages of simple structure, convenient operation, repeated use, difficult to damage, and so on.

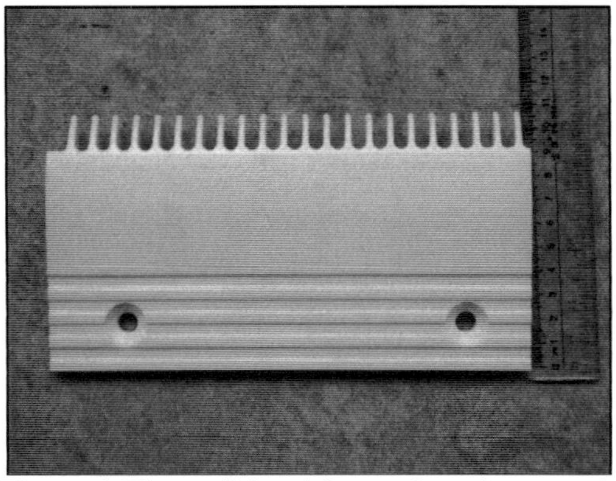

Figure. 1 The traditional form of escalator comb

## 2. The structure of the escalator comb

This article researches an escalator comb device that can turn upward, as shown in Figure 2. The device can separate the front section of the comb from the step under the state of impact and extrusion. When foreign matters enter the comb teeth, the comb plate will be pushed by the steps. Under the action of thrust, the comb plate can be turned upward to be separated from the step. After the escalator stops, the foreign matter can be quickly removed. The specific structure of the escalator comb is as follows:

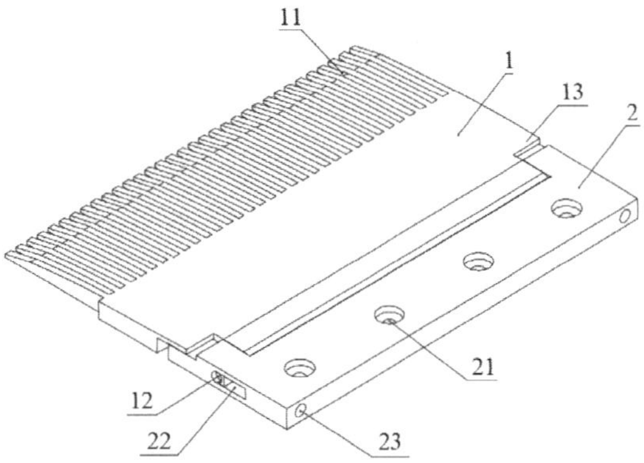


Figure. 2 The structure diagram of the escalator comb
1. Comb plate 2. Mounting base 11. Comb teeth
12. Rotating shaft 13. Upper baffle 21. Mounting hole
22. Long hole 23. Long hole

The device comprises a comb plate and a mounting seat. The front end of the comb plate is provided with a row of comb teeth. The rear two sides of the comb plate are provided with a rotating shaft. A chute which can make the comb plate move horizontally is in front of each rotating shaft. The bottom of the front end of the chute is provided with a notch connected with the chute, as shown in Figure 3.

The rear part of the mounting base is provided with several mounting holes. In this article, the mounting holes are countersunk holes, as shown in Figure 4. Both sides of the front end of the mounting base extend in the direction of the comb plate to form a lower baffle. Both sides of the mounting base are respectively provided with a long hole. Each long hole is provided with an arc push block. The inner arc surface of the arc push block is in contact with the rotating shaft, and the outer surface is vertically fixed with a guide rod. The guide rod is inserted into the inside of the guide hole. There is a spring sleeved on the outer surface of the guide rod. The two ends of the spring are respectively against the arc push block and the mounting base.

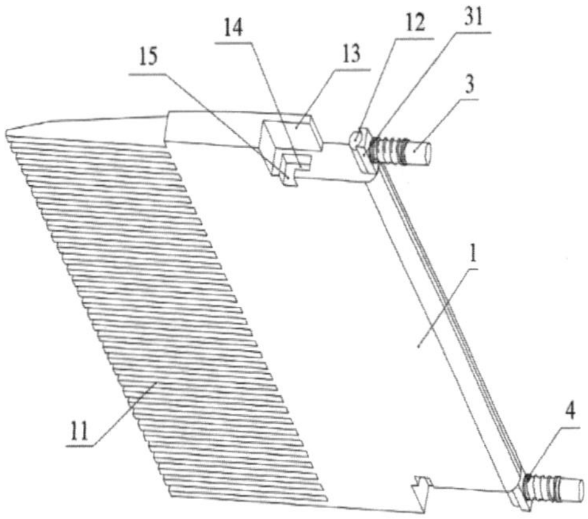


Figure. 3 The structure diagram of the comb plate

1. Comb plate 3. Guide rod 4. Spring
11. Comb teeth 12. Rotating shaft
13. Upper baffle 14. Chute
15. Notch 31. Arc push block

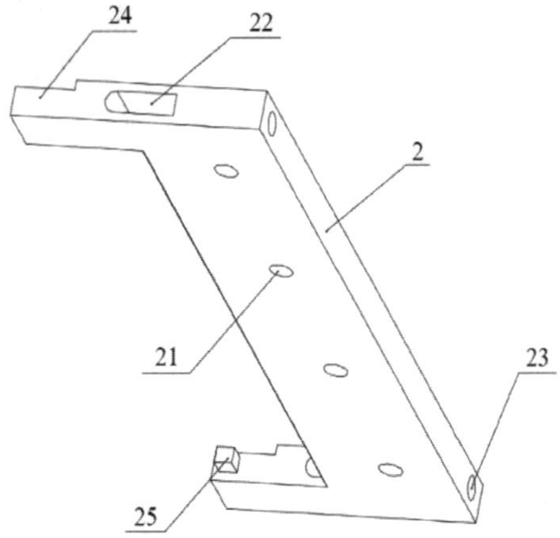


Figure. 4 The structure diagram of the mounting base

2. Mounting base 21. Mounting hole
22. Long hole 23. Guide hole
24. Lower baffle 25. Positioning block

The inner sides of the two lower baffles of the mounting base are respectively provided with positioning blocks extending inward. Under normal conditions, the positioning block is located inside the chute under the pre tightening and pushing action of the spring. At this time, the comb plate will not rotate relative to the mounting base. The comb plate and the mounting base are equivalent to a stationary entirety. When the mounting base is installed, the comb teeth is inserted into the groove of the step. When foreign matters are caught in the gap between the comb teeth and the step, the spring is compressed under the thrust of the step. The rotating shaft of the comb plate pushes the arc push block along the long hole to the direction of the mounting base until the positioning block is directly opposite to the notch. At this time, the positioning block comes out of the notch, and the comb plate can rotate around the rotating shaft and turn upward, as shown in Figure 5. In this way, the jammed foreign matter can be taken out and the comb plate can be reset by pushing hard. The positioning block will enter the chute again from the notch and return to the initial position inside the chute under the thrust of the spring.

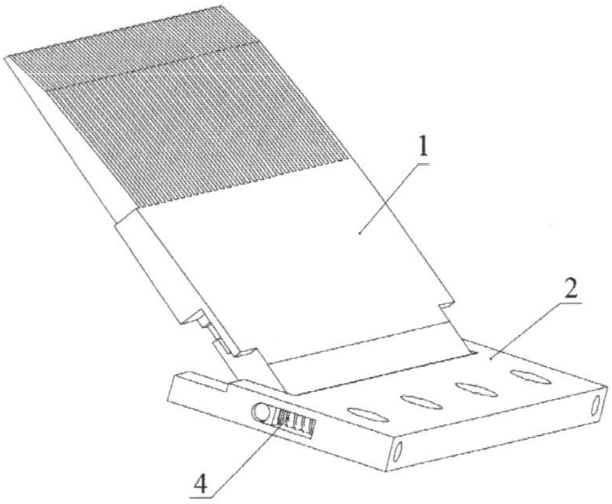


Figure. 5 The structure diagram of comb plate turning upward

1. Comb plate 2. Mounting base 4. Spring

## 3. The feature of the upward turning comb

The comb plate researched in this article can be turned upward relative to the mounting base. When foreign matters enter the comb plate, the comb plate will receive the thrust of the step. Under the action of the thrust, the comb plate can be turned upward to separate it from the step, and the foreign matters can be quickly taken out. The comb can be reused for many times. After turning upward, it can also be reset and used again.

## 4. Conclusion

With the continuous improvement of economic level, the rapid development of construction industry and the steady progress of urbanization, the number of escalators is gradually increasing. As a convenient mobile transportation equipment, escalator is widely used in stations, airports, shopping malls and other occasions. However, there are more and more escalator accidents at the same time. As an important safety component of the escalator, the comb facilitates the transportation of passengers and prevents the escalator from being damaged by foreign matters to a certain extent.

This article researches a special structure of the comb that the teeth can turn up in order to overcome the defects of the above existing technology. The device comprises comb plate, mounting base, guide rod, spring and so on. This device can effectively reduce the clamping accident of escalator caused by comb plate and ensure the safety of passengers. The comb designed in this article has the following advantages:

- Better protection to passengers. The traditional comb adopts a fixed structure and is rigidly connected with the entrance and exitance support. This makes it easy to clip the passenger's limbs in. In recent years, it is common for children to be injured by comb [7-8]. The new comb researched in this article adopts a reversible structure. Under the action of thrust, the comb teeth can be turned upward and separated from the steps, so that the clamped passengers can quickly disengage and avoid more serious injury.
- Reduction of the repair cost of comb. The traditional comb plate is made of easily broken plastic or aluminum material to ensure that the object can be released when clamped. This causes the comb to be constantly replaced [9]. Moreover, if the comb is not found damaged due to the negligence of the maintenance personnel, there will be potential safety hazards in the escalator. The new comb researched in this article adopts a reversible structure. First, the probability of comb damage is reduced. Second, the comb teeth turn upward, and the missing teeth are easier to be found. Third, if the comb is broken, only the teeth part of the comb needs to be replaced.

**Acknowledgments**
This research was supported by The Youth Project of Shanghai Institute of Special Equipment Inspection and Technical Research (2021-17).

**References**
[1] Huang Zhen, Peng Langcao, Peng Guang. Common risk points and preventive measures of escalator [J]. China Elevator, 2021,32 (17): 44-47.
[2] Huang Jianliang, Yu Ying, Lv Xiaodong. On the inspection method of comb plate of escalator and moving sidewalk [J]. China Plant Engineering, 2021 (09): 164-166.
[3] Tong Mengqi Protection principle and improved design of escalator safety device [J]. Design, 2018 (01): 92-93.
[4] Bi Chenshuai, Xie Ran, Dai Guangyu. Discussion and Reflection on the protection of comb plate of escalator and the prevention of foreign matter inclusion [J]. China Elevator, 2019,30 (01): 56-59.
[5] Jiang Yefeng, Chen Weigang, Bai Zhiguo. A new detection device and method for foreign matter protection device of escalator comb plate [J]. China Elevator, 2018,29 (21): 65-66 + 68.
[6] Wang Yifan, Tang Zhihao, Cao Hailin, ye Zijun, Wang Mincheng, Wang Yulin. Safety detection device for escalator comb plate [J]. Mechanical Engineer, 2018 (11): 54-56.
[7] Huo Jide, Chen Yinhui, Shi Zilu, Zhang Shuai, Yang Yafei, Wang Yulin. A safety detection device for escalator comb plate [J]. Mechanical Engineering & Automation, 2021 (05): 124-126.
[8] Huang Jianliang, Yu Ying, LV Xiaodong. On the inspection method of comb plate of escalator and moving sidewalk [J]. China Plant Engineering, 2021 (09): 164-166.
[9] Qi Zhengwu, Liang Minjian, Wang Kui. Research on dynamic measurement and hidden danger location method of safety gap between escalator steps and apron [J]. Hoisting and Conveying Machinery, 2017 (07): 108-110.

# Design of leakage prevention and diversion alarm system based on wireless control

**Zhengdong Li[1], Yulin Wang[1], Xuanyan Wu[1], Zhao Jin[1], Wen Lu[1], and Xiuling Li[2*]**

[1] College of Intelligent Transportation, Chongqing Vocational College of Transportation, Chongqing, 402247, China

[2] Communication NCO Academy, Army Engineering University, Chongqing, 400035, China

[*]Corresponding author's e-mail: xlli@mail.ustc.edu.cn

**Abstract.** To solve the leakage of contrast agents due to high pressure and fast speed of the high pressure injector, an intelligent anti-leakage system is designed. The pressure sensor MS5837 detects the pressure signal in real time and transmits the pressure signal to MSP430F5529. Once the micro controller detects that the pressure exceeds the threshold value, the leakage of the contrast agent is judged. The controller immediately starts the sound and light alarm and diversion switch function, and informs the medical staff of the clinic to deal with the leakage wound in time through the wireless transmission chip NRF24L01+. This system overcomes the disadvantages of previous mechanical design , such as easy loosening, no diversion. It adds diversion switching function and wireless communication module, which ensures that medical stuff can deal with the wound in the first time without radiation and reduce the leakage damage to the minimum. The results of animal experiments show that the intelligent diversion alarm system is stable, reliable and highly sensitive, which can effectively reduce the leakage of the contrast agents of the high pressure injector and relieve the pain of the patient.

## 1. Introduction

With the development of X-ray, image enhancement and artificial contrast technology, CT enhanced contrast-enhanced high pressure syringe is more and more used in radiology diagnosis and treatment[1-3]. However, in clinical application, because of the high injection speed and pressure of high pressure syringe when injecting contrast agent, it is easy to cause the pinhole to break away from the blood vessel, making the contrast agent penetrate into the skin and produce swelling[4-8]. Even though some high pressure syringes have the function of pressure detection, they measure the internal pressure of the high pressure syringe. Clinical trials show that this indirect monitoring method often can't detect the leakage phenomenon. Sometimes the patient's arm has developed severe pimples, but the high pressure syringe is still injecting contrast medium. In addition, because there is radioactivity in CT scan, in order to avoid the harm of radioactivity to medical personnel, the high pressure syringe must be operated in separate rooms. Therefore, scholars have also designed a mechanical device to monitor the arm pressure of patients, but there are still problems such as low sensitivity, no diversion device, no compartment alarm. In order to solve the above problems, a wireless intelligent anti leakage system is designed by using MSP430F5529[9-13], pressure sensor MS5837[14], wireless transmission chip NRF24L01+[15-16]. The system can monitor the high pressure injection process in real time.

Once the leakage of the contrast agent is detected, the wireless alarm and diversion switch can be started immediately. The system uses wireless transmission mode to alarm, which not only does not produce radiation to the medical staff, but also can timely inform the medical staff to close the high pressure syringe and timely dispose the leakage wound, which can effectively reduce the damage caused by the leakage of contrast agent to the patient.

## 2. Materials and methods

### *2.1. Design of system architecture*

The high pressure injection intelligent anti leakage system mainly consists of two parts: the main control system (Figure 1) and the terminal system (Figure 2). The main control system is installed in the high pressure syringe injection room, and the terminal system is installed in the medical personnel monitoring room. The main control system consists of pressure sensor module, micro controller module, sound light alarm module, diversion switch module and wireless communication module. The terminal system is composed of micro controller module, wireless communication module and sound light alarm module. The microcontroller module of the main control system can detect the pressure value of the injection point of the patient's arm in real time, and transmit the pressure information to the terminal system through the wireless communication module. The display screen of the main control system and the terminal system will simultaneously display the current pressure signal in real time. In case of leakage of contrast medium, the pressure value of pressure sensor will increase rapidly and exceed the preset voltage threshold. The micro control module will start the diversion module immediately and send the information to the terminal system in time through the wireless transmission module. After receiving the leakage information, the terminal system responds to the sound and light alarm quickly. The medical staff in the monitoring room will immediately close the high pressure syringe and deal with the patient's wound in time.

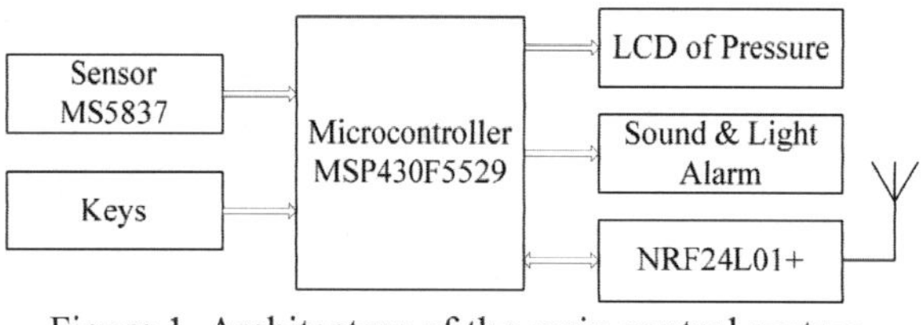


Figure 1. Architecture of the main control system.

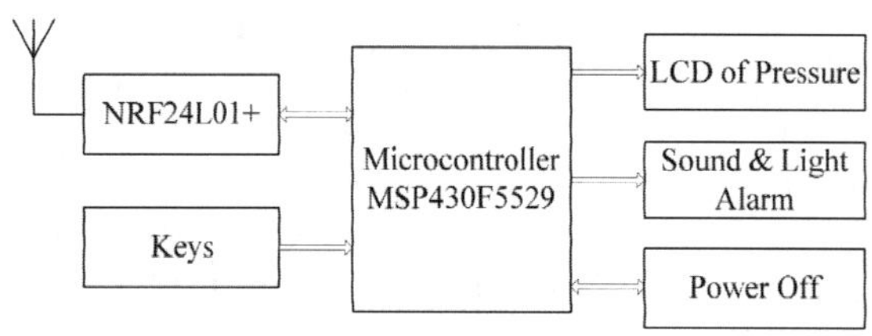


Figure 2. Architecture of the terminal control system.

### *2.2. Hardware platform of system*

*2.2.1. Module of pressure sensor* MS5837 is a new type of high linear digital pressure sensor designed, which is developed by TE Connectivity. It is mainly composed of piezoresistive sensor and interface conversion circuit. MS5837 can convert the uncompensated analog output voltage from piezoresistive pressure sensor to 24 bit digital voltage value, and then transmit the digital voltage signal to microcontroller for analysis and processing through I2C bus. The supply voltage range of MS5837 is 1.5 ~ 3.6 V, and the pressure detection range is 10 ~ 2000 mbar. In addition, MS5837 has the advantages of small size, low power consumption and strong anti-interference ability. This system uses MS5837 pressure sensor as the pressure detection module of the front end of the system. The sensor is directly fixed on the injection needle of the patient's arm in the form of cuff, which can monitor the swelling pressure caused by the leakage of contrast agent in real time, and transmit the pressure signal to the microprocessor for analysis in real time.

*2.2.2. Module of Microcontroller* MSP430F5529 chip of Texas Instruments is used as the microcontroller of the main control system and terminal system. In the main control system, the microcontroller receives the voltage signal from the pressure sensor in real time through I2C bus, and compares it with the preset threshold voltage. When MSP430F5529 detects that the detection voltage of the pressure sensor exceeds the original threshold voltage, it immediately starts the wireless transmission module to send an audible and visual alarm signal to the terminal system, and triggers the servo motor to conduct diversion switching, and diverts the remaining contrast agent in the high pressure syringe into the sterile bottle. At the same time, after receiving the alarm signal from the main control system, the terminal system immediately starts the audible and visual alarm to remind the medical staff in the monitoring room to close the high pressure syringe and deal with the patient's wound in time, so as to relieve the patient's pain caused by the leakage of contrast medium. In the process of contrast medium injection, the main control system and the terminal system will display the pressure data of the current patient's arm sensor and the preset threshold voltage in real time. In clinical application, medical staff can fine tune the initial threshold voltage in the main control system according to different individual characteristics of patients, so as to further improve the accuracy of contrast agent leakage prediction.

*2.2.3. Module of sound & light alarm* The main control system and terminal system are designed with sound and light alarm module. The alarm signal is transmitted by wireless way. In case of leakage of contrast medium, the medical staff in the monitoring room can receive the leakage alarm of high pressure syringe without going to the injection room, and then quickly close the syringe and deal with the patient's wound, so as to avoid the risk of radiation to the medical staff.

*2.2.4. Module of wireless communication* NRF24L01+ is a single chip wireless transceiver chip produced by Nordic semiconductor company. Its working channel is from 2.4 GHz to 2.5 GHz. NRF24L01+ chip's working voltage range is 1.9 ~ 3.6 V, the maximum input voltage value of input pin is 5 v. The data transmission rate is 1 Mbps, 2 Mbps. NRF24L01+ has 126 communication channels and 6 data channels, which can realize multi-point communication and frequency modulation. The data communication between NRF24L01+ wireless communication module and MSP430F5529 microcontroller is carried out through SPI bus .

*2.2.5. Module of diversion switch* The function of the diversion switch module is to divert the remaining contrast agent in the high pressure syringe into the sterile bottle when leakage occurs, and no longer continue to inject to the patient, so as to reduce the damage of leakage to the patient. Through analysis and test, the response speed and torque of servo motor with speed of 0.18 ~ 0.16 sec / 60 °, torque of 15 kg and working voltage of 5 ~ 7.2 V meet the design requirements of the system.

### *2.3. Software platform of system*

The system software is designed in the IAR embedded workbench development environment. After the system is powered on, the program completes initialization, sets timer, comparator and register, resets initial threshold voltage, and outputs PWM wave to control servo motor reset. When normal wireless communication is established between the main control system and the terminal system, the threshold voltage of reference voltage can be adjusted according to the individual characteristics of the patient. The microcontroller of the main control system detects the voltage value of the pressure sensor every 10 ms and compares it with the threshold voltage. If the output voltage of the pressure sensor exceeds the initial threshold voltage of the system, the main control system will start the audible and visual alarm, and control the servo motor to complete the diversion switching. At the same time, the alarm is sent to the terminal system through the wireless communication module, and the terminal system will start the audible and visual alarm to inform the medical staff in the monitoring room to deal with the leakage (Figure 3, Figure 4).

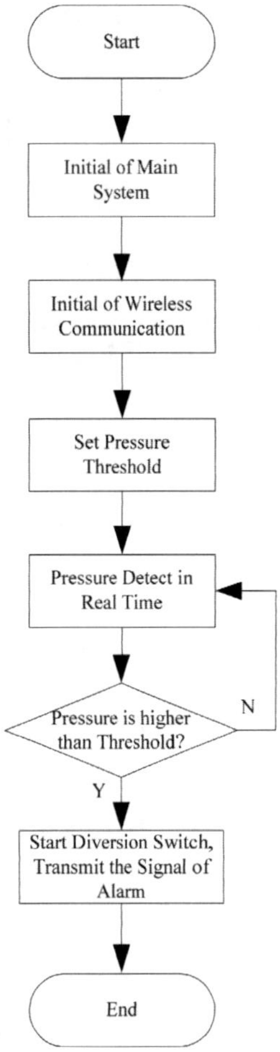


Figure 3. Program architecture of main system

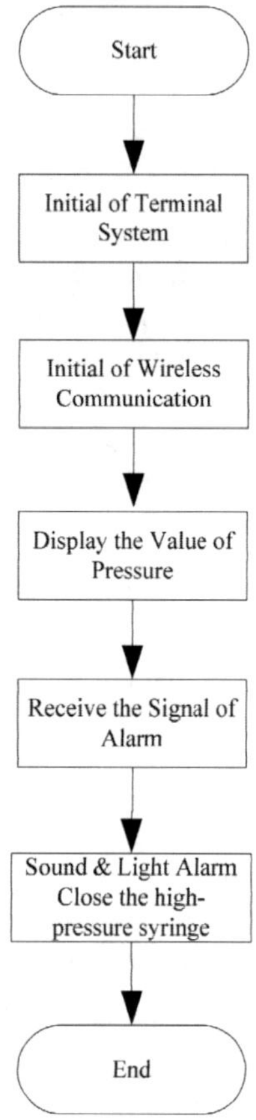


Figure 4. Program architecture of terminal system.

## 3. Results and discussion

### *3.1. Results*

Three male rabbits are selected with a body weight of 2.5 ± 0.5 kg. The experimental instruments include the main control system and terminal system of high pressure syringe intelligent anti leakage system, medical syringe, medical tape, scalp needle, rabbits, operating table . First, the rabbits are anesthetized with 30% Pentobarbital Sodium through the ear vein. After the domestic rabbit is anesthetized, it is fixed by prone position and its limbs are prepared. Then, the syringe is aspirated with normal saline and connected with scalp needle. Then the scalp is needled under the skin of the experimental limb (forelimb / hindlimb). At 0.5 cm in front of the needle, the pressure sensor of the main control system and the experimental limb are wound and fixed with medical tape. The fixed degree of tightness is subject to the initial pressure reaching the threshold value. At last, the normal saline is injected with syringe at different injection speed to make the experimental limb rapidly form a pimple. At this time, the main control system would display the current pressure value in real time and transmit it to the terminal system through wireless transmission. The experiment recorded whether the system can start the wireless alarm and diversion function in time. Through 12 experiments (3 × 4 limbs), the results show that the system can start the acousto-optic alarm and the diversion switch before the mound expansion diameter is 1.8 cm. The wireless transmission function is normal also.

### *3.2. Discussion*

At present, the reported devices also adopt this idea, but there are great differences in the specific implementation. In previous designs, the method was to detect the pressure through the pressure sensor and closes the high-pressure syringe by wireless method. However, its pressure sensor pressure threshold is set in manual rotation mode. This method is greatly influenced by subjective factors and its accuracy is not high. At the same time, the high-pressure syringe is closed by transmitting wireless signal. In reference [17], an anti-leakage device of high-pressure syringe is designed. The pressure signal is detected by the pressure sensor, but there is no wireless alarm function. In case of leakage, although the diversion function can be switched automatically, the medical staff can't be reminded to deal with the leakage at the first time, thus increasing the sufferings of patients.

In this paper, the skin is monitored intuitively and the sensory pressure is digitized, so it is easier to control the use threshold under different tissues. It can stop the source of injection and prevent the hypertonic liquid from entering the patient's body to cause extravasation. The device has the advantages of simple manufacture, low price and convenient use. It can be connected with various types of high-pressure syringes without modifying the internal software according to different high-pressure syringes. The control part has good compatibility and practicability, so it has a good prospect of clinical application. The application of this alarm device can avoid the harm of X-ray to non-essential personnel and control the extravasation of injection timely and effectively.

## 4. Conclusions

In order to solve the problem of liquid leakage in CT enhanced scanning of high pressure syringe during injection of contrast medium, an intelligent anti leakage system is designed based on MSP430F5529 micro control chip of Texas Instruments, MS5837 pressure sensor of TE Connectivity, NRF24L01+ wireless transmission chip of Nordic semiconductor. The system can monitor the swelling pressure of skin caused by the leakage of contrast agent in real time. If the leakage is found, the wireless alarm and diversion switch will be started immediately. The alarm signal is transmitted by wireless mode, which can not only avoid the radiation risk, but also grasp the leakage in the process of contrast agent injection. The results of rabbit experiment show that the system has good security and high reliability, and can effectively reduce the leakage of contrast agent, which has a certain practical application value. Although the basic functions of the system can be realized, there are still some areas to be improved, such as the inconvenient winding operation of sensors and limbs with tape, the system is not small enough, etc., which need to be further optimized later.

## Acknowledgments

This work was supported by the Research Project on Science and Technology of Chongqing Education Committee in 2020(KJQN202005803, KJQN202005801), the General Project of "14th five" Chongqing Educational Science in 2021(2021-JZ-024), the Seventh Batch of Pilot Projects of Comprehensive Education Reform in Chongqing in 2021(21JGS63), the Scientific Research Project of Chongqing Vocational College of Public Transportation (YSKY2021-02).

## References

[1] Liang, H.D., Yan, X., Huang, J. (2018) Application of iodinated contrast agent by the intravenous infusion and nursing in CT high pressure injector. Journal of Medical Imaging, 28(3):481-483.

[2] Ren, H.H. (2018)Nursing study on extravasation of contrast agent in high pressure injector with enhanced magnetic resonance scanning. Chinese Remedies & Clinics, 18:122-124.

[3] Xue, Y.C. (2016)The Prevention and nursing analysis of the leakage of contrast agent in CT enhanced scanning. Psychological Doctor, 22(1):178-179.
[4] Kong, J.Y. (2019)Application of risk management in the prevention of leakage and complications of contrast medium in high pressure syringe CT Enhanced Scan. Cardiovascular Disease Journal of Integrated Traditional Chinese and Western Medicine, 7(22): 72.
[5] Li, Y.L. (2018)Nursing method and effect analysis of intravenous cannula on prevention of extravasation of contrast agent for high pressure injector. Journal of Imaging Research and Medical Applications, V2(19):177-178.
[6] Liang, J.L. (2016) Analysis on the causes of contrast agent extravasation in enhanced CT scan and nursing intervention. Nursing of Integrated Traditional Chinese and Western Medicine, 2(3):26-28.
[7] Meng, J.Y. (2018) New progress in prevention and nursing of contrast agent extravasation. Nursing Research of China, 32(8):1193-1195.
[8] Jiang, X.H. (2017)Predictive nursing intervention for the prevention of CT contrast extravasation. China & Foreign Medical Treatment, 29: 162-164.
[9] TEXAS INSTRUMENTS. (2018) MSP430F552x, MSP430F551x Mixed-Signal Microcontrollers [EB/OL]. http://www.ti.com.
[10] TEXAS INSTRUMENTS. (2018) MSP430x5xx and MSP430x6xx Family User's Guide [EB/OL]. http://www.ti.com.
[11] Qu, Z. G. (2019) Design and Implementation of Weak Signal Detection System Based on MSP430. Computer Technology and Development,29(9), 146-151.
[12] Chen, Z., et al. (2018) Design of Remote DC Motor Control System Based on MSP430 MCU. Journal of Yancheng Institute of Technology(Natural Science Edition), 31(3),34-39.
[13] Zhao, M.J. (2017)Portable asthma monitoring system based on MSP430. Application of Electronic Technique, 43(9), 68-71,75.
[14] TE Connectivity. (2019) MS5837-02BA Ultra-Small Gel Filled Pressure Sensor [EB/OL]. http://www.te.com.cn.
[15] Wu, Q. (2018)The Design of Multi Hop Communication Protocol Based on nRF2401. Journal of Jiangxi Normal University (Natural Sciences Edition), 42(2), 149-154.
[16] Han, M.L. (2019)Design of intelligent temperature monitoring system for wireless bearing lubricating oil based on Arduino and Nrf24l01. Journal of Chinese Agricultural Mechanization, 40(5):167-171.
[17] Li, Z.D., et al. (2011)Design on Intelligent Diversion & Alarm System of High Pressure Syringe, Modern Electronics Technique, 34(16), 164-165,168.

# Analysis of stator rotor of rotating traveling wave ultrasonic motor

**Yi Cai[1*], Jingzhen Han[1], Yongfu Li[1], Yang Liu[1], Guangteng He[1]**

[1] Department of School of Mechanical and Electrical Engineering, North China Institute of Aerospace Engineering, Langfang 065000, China

[*]Corresponding author's e-mail: caiyi@nciae.edu.cn; 1399810399@qq.com

**Abstract.** As a new type of energy exchange device, the operation characteristics of ultrasonic motor is a hot issue in current research. In this paper, the thermo-electro-mechanical coupling model of the stator and rotor of the rotary traveling wave ultrasonic motor is constructed by finite element software. The frequency and amplitude of the working mode of the stator and rotor of the ultrasonic motor under different temperature conditions are studied, and the influence of the piezoelectric ceramic lamination on the working state of the ultrasonic motor is discussed. The following conclusions are drawn: with the increase of temperature, the resonant frequency of the working mode of the ultrasonic motor B09 decreases and the amplitude increases slightly; piezoelectric ceramic delamination can improve the amplitude-frequency characteristics of stator and rotor. Considering comprehensively, the superposition of two layers of piezoelectric ceramic plates is the most suitable structure.

## 1. Introduction

As a new type of energy exchange device, ultrasonic motor has many characteristics, such as fast response (millisecond level), free from magnetic field interference, no noise, small inertia, and power-off self-locking. Its development has attracted wide attention. However, due to the unstable operation performance, short service life and high processing cost, the industrialization process is limited [1]. One of the most important reasons for the above shortcomings is the high temperature generated by the friction between the stator and rotor during the operation of the ultrasonic motor. The increase of the body temperature will change the resonance frequency of the stator elastomer and the physical parameters of the friction material, resulting in the instability of the operation characteristics of the ultrasonic motor, and the high body temperature will also lead to the depolarization of the piezoelectric ceramics [2,3]. Therefore, it is necessary to consider the influence of temperature rise when studying the operating characteristics of ultrasonic motor. Many scholars have studied the operating characteristics of ultrasonic motors. Hou [4] carried out the vibration modal and harmonic response analysis of the ultrasonic motor stator through the finite element software, and preliminarily verified the finite element results with the dynamic signal analyzer. Liu [5] analyzed the contact stress between stator and rotor of traveling wave rotary ultrasonic motor, and optimized the contact state by adjusting the structural parameters of stator and rotor. Han [6] analyzed the electromechanical coupling of stator and rotor of traveling wave rotary ultrasonic motor, and optimized the structural parameters of stator elastomer. The above studies did not consider the influence of temperature rise conditions on the operating characteristics of ultrasonic motors. Zhou [7] established the theoretical model and heat source experiment of the heating capacity of the ultrasonic motor, and determined that the heating of the motor mainly came from the heating of piezoelectric ceramics and the heating of friction, but did not discuss the influence of the heating of the ultrasonic motor. Lu [8] studied the stator characteristics of ultrasonic motor under thermo-electro-mechanical coupling field, and obtained the variation law of stator vibration under temperature conditions. But in the actual work of ultrasonic motor, synchronous interaction between stator and rotor, the stator and rotor are discussed separately, will inevitably ignore the pre-pressure effect of stator tooth surface.

In view of this, this paper takes the stator and rotor of rotary ultrasonic motor as the research object, carries on the thermo-electro-mechanical coupling analysis, studies its operation characteristics under different temperature conditions, and discusses the influence of laminated piezoelectric ceramic plate at the bottom of the stator on the resonant frequency and stator tooth amplitude, which lays the foundation for the research and design of ultrasonic motor.

## 2. Thermo-electro-mechanical coupling model

### 2.1. Structure of stator and rotor

The schematic diagram of the stator and rotor structure of the rotary traveling wave ultrasonic motor is shown in Fig.1, which is mainly composed of piezoelectric ceramics, stator, rotor and friction layer. Piezoelectric ceramic pieces adhere to the bottom edge of the stator according to the polarization direction, forming a piezoelectric ceramic ring. The rotor is closely bonded with the stator tooth through a certain pre-pressure. In order to increase the friction between the stator tooth and the rotor, a friction layer is set on the contact surface of the rotor, and the diameter of the stator is slightly larger than that of the rotor, which ensures the complete contact between the friction surfaces. The structure size of stator and rotor is shown in table 1.

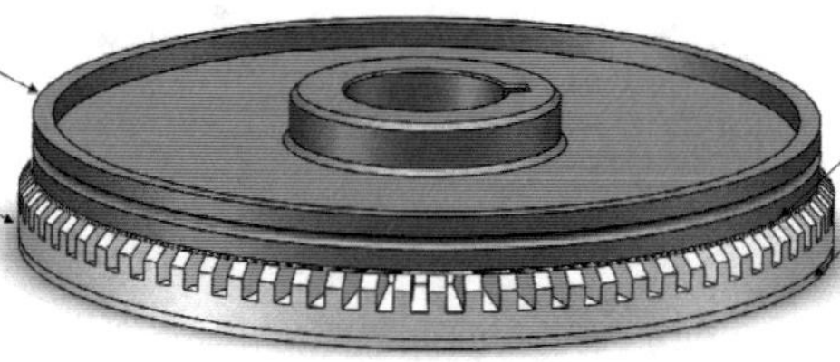


Figure 1. Diagram of stator and rotor structure

Table 1. Structure parameters of stator and rotor

| Parts | outside diameter /mm | inside diameter /mm | thickness /mm | tooth number |
|---|---|---|---|---|
| Stator | 60 | 18 | 4.5 | 72 |
| piezoelectric ceramics | 60 | 44 | 0.5 | / |
| Rotor | 58 | 20 | 4.6 | / |

The operation principle of ultrasonic motor is that when the piezoelectric ceramic is subjected to an external electric field with opposite polarization direction, the piezoelectric ceramic sheet will shrink, and vice versa, the piezoelectric ceramic sheet will elongate. Using this characteristic of piezoelectric ceramics, under the action of high frequency continuous transformation of external electric field, the piezoelectric ceramic ring drives the stator reciprocating bending deformation. At this time, any particle on the stator surface moves elliptically. Due to the adhesion between the rotor and the stator tooth surface, the elliptical motion of the stator particle makes the rotor rotate through friction [9].

### 2.2. Theoretical model

Stator kinetic energy can be expressed as:

$$T_s = \frac{1}{2}\dot{x}^T M \dot{x} \tag{1}$$

In formula (1), $\dot{x}$is the derivative of modal coordinate $x$ to time, $M$ is the mass matrix of stator.
The elastic potential energy of stator can be expressed as:

$$V_s = \frac{1}{2}x^T K x \tag{2}$$

In formula (2), $K$ is the stator modal stiffness.
The potential energy of piezoelectric ceramics can be expressed as:

$$W_c = \frac{1}{2}v^T C v + \frac{1}{2}v^T \psi x + \frac{1}{2}v^T \xi x \tag{3}$$

In formula (3), $C$ is the capacitance matrix, $\xi$ is the thermoelectric matrix, and $\psi$ is the modal excitation force.

Elastic potential energy of piezoelectric ceramics can be obtained from the second kind of nonadiabatic piezoelectric equation.

$$V_c = \frac{1}{2}x^T K_p x + \frac{1}{2}x^T \psi v + \frac{1}{2}x^T \theta T \tag{4}$$

In formula (4), $K_p$ is modal stiffness, and $\theta$ is modal thermal stress.

The above energy expression is used to derive the thermo-electro-mechanical coupling equation of stator and rotor by Hamilton variational principle.

$$\begin{cases} M\ddot{x} + Kx = \psi v + \frac{1}{2}\theta T \\ \psi v + Cv = Q - \frac{1}{2}\xi T \end{cases} \tag{5}$$

In formula (5), $Q$ is the virtual work of piezoelectric ceramic charge in the electric field.

## 3. Finite element analysis

The finite element model is shown in Fig. 2. The stator and rotor structure adopts tetrahedral mesh, while the piezoelectric ceramic and friction layer adopts hexahedral mesh, with a total of 10390 meshes. Considering that the piezoelectric ceramic plate is subjected to the coupling effect of thermal field, electric field and stress field, solid5 element is adopted, while solid187 element is adopted for other components. Setting the contact between the stator tooth surface and the friction layer as sliding friction, and defining the friction coefficient as 0.17. Binding constraints are adopted between stator and piezoelectric ceramic, ignoring the influence of adhesive layer. Material parameters are shown in Table 2.

Table 2. Material parameters

| Parts | Material | Density g/cm$^3$ | elastic modulus/Pa | poisson ratio | thermal expansion coefficient /°C |
|---|---|---|---|---|---|
| Stator | Tin bronzes | 8400 | $1.13\times10^{10}$ | 0.3 | $1.8\times10^{-5}$ |
| Rotor | Duralumin | 2770 | $7.1\times10^{10}$ | 0.33 | $2.3\times10^{-5}$ |
| Friction layer | PTFE | 2500 | $1.2\times10^{10}$ | 0.37 | $11.6\times10^{-5}$ |
| piezoelectric ceramics | PZT-4 | 7500 | $7.65\times10^{10}$ | 0.32 | $2.1\times10^{-5}$ |

In this case, 200N pre-pressure is applied on the middle convex surface of the rotor, and three bolt holes of the stator are constrained with full freedom. Since the polarization direction of the adjacent piezoelectric ceramics is opposite, it is necessary to re-establish the coordinate system opposite to the global coordinate system, so as to define the piezoelectric ceramics with opposite polarization direction in this coordinate system, and define the bond surface between the piezoelectric ceramics and the stator as the zero potential surface by default. The temperature conditions of stator and rotor contact surface are set as 30°C, 50°C, and 70°C respectively to simulate the temperature rise of ultrasonic motor during operation. The solution range is limited to the first 40 modes of 20kHz ~50kHz.

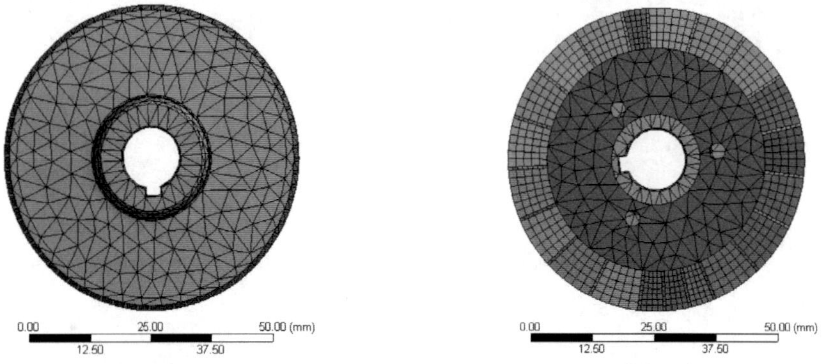


(a) front (b)back

Figure 2. Finite element model of stator and rotor

## 4. Result and discussion

Taking the modal results at 50°C as an example (see fig.3), the bending vibration modes of B08 and B09 appear in the first 40 modes of 20 ~ 50kHz frequency, which represent the vibration modes of 8 and 9 bending waves on the stator surface, respectively. The resonant frequencies of B08 are 23637Hz and 23653Hz, and the resonant modes of B09 are 29172Hz and 29177Hz, and the frequencies of the two degenerate modes are similar. However, from the point of view that the two-phase symmetry can improve the elastic excitable region, the traveling wave ultrasonic motor generally uses odd mode. Therefore, it is appropriate to choose B09 mode as the working mode of the ultrasonic motor.

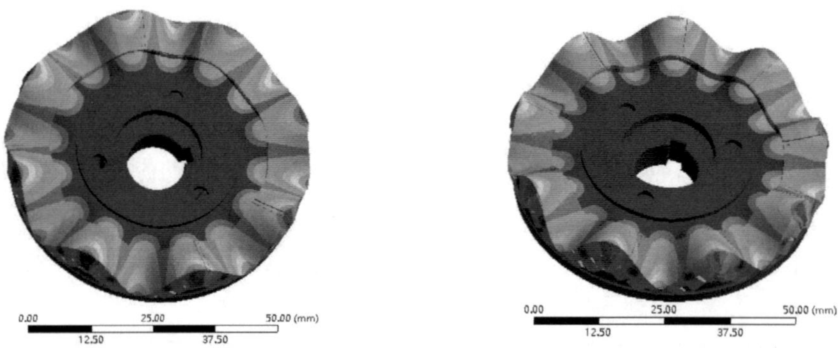


(a) B08 vibration mode (b) B09 vibration mode

Figure 3. B08 and B09 vibration modes of stator and rotor at 50°C

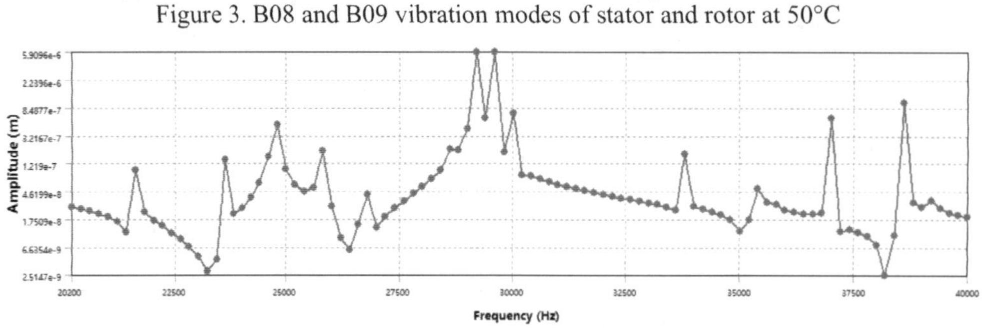


Figure 4. Harmonic response analysis of stator at 50°C

Modal analysis cannot determine whether the stator and rotor structure is applicable, and the response amplitude of the stator and rotor structure at a specific frequency should be determined through harmonic response analysis. A point on the surface of the sub-tooth is selected to calculate the amplitude-frequency characteristics of the point under a given excitation. As shown in Fig.4, when the thermal condition is 50°C, the amplitude of the stator tooth is the maximum at 29200 Hz, which is close to the resonant frequency of the B09 mode, and the maximum is 5.34μm.

Fig.5 shows the comparison of B09 modal frequencies and amplitudes of stator and rotor structures at 30°C, 50°C and 70°C. It can be seen from the figure that with the increase of temperature conditions, the resonant frequency of B09 mode of the stator and rotor structure decreases by 750Hz, and the amplitude increases by 0.09μm. The reason is that the rotor is attached with a friction layer, and the thermal conductivity of the friction layer is poor, so most of the heat generated by the friction between the friction layer and the surface of the stator tooth is transferred to the stator. The stator elastomer is subjected to thermal stress, and the material properties also change after the temperature increases. So, the resonance frequency is reduced and the amplitude is slightly larger.

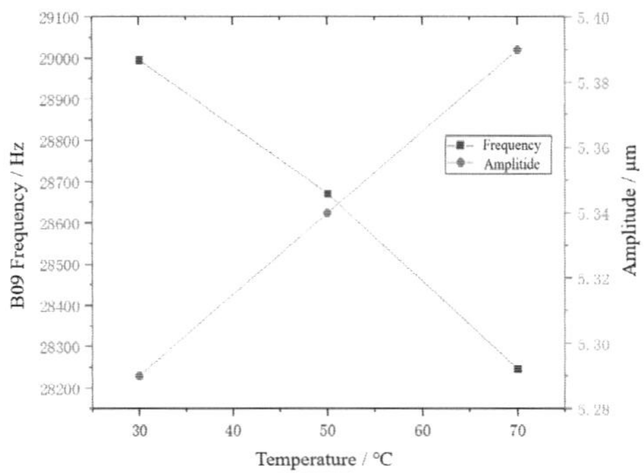


Figure 5. Frequency amplitude comparison under different temperature conditions

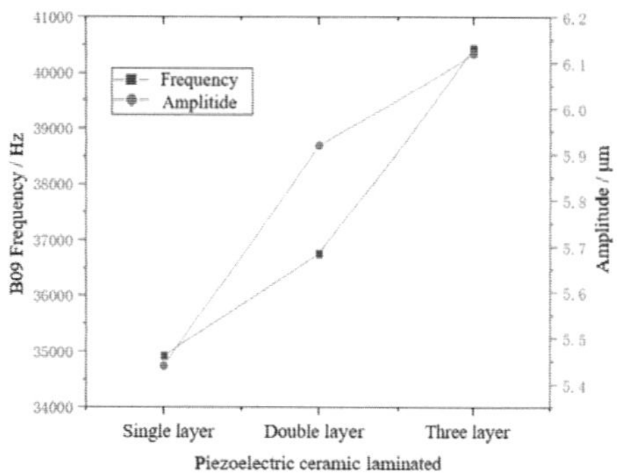


Figure 6. Frequency amplitude comparison of different stacking structures

## 5. Influence of laminated piezoelectric ceramics

Piezoelectric ceramic plate is the main energy exchange component of ultrasonic motor and the core component of ultrasonic motor. Its performance directly affects the vibration characteristics of the stator. Therefore, in view of the current single layer piezoelectric ceramic plate can affect the degree of stator is limited, the structure of laminated piezoelectric ceramic plate is proposed. When piezoelectric ceramics are designed, they tend to be thin, but piezoelectric ceramics cannot be too thin, too thin piezoelectric ceramics are prone to brittle fracture under high frequency vibration; the piezoelectric ceramic plate is not too thick, and the influence of too thick piezoelectric ceramic plate on the characteristics of ultrasonic motor is not obvious. The 1mm, 0.5mm and 0.33mm piezoelectric ceramics were selected for layered stacking. The 0.33mm piezoelectric ceramics were stacked into three layers, 0.5mm stacked into two layers, and 1mm was a single layer. Then the vibration frequency and amplitude under the working mode of B09 were analyzed.

Fig.6 shows that the resonance frequency and amplitude of B09 mode change with the change of piezoelectric ceramic stack structure. When the piezoelectric ceramics are stacked in double layers, the resonant frequency of B09 mode increases by 1840Hz compared with that of the single-layer piezoelectric ceramics, and the amplitude increases by 0.48μm. When the piezoelectric ceramic plate is stacked with three layers, the resonant frequency of B09 mode is 3682Hz higher than that of the double-layer stacked piezoelectric ceramic plate, and the amplitude increases by 0.2μm. With the increase of the piezoelectric ceramic layer, the resonant frequency increases significantly, which is because the same piezoelectric ceramic layer is stacked layer by layer which is prone to local resonance. With the increase in the number of piezoelectric ceramic slices, the resonant frequency increases significantly, because the same piezoelectric ceramic slices are stacked in layers, which is prone to local resonance. If the piezoelectric ceramic slice is superimposed by three layers, the process difficulty will be greatly increased, and the amplitude increase is not obvious. In summary, the piezoelectric ceramic plate can be designed by two layers of superposition structure. If the piezoelectric ceramic slice is superimposed by three layers, the process difficulty will increase greatly, and the amplitude increase is not obvious. In summary, the piezoelectric ceramic plate can be designed by two layers of superposition structure.

## 6. Conclusion

In this paper, ANSYS software is used to analyze the thermo-electro-mechanical coupling of stator and rotor of ultrasonic motor, and the working mode and maximum amplitude of stator under different temperature conditions are calculated. Through comparison, it is found that the resonant frequency of stator working mode decreases and the amplitude increases slightly with the increase of temperature. Limited to the single-layer piezoelectric ceramic plate, it is also proposed to use the laminated structure of piezoelectric ceramic plate to improve the frequency and amplitude characteristics of the stator. Through the finite element analysis, it can be seen that the laminated structure of piezoelectric ceramic can improve the amplitude and frequency characteristics of the stator and rotor. However, considering comprehensively, the superposition of two layers of piezoelectric ceramic plates with the same thickness is the most suitable.

## Acknowledgments

This work was financially supported by the Graduate Student Innovation Funding Project of North China Institute of Aerospace Engineering (YKY-2020-06).

## References

[1] Zhou, P., Yao, Z. (2009) Application of finite element method in temperature field simulation of motor. WORLD SCI-TECH R&D, 31(6): 1095-1097.
[2] Chen, C., Zeng, J., Zhu, H. (2009) Parameter optimization design of stator of miniature traveling wave ultrasonic motor. China Mechanical Engineering, 20(5): 568-572.
[3] Zhou, P., Yao, Z., Zhou, F Z. (2009) Estimation method of traveling wave ultrasonic motor calorific value. Piezoelectrics & Acoustooptics, 31(1): 132-135.
[4] Hou, F., Ji, Y B. (2012) Vibration modal and harmonic response analysis of stator of ring ultrasonic motor. Mechanical & Electrical Technology, 35(5): 85-88.
[5] Liu, J., Ji, Y B., Zhao, B W. (2019) Contact analysis of stator and rotor of traveling wave rotating ultrasonic motor[J]. Journal of Xiamen University (Natural Science), 58(1): 145-149.
[6] Han, X., Han, J C., Cai, Y., et.al. (2019) Finite element analysis of fixed rotor electromechanical coupling for rotary traveling wave ultrasonic motor. Modular Machine Tool & Automatic Manufacturing Technique, 543(5): 55-58, 67.
[7] Zhou, P. (2016) Performance test and analysis of traveling wave ultrasonic motor. New Technology & New Process, 339(3): 114-117.
[8] Lu, X L., Zhao, C S. (2011) Thermo-electromechanical coupling characteristics of ultrasonic motor stator. Journal of Vibration Engineering, 24(5): 522-528.
[9] Su, G B., Han, J C., Qi, X Z., Cai, Y. (2017) Design and finite element analysis of fixed rotor of traveling wave rotary ultrasonic Motor. Electric Machines and Control Application, 44(12):100-106.

 IOP Publishing
**2264** (2022) 012008 doi:10.1088/1742-6596/2264/1/012008

# Taking 2oo3 as an example to study the improvement of Markov model considering periodic function test

**Peng Chen** [1], **Qinbei Lin** [2], **Xuefeng Han**[3]*

[1] College of Safety Science and Engineering, Nanjing Tech university, Nanjing 211816, China

[2] College of Safety Science and Engineering, Nanjing Tech university, Nanjing 211816, China

[3] College of Safety Science and Engineering, Nanjing Tech university, Nanjing 211816, China

*Corresponding author's e-mail:safety@njtech.edu.cn * Xuefeng Han

**Abstract:** In order to make the safety instrumented system meet the functional safety requirements of the classification in a longer period of time, and considering the model incompleteness of the original Markov model, we propose to add the consideration of periodic functional testing to the original Markov model for modeling. The results show that the PFDavg of the safety instrumented system with periodic functional testing is lower than that of the safety instrumented system without periodic functional testing, and the Markov model with periodic functional testing is more comprehensive and more consistent with the engineering reality.

## 1. Introduction

In recent years, with the rapid economic development and the increasing demand for living and production, new coal chemical, petrochemical, metallurgical, and electric power industries have been expanding their enterprise production scale in order to improve their profitability [1]. These industries often involve flammable and explosive, toxic and hazardous materials and products, accompanied by special process requirements such as high temperature and pressure, which can bring great danger to the enterprises and surrounding communities in case of fire, explosion or leakage. In order to reduce the risk of these industrial processes, safety instrumentation systems were created [2]. The role of safety instrumentation systems is to monitor the status of the production process and, when a hazardous condition is detected, to immediately and automatically perform its defined safety functions to convert the production process to a safe state. In case of device or system failure, it mitigates the consequences of accidents and the impact caused by hazardous events [3].As early as the 1990s, scholars in Europe and North America began to study the functional safety assessment of safety instrumentation systems, and now a relatively complete functional safety assessment system for safety instrumentation systems has been formed. China's research in this area started late, and problems such as imperfect risk analysis, improper design selection, and unreasonable setting of redundant structures are still prevalent when functional safety assessment of safety instrumentation systems in the design stage [4].

In addition, even though safety instrumentation systems can significantly reduce the frequency of accidents in industrial processes, they cannot guarantee that the system will always be safe. When the safety instrumentation system itself fails, it will not respond properly and perform the required safety functions, thus causing accidents. At the same time, the "Guidance of the State Administration of Safety Supervision on Strengthening the Management of Chemical Safety Instrumentation Systems", the General Administration of Safety Supervision No. 3 (2014) 116, proposed that since 2018, all new chemical installations and hazardous chemical storage facilities involving key regulated hazardous chemicals, key regulated hazardous chemical processes and major sources of danger should be designed with safety instrumentation systems [5].

Therefore, it is important to conduct safety assessment research on the functional safety of safety instrumentation systems and improve the methods for grading and verifying their safety integrity levels to ensure the safety of production processes in petrochemical, new coal chemical, power, metallurgical and other industries.

## 2. Safety integrity classification method

In this paper, an FTA-BN method is proposed as a method for safety integrity level grading. the FTA-BN method is to first calculate the probability of occurrence of event consequences quantitatively using a fault tree after performing HAZOP analysis [6]. Considering the subjectivity and uncertainty of the reliability data used in the quantitative calculation, the fault tree is mapped to a Bayesian network, and the fault tree analysis is corrected using the Bayesian network to reduce the subjectivity and uncertainty of the initial data. At the same time, considering that the fault tree analysis method is based on the independence between events to calculate the top event occurrence probability, which is not in line with the actual situation that the equipment may interact with each other in the industrial production process, i.e., the hazardous events will affect each other, the calculation results are corrected using the conditional probability table of Bayesian network to make the calculation results more consistent with the actual engineering applications [7]. Finally, the confirmation of PFDavg is performed to determine whether the safety instrumentation system needs to be set up and the safety integrity level is calculated.

## 3. Security integrity level verification study

### *3.1. Markov model considering cyclic functional testing*

The Markov model represents the system through state changes and the normal and failure states of the system can be established simultaneously in the model. It is able to include a wide range of factors affecting reliability, is not affected by the subordination relationships between devices, and has high computational accuracy [8].

Safety instrumented systems are passive systems, and to detect hidden failures in safety instrumented systems in a timely manner, it is necessary to perform cyclic functional tests. By performing periodic function tests, dangerous failures in the system that are not detected can be found and repaired, effectively preventing the safety instrumented system from failing to perform its safety functions correctly in the event of dangerous conditions. That is, periodic functional testing can ensure to a greater extent that the safety instrumented system maintains a low failure rate over a long period of operation and meets the SIL rating required for safety integrity level classification [9].Therefore, Markov model is selected as the method for safety integrity level verification, and on the basis of considering failure mode, redundant voting structure, common cause failure, and diagnostic coverage, a more complete Markov model is established by adding consideration of cyclic functional testing. In this paper, we will study from the perspective of model integrity and take 2oo3 as an example to establish a Markov model considering cyclic functional testing [10].

### *3.2. A 2oo3 redundant structural model considering periodic functional testing*

The Markov model of the 2oo3 redundant structure considering periodic functional tests is shown in Figure 1.

In the 2oo3 redundant structure, the system can perform safety functions as long as two of all channels are in the normal state. The 12 operating states of the structure are shown in Table 1.

Table 1.Working status of 2oo3 redundant structure

| Status | Description |
|---|---|
| 0 | Normal State |
| 1 | Detected security failure on one of the channels |
| 2 | Undetected safety failure on one of the channels |
| 3 | Dangerous failure detected in one of the channels |
| 4 | Undetected hazardous failure on one of the channels |
| 5 | Detected safety failure on one of the channels and detected hazardous failure on one of the other two channels |
| 6 | Detected safety failure on one of the channels and undetected hazardous failure on one of the other two channels |
| 7 | Undetected safety failure in one channel, detected hazardous failure in one of the other two channels |
| 8 | Undetected safety failure on one channel, undetected hazardous failure on one of the other two channels |
| 9 | A safety failure occurs in the system |
| 10 | A detected hazardous failure occurs in the system |
| 11 | System has an undetected hazardous failure |

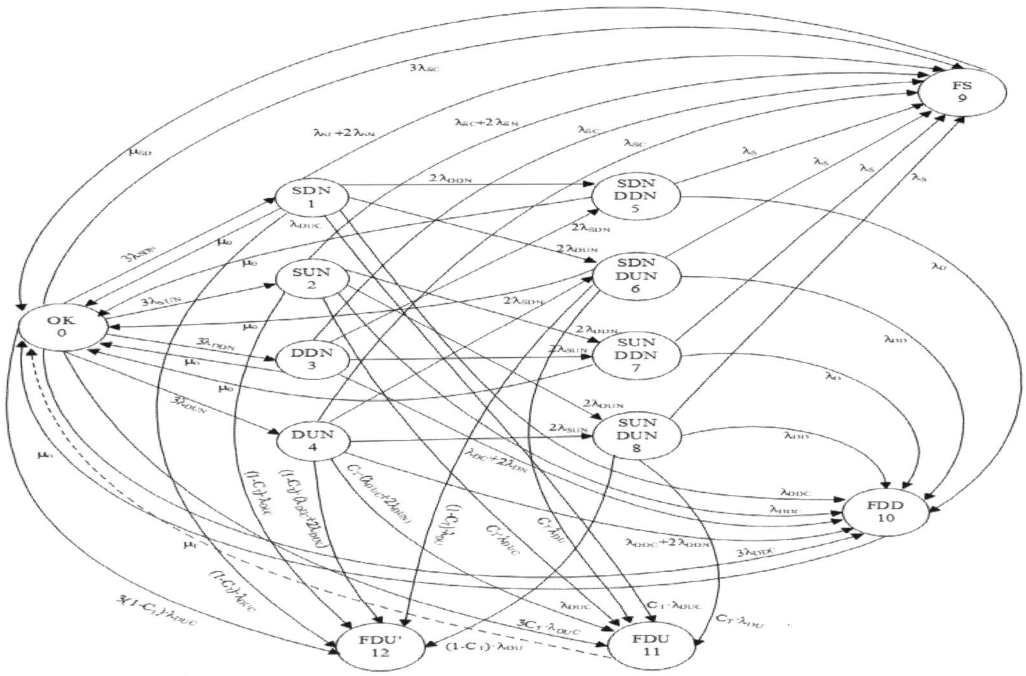


Figur 1. 2oo3 Markov model considering periodic functional test

The 2oo3 redundant structural Markov model state transfer matrix P considering periodic functional tests is shown in equation (1).

$$
P=\begin{bmatrix}
1-\Sigma & 3\lambda_{SDN} & 3\lambda_{SUN} & 3\lambda_{DDN} & 3\lambda_{DUN} & 0 & 0 & 0 & 0 & 3\lambda_{SC} & 3\lambda_{DDC} & 3C_T\cdot\lambda_{DUC} & 3(1-C_T)\cdot\lambda_{DUC}\\
\mu_0 & 1-\Sigma & 0 & 0 & 0 & 2\lambda_{DDN} & 2\lambda_{DUN} & 0 & 0 & \lambda_{SC}+2\lambda_{SN} & \lambda_{DDC} & C_T\cdot\lambda_{DUC} & (1-C_T)\cdot\lambda_{DUC}\\
0 & 0 & 1-\Sigma & 0 & 0 & 0 & 0 & 2\lambda_{DDN} & 2\lambda_{DUN} & \lambda_{SC}+2\lambda_{SN} & \lambda_{DDC} & C_T\cdot\lambda_{DUC} & (1-C_T)\cdot\lambda_{DUC}\\
\mu_0 & 0 & 0 & 1-\Sigma & 0 & 2\lambda_{SDN} & 0 & 2\lambda_{SUN} & 0 & \lambda_{SC} & \lambda_{DC}+2\lambda_{DN} & 0 & 0\\
0 & 0 & 0 & 0 & 1-\Sigma & 0 & 2\lambda_{SDN} & 0 & 2\lambda_{SUN} & \lambda_{SC} & \lambda_{DDC}+2\lambda_{DDN} & C_T\cdot(\lambda_{DUC}+2\lambda_{DUN}) & (1-C_T)\cdot(\lambda_{DUC}+2\lambda_{DUN})\\
\mu_0 & 0 & 0 & 0 & 0 & 1-\Sigma & 0 & 0 & 0 & \lambda_S & \lambda_D & 0 & 0\\
\mu_0 & 0 & 0 & 0 & 0 & 0 & 1-\Sigma & 0 & 0 & \lambda_S & \lambda_{DD} & C_T\cdot\lambda_{DU} & (1-C_T)\cdot\lambda_{DU}\\
\mu_0 & 0 & 0 & 0 & 0 & 0 & 0 & 1-\Sigma & 0 & \lambda_S & \lambda_D & 0 & 0\\
0 & 0 & 0 & 0 & 0 & 0 & 0 & 0 & 1-\Sigma & \lambda_S & \lambda_{DD} & C_T\cdot\lambda_{DU} & (1-C_T)\cdot\lambda_{DU}\\
\mu_{SD} & 0 & 0 & 0 & 0 & 0 & 0 & 0 & 0 & 1-\mu_{SD} & 0 & 0 & 0\\
\mu_0 & 0 & 0 & 0 & 0 & 0 & 0 & 0 & 0 & 0 & 1-\mu_0 & 0 & 0\\
0 & 0 & 0 & 0 & 0 & 0 & 0 & 0 & 0 & 0 & 0 & 1 & 0\\
0 & 0 & 0 & 0 & 0 & 0 & 0 & 0 & 0 & 0 & 0 & 0 & 1
\end{bmatrix} \quad (1)
$$

## 4. 2oo3Comparison of redundant structure models

In this section, the original model and the improved model of 2oo3 redundancy structure are analyzed computationally. The time TD for online repair of instrument failure is 8 hours, i.e., MTTR (mean time to repair) is 8, and the repair rate of online repair is; the time for system restart after one failure-free stop is 24 hours, i.e., SD is 24, then. Let the periodic functional test coverage CT be 90%, and the common cause failure correction parameter C is taken as 2.4. 2oo3 failure data are shown in Table 2.

Table 2. 2oo3 Failure data

| Failure Type | 2oo3 |
|---|---|
| Detectable safety failure rate$\lambda_{SD}$ | $2.00\times10^{-7}$ |
| Undetected safety failure rate$\lambda_{SU}$ | $3.00\times10^{-7}$ |
| Detectable Hazardous Failure Rate$\lambda_{DD}$ | $5.00\times10^{-7}$ |
| Undetected hazardous failure rate$\lambda_{DU}$ | $3.00\times10^{-7}$ |

The PFDavg was calculated by Matlab for each year within 20 years, and the PFDavg of the original model and the improved model were compared for each year, and the PFDavg data comparison is shown in Table 3, and the 2oo3 comparison results are shown in Figure 3.

Table 3. Comparison of PFDavg data

| | 1st year | 2nd year | 3rd year | 4th year | 5th year | 6th year | 7th year | 8th year | 9th year | 10th year |
|---|---|---|---|---|---|---|---|---|---|---|
| Original | 4.78E-4 | 9.62E-4 | 1.453 E-3 | 1.952 E-3 | 2.456 E-3 | 2.968 E-3 | 3.486 E-3 | 4.010 E-3 | 4.540 E-3 | 5.076 E-3 |
| Improvements | 4.768 E-4 | 4.926 E-4 | 5.083 E-4 | 5.241 E-4 | 5.398 E-4 | 5.556 E-4 | 5.714 E-4 | 5.871 E-4 | 6.029 E-4 | 6.186 E-4 |
| | 11th year | 12th year | 13th year | 14th year | 15th year | 16th year | 17th year | 18th year | 19th year | 20th year |
| Original | 5.618 E-3 | 6.166 E-3 | 6.719e -3 | 7.277e -3 | 7.841e -3 | 8.410 E-3 | 8.984 E-3 | 9.563 E-3 | 1.015 E-2 | 1.073 E-2 |
| Improvements | 6.344 E-4 | 6.502 E-4 | 6.659 E-4 | 6.817 E-4 | 6.974 E-4 | 7.132 E-4 | 7.290 E-4 | 7.447 E-4 | 7.605 E-4 | 7.762 E-4 |

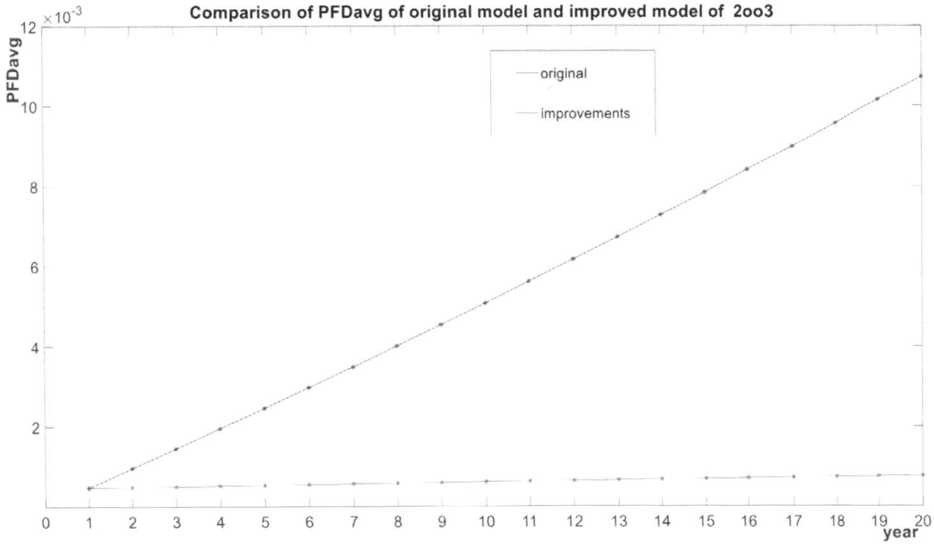


Figure 2. Comparison of PFDavg of original model and improved model of 2oo3

## 5. Comparison of original Markov model and Markov model safety integrity level verification considering cyclic functional testing

Let the time TD for online repair of instrument failure be 8 hours, the time for system restart after one trouble-free stop be 24 hours, the periodicity of functional testing be 1 year, the coverage of functional testing CT be 90%, and the system reach the limit state in 10 years. According to the original Markov model, the PFDavg of each component of the safety instrumented system can be found in 10 years. in the Markov model considering periodic functional testing, the failure data of the equipment is re-substituted and calculated [11]. the comparison of the PFDavg calculation results for 2oo3 is shown in Table 3.

Table 4. 2oo3 Calculation results

| 2oo3PFDavg | | | | | | | | | | |
|---|---|---|---|---|---|---|---|---|---|---|
| Time | 1 | 2 | 3 | 4 | 5 | 6 | 7 | 8 | 9 | 10 |
| Original PFDavg (×10-4) | 4.61 | 9.15 | 13.7 | 18.3 | 23.0 | 27.6 | 32.3 | 37.0 | 41.7 | 46.4 |
| Improvements PFDavg (×10-4) | 4.48 | 4.63 | 4.77 | 4.92 | 5.07 | 5.22 | 5.37 | 5.51 | 5.66 | 5.81 |

The calculation results show that the PFDavg of the safety instrumented system with cyclic functional testing is lower than that of the safety instrumented system without cyclic functional testing, which means that the PFDavg of the safety instrumented system can be significantly reduced by performing cyclic functional testing, and because even with cyclic functional testing, there are still some failures that are not detected and cannot be repaired, so the PFDavg of the safety instrumented system with cyclic functional testing still grows gradually. Therefore, the PFDavg of a safety instrumented system with cyclic functional testing will still grow gradually. The Markov model that considers periodic functional testing is more comprehensive and more consistent with the engineering reality.

## 6. Conclusion

In this paper, the Markov model with the highest computational accuracy is selected as the method for safety integrity level verification. It is pointed out that previous research on Markov model has less consideration of model integrity, and it is proposed that Markov model is constructed by considering periodic functional tests, and models of 1oo1, 1oo2 and 2oo3 redundant structures are constructed. From the validation results, it can be seen that the Markov model considering periodic functional tests is more comprehensive and more in line with engineering reality.

## Acknowledgements

I would like to thank my supervisor, Mr. Han Xuefeng, for teaching me and urging me to complete this dissertation. In addition, I would like to thank my sisters and brothers who taught me to learn matlab. I would also like to thank Sister Lin and Brother Huang for helping me with the outline guidance and the final touches of the paper. I am lucky to have the opportunity to accomplish such a thing at the graduate level, which means a lot to my life.

## References

[1] Yu Zeke,Li Bin,Yao Yutian. Design of petrochemical safety instrumentation system[J]. ChinaNewTechnologyNewProducts,2017(09):27-28.DOI:10.13612/j.cnki.cntp.2017.09.016.
[2] Jin Jianghong, Wu Zongzhi, Hu Cellophane. A study of IEC61508, the basic standard for functional safety [J]. China Safety Production Science and Technology, 2009, 5(2): 71~75.
[3] Wu, N. N. Research on the reliability modeling method of safety instrumentation system based on Markov model[D]. Zhejiang.
[4] Shi Jianfeng. The necessity of safety integrity level (SIL) assessment verification for safety instrumented systems (SIS)[J]. China Chlor-Alkali,2019(05):37-38.
[5] Guidance from the State Administration of Safety Supervision on strengthening the management of chemical safety instrumentation systems [J]. Total corrosion control,2014,28(12):14-15+43.
[6] Shen Xiaoyan,Xiao Danlei,Wei Shanshan,Sui Xinyu,Zhu Zigeshuo. Accident analysis of dangerous goods road transport based on FTA-BN [J/OL]. Journal of Safety and Environment:1-9[2021-11-19]. https://doi.org/10.13637/j.issn.1009-6094.2021.0821.
[7] Haifeng Bian,Jun Zhang,Ruixue Li,Huanhuan Zhao,Xuexue Wang,Yiping Bai. risk analysis of tripping accidents of power grid caused by typical natural hazards based on FTA-BN model[J]. Natural Hazards,2021(prepublish):
[8] Sassi Imad,Anter Samir,Bekkhoucha Abdelkrim. A graph-based big data optimization approach using hidden Markov model and constraint satisfaction problem[J]. Journal of Big Data,2021,8(1):
[9] Ji K, Wang L, Wang MF, Wang Xianzhao, Xu P, Liu XL. Markov simplified model for safety evaluation of safety instrumentation system based on inspection test[J]. Safety and Environmental Engineering,2021,28(05):18-26.
[10] Wang Huifeng, Zhang Heng, Tang Chenhui, et al. Failure rate analysis of diagnostic module based on Markov model in SIS[J]. Chemical Automation and Instrumentation, 2013, (02): 68-71.
[11] Wei Xueyong. Markov model-based security attack and defense strategies for smart campus networks[J]. Electronic design engineering, 2021, 29(15):72-76.

# Performance Analysis of Unmanned Aerial Vehicle Assisted Fiber-based Visible Light Communication System

**Lisu Yu, Chaoliang Liu, Jiajia Qian, Yuhao Wang*, and Zhenghai Wang***

School of Information Engineering, Nanchang University, Nanchang, Jiangxi, 330031, China

*Corresponding author's e-mail: wangyuhao@ncu.edu.cn; wangzhhai@126.com

**Abstract:** Employing unmanned aerial vehicle (UAV) in fiber-based visible light communication (FVLC) can not only enhance the flexibility of visible light communication, but also provide users with sufficient lighting conditions. In the downlink UAV-assisted FVLC (UFVLC) system based on power domain non-orthogonal multiple access (PD-NOMA), the outdoor weak-turbulence channel environment is considered in this paper, where the UAV is set as a communication relay. We have analysed the distribution function of received signal and system noise in the actual environment, derived the calculation formulas of noise and signal-to-noise ratio (SNR). The theoretical bit error rate (BER) expression of the UFVLC system is derived by analysing the influence of UAV and receiver position. Moreover, numerical results demonstrate that the theoretical analysis is consistent with the simulation results, especially in the case of high SNR.

## 1. Introduction

Visible light communication (VLC) is a wireless communication technology using visible light as the signal transmission carrier. Compared with other wireless communication technologies, visible light communication has many advantages, such as high security, good confidentiality, high communication rate, integrating communication and lighting functions, which is expected to become one of key technologies in the next generation of mobile communication system [1-3]. Non-orthogonal multiple access (NOMA) technology can enable multiple users to share spectrum resource, which can greatly improve the spectrum efficiency and communication capacity. Successive interference cancellation (SIC) technology is used to eliminate the interference between different users [4-5]. The application of NOMA technology to the downlink visible light communication system can further improve the traffic capacity of the system and support more users for communication connection [6].

At present, unmanned aerial vehicle (UAV) has been widely used in more and more fields because of its convenient deployment and strong mobility, such as agriculture, communication, monitoring and auxiliary communication network reconstruction in disaster areas [7]. UAV can interact with high-speed data. The traditional radio frequency (RF) communication link cannot meet the increasing high-speed communication requirements. The traditional link is limited by factors such as capacity, interference, confidentiality and spectrum resources, and cannot fully meet the communication requirements of heterogeneous systems of air to ground UAV [8]. However, the existing VLC is limited by the large attenuation of communication link and is not convenient for flexible deployment, especially in some mountainous areas and areas with weak signal. The combination of VLC, NOMA, and UAV technology can not only improve the flexibility of visible light communication, but also meet the requirements of high-speed and large capacity communication.

The research on UAV-assisted visible light communication has also attracted more and more researchers' attention. The authors of [9] studied a new method of combining VLC with UAV to provide flexible communication and lighting at the same time, the location and unit Association of UAV are optimized under the constraints of illumination and communication. The authors of [proposed a UAV-assisted VLC system using NOMA, considering the joint optimization of power allocation and UAV layout to maximize the total rate of all users. The authors of [11] worked on the UAV deployments of UAV network based on VLC, in which a federated learning framework based on convolutional auto encoder (CAE) machine learning algorithm is proposed.

The research on visible light communication is inseparable from the analysis of visible light channel noise. The signal-to-noise ratio of VLC system in visible light channel environment will be very different in different scenes during the day and at night [12]. The received signal-to-noise ratio (SNR) of visible light communication system will also be very different under different light source layout [13]. Some researchers have studied the analysis of optical power distribution and received SNR of indoor visible light communication system [14]. In [15], the authors studied the statistical characteristics of mobile users' SNR in random direction in indoor VLC system, obtained the new closed-form expressions of probability density function and cumulative distribution function of instantaneous received SNR, and verified the accuracy of the derived expressions through simulations.

For the research of indoor VLC system, in most cases, the signal intensity of line of sight (LoS) is significantly greater than that of non-line of sight (NLoS). Thus, the signal of non-direct path can be ignored. At this time, the signal light is equivalent to static direct signal, and the channel characteristics here are often static. However, in some indoor visible light communication environments, the signal light of NLOS cannot be completely ignored. There are many factors affecting the performance of indoor VLC system, such as multipath attenuation, path loss, occlusion and environmental noise [16]. Lambert radiation model is often used to describe static or quasi-static VLC channel characteristics [17-19]. Compared with the research of indoor VLC, the channel characteristics of outdoor VLC often need to consider the influence of turbulence. Turbulence can constantly change the refractive index of the atmosphere, which will lead to the irregular change of the refractive index of visible light signal, thus will have an adverse impact on the performance of VLC system. Now, in outdoor VLC system, the commonly used light intensity attenuation models are negative exponential model [20], gamma-gamma model [21] and log normal model [22], which correspond to strong turbulence, medium turbulence and weak turbulence respectively. In this paper, only the log normal intensity attenuation in weak turbulence state is analyzed.

This paper describes a UAV-assisted outdoor VLC system, which can provide high-speed and high-capacity communication for some users who are blocked by natural disasters or mountains. Meanwhile, it can also meet certain lighting conditions. The logarithm normal distribution channel model of UAV-assisted VLC system based on power domain non-orthogonal multiple access (PD-NOMA) in outdoor weak turbulence environment is studied. The distribution function of the received signal and the noise of the system in the actual environment are analyzed. The calculation formulas of noise and SNR and the theoretical bit error rate (BER) expression are derived.

## 2. System Model

### *2.1. Scene Description*

The scene of UAV-assisted fiber-based visible light communication (FVLC) system [3] described in this paper is shown in figure 1, which is mainly composed of four parts: base station (BS), UAV, light emitting diode (LED), and ground user equipment (UE). Among them, the UAV acts as the mobile communication relay of the system, while the flight path of the UAV is fixed. In the whole communication process, the base station first communicates with the UAV and transmits the information to the UAV relay. Then, the UAV sends the information received from the base station to the LED, and finally transmits the signal to the ground UEs using the signal converter. It is assumed that the flight altitude of the UAV is fixed at H_UAV, and the UAV has a message buffer that can store the information received from the base station [23]. In addition, for convenience, it is assumed that the Doppler effect of the communication system is fully compensated in the whole communication process.

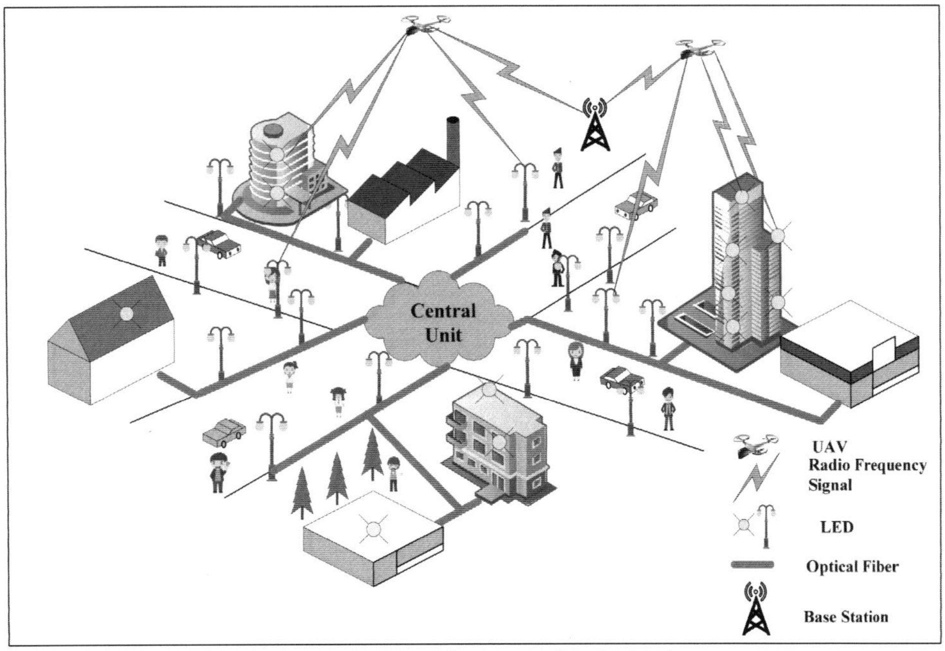


Figure 1. The scene of UAV-assisted FVLC system.

According to the scenario of a UAV-assisted FVLC (UFVLC) system shown in figure 1, a three-dimensional (3D) coordinate system as shown in figure 2 is established. Considering that the distance between the base station and the UAV is much greater than that between the UAV and the ground user, only the signal transmitted from the UAV to the LED and the communication link from LED to the ground UE are considered here. In general, considering that there are only two ground UEs, the connecting line of UE U1 and U2 is taken as the axis of the rectangular coordinate system, the axis perpendicular to the axis on the horizontal plane is taken as the axis of the coordinate system, and the axis is the direction perpendicular to the horizontal ground. Here assuming that flight time is T, when the time is t, the position coordinate of the flying UAV in the 3D coordinate system is $\left(x(t), y(t), H_{\_\mathrm{UAV}}\right), 0 \le t \le T$. It is known that the UAV flies at a uniform speed of speed V in the whole communication process.

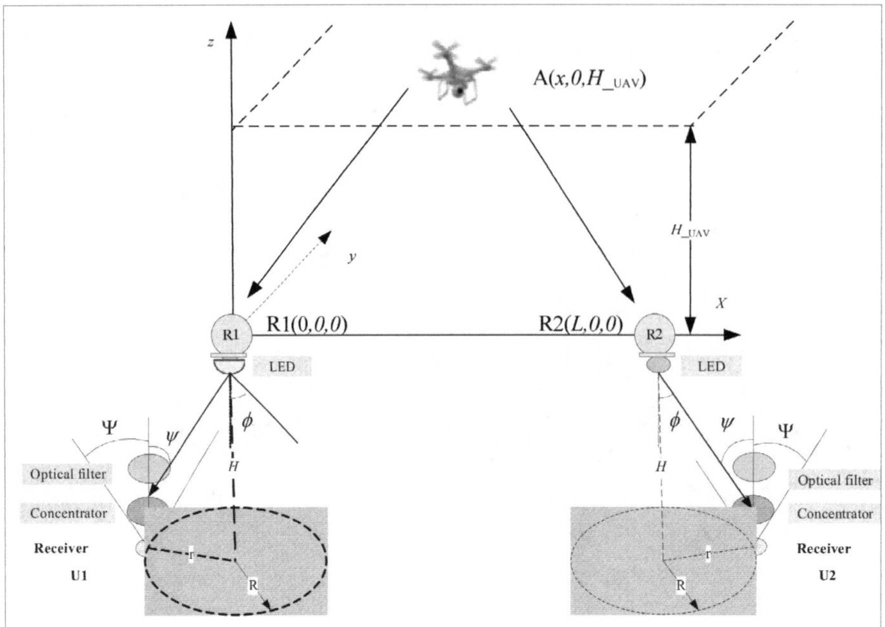


Figure 2. Channel model of UAV-assisted VLC system.

### 2.2. Communication Model

*2.2.1. Channel Model.* Considering that the distance from the base station to the UAV is much greater than that from the UAV to the ground users, it can be approximately considered that the signal strength from the base station to the UAV remains unchanged during the flight of the UAV. Therefore, only the channel model from UAV to ground UEs is considered here. In the n-th communication time slot of UFVLC system, the channel gain of information from UAV A to the LED receiver R as [24]

$$h_{ar}[n] = \beta d_{ar}^{-2}[n], \tag{1}$$

where $\beta$ is the channel power per unit distance $(d_0 = 1\text{ m})$, $d_{ar}[n]$ is the distance from UAV A to the LED receiver R. In this process, the information is transmitted through the direct path, regardless of fading.

In the $n$-th communication time slot, it is assumed that the channel model from LED receiver R to ground user $U_i$ is the same, which is

$$h = \frac{(m+1)}{2\pi}\frac{n_c^2 A_{rx} T_s}{\sin(\Psi)^2}\cos(\phi)^m \frac{\cos(\psi)}{d^2} rect(\frac{\psi}{\Psi}), \tag{2}$$

where $m$ is the Lambert radiation order, m= $-\text{In}(2)/\text{In}(\cos(\phi_{1/2}))$, $\phi_{1/2}$ is the half power angle emitted by the LED, $A_{rx}$ is the receiving area of the photoelectric detector (PD), $n_c$ is the refractive index of the receiving lens, $\Psi$ is the field angle of view of the PD, $d$ is the direct link distance between the LED and the PD receiver, $\phi$ is the radiation angle of the LED, $\psi$ is the incident angle of light, and $rect(\cdot/\Psi)$ is the rectangular function between $[0,\Psi]$, $T_s$ is the gain coefficient of the optical filter.

As shown in figure 2, in the LoS channel, we have $\cos(\phi) = H/d$, and $d = \sqrt{r^2 + H^2}$, where r represents the receiving radius of PD, so h can be further expressed as

$$h=\frac{(m+1)}{2\pi}\frac{n_c^2 A_{rx} T_s}{\sin(\Psi)^2}H^m\frac{\cos(\psi)}{\left(r^2+H^2\right)^{\frac{(m+2)}{2}}}rect(\frac{\psi}{\Psi}). \tag{3}$$

In the communication link between LED receiver R and ground UEs, it is affected by outdoor atmospheric turbulence. Here, only the influence of outdoor weak turbulence is considered. In the outdoor weak turbulence channel environment, the fluctuation of channel gain h follows log normal (LN) distribution, and the channel gain from can be expressed as

$$h^{LN}=hh^{'}, \tag{4}$$

where $h^{'}$ represents the outdoor turbulence interference gain.

The probability density function distribution of the received SNR at the receiving end can be expressed as [25]

$$\text{pdf}_{\gamma}^{LN}(\gamma)=\frac{10/\text{In}10}{\sqrt{2\pi\sigma_L^2}\gamma}\exp\left(-\frac{(10\lg(\gamma/\overline{\gamma}))^2}{2\sigma_L^2}\right), \tag{5}$$

where $\gamma$ is the instantaneous received SNR, $\overline{\gamma}$ is the average received SNR, $\overline{\gamma}=\left(\rho P_r\right)^2/\sigma_{total}^2$, $\rho$ is the photoelectric response of PD and $\sigma_L^2$ is the disturbance variance of light intensity. In the weak turbulence environment, let the $\sigma_L$ is set as 0.2 [25].

*2.2.2. Signal Transmission Model.* In the UFVLC system, the UAV superimposes and forwards the data stream received from the base station in the way of PD-NOMA, and performs Mk-QAM mapping modulation on the data sent by the user, where k represents the k-th user and Mk represents the modulation order of the k-th user. Without loss of generality, only the PD-NOMA communication of two users is considered here k$\in$ {1,2}, Then, the power is allocated to each user[26]. Next, the signals sent by the two users are mixed together and transmitted in the form of radio frequency (RF) signal. In the n-th communication time slot, the signal model transmitted from LED receiver R to the ground UE is

$$y_i[n]=h_{i,ar}[n]\sum_{k=1}^{2}\sqrt{P_k}x_k^{ar}[n]+\omega[n]\ ,\ i=1,2, \tag{6}$$

where $\sum_{k=1}^{2}\sqrt{P_k}x_k^{ar}[n]$ represents the superposition information transmitted from UAV A to LED receiver R, $P_k$ represents the power allocation ratio of the k-th user satisfying $P_1+P_2=1$, $h_{i,ar}[n]$ is the channel gain from UAV A to the i-th LED receiver R, and $\omega[n]\sim C\left(0,\sigma^2\right)$ is the noise in the transmission process.

In the *n*-th communication time slot, the signal model transmitted from the LED receiver R to the ground UE is

$$y_i[n]=h_i^{LN}[n]\lambda_i\left[h_{i,ar}[n]\sum_{k=1}^{2}\sqrt{P_k}x_k^{ar}[n]+\omega[n]\right]+\tilde{\omega}[n],\quad i=1,2, \tag{7}$$

where $\lambda_i$ represents the amplification gain of the i-th LED, and $\tilde{\omega}[n]\sim C\left(0,\sigma_0^2\right)$ is the noise in the receiving process. The expression of $\lambda_i$ is

$$\lambda_i=\sqrt{\frac{P_t}{P\left|h_{i,ar}\right|^2+\sigma^2}}, \tag{8}$$

where $P_t$ indicates that LED transmits the amplified signal to the destination node with power $P_t$, and $P$ indicates the total power of the signal sent by the UAV.

## 3. Performance Analysis

### 3.1. Noise Analysis of Visible Light Channel

The total noise variance of the visible light channel can be expressed as

$$\sigma_0^2 = \sigma_{\text{shot}}^2 + \sigma_{\text{thermal}}^2, \tag{9}$$

where $\sigma_{\text{shot}}^2$ represents the variance of shot noise, which can be expressed as

$$\sigma_{\text{shot}}^2 = 2q\rho P_r B + 2q\rho P_{am} I_2 B, \tag{10}$$

where $q$ represents the electronic charge, $P_r = hP_t$, $P_r$ represents the optical power received by the user, $h$ represents the channel gain of visible light communication, $B$ represents the noise bandwidth, $P_{am}$ represents the ambient noise power, and $I_2$ represents the noise bandwidth factor that accounts for a rectangular transmitter pulse shape [12], $\sigma_{\text{thermal}}^2$ represents the variance of thermal noise, which can be expressed as

$$\sigma_{\text{thermal}}^2 = \frac{8\pi kT_A}{G}\eta A_{rx} I_2 B^2 + \frac{16\pi^2 kT_A\Gamma}{g_m}\eta^2 A_{rx}^2 I_3 B^3, \tag{11}$$

where $T_A$ represents the absolute temperature of the surrounding environment, G represents the open-loop voltage gain, $\eta$ represents the capacitance per unit area of PD, $\Gamma$ represents the FET (field effect transistor) channel noise coefficient, $g_m$ is the FET transconductance, $I_3$ represents the fully raised cosine equalization pulse forming noise bandwidth factor, and $k$ represents the Boltzmann constant [12].

### 3.2. SNR Analysis of The Whole System

With the continuous change of UAV position, the SNR received by the ground UE is also changed. The SNR received by the i-th ground UE as

$$SNR_i = \frac{\left(\rho h_i^{LN}\right)^2 \lambda_i^2 h_{i,ar}^2 P}{\left(\rho h_i^{LN}\right)^2 \lambda_i^2 \sigma^2 + \sigma_0^2}. \tag{12}$$

In the previous noise analysis of VLC system, the impact of environmental noise transformation on system performance is often not considered. However, in actual visible light communication system, environmental noise will have a great impact on the system, especially when NOMA technology is applied. After adding the environmental noise, the new SNR expression is

$$SNR'_{i,U} = \frac{\left(\rho h_i^{LN}\right)^2 \lambda_i^2 h_{i,ar}^2 P}{\left(\rho h_i^{LN}\right)^2 \lambda_i^2 \sigma^2 + \sigma'^2_0 + \sigma_{am}^2}, \tag{13}$$

where $\sigma_{am}^2$ represent ambient noise variance and $\sigma_{am}^2 = 2q\rho P_{am} I_2 B$, the $\sigma'^2_0$ can be expressed as

$$\sigma'^2_0 = 2q\rho P_r B + \sigma_{\text{thermal}}^2. \tag{14}$$

In the absence of ambient noise, the SNR can be expressed as

$$SNR_{i,U} = \frac{\left(\rho h_i^{LN}\right)^2 \lambda_i^2 h_{i,ar}^2 P}{\left(\rho h_i^{LN}\right)^2 \lambda_i^2 \sigma^2 + \sigma_0'^2}. \tag{15}$$

### 3.3. BER Analysis

Under the influence of outdoor weak turbulence, the signal received by the i-th PD is

$$y_i = h_i^{LN} \lambda_i \left[ h_{i,ar} \sum_{k=1}^{2} \sqrt{P_k} x_k^{ar} + \omega \right] + \tilde{\omega}, \quad i=1,2, \tag{16}$$

where the distribution of SNR at the receiving end follows Log-Normal distribution. According to the decoding order of SIC, the signal of strong user is detected at first. Here, the receiving $U_1$ is regarded as a strong user, and greater power is allocated. Without loss of generality, the received signal is demodulated by $M_1$-QAM, thus, the BER expression of $U_1$ received signal is [26] [27, Eq. (5.1)]

$$BER_{U_1}^{LN} = \frac{1}{\log 2(M_1)} \frac{4}{M_1 M_2} \sum_{i=1}^{\sqrt{M_2}} \sum_{j=1}^{\sqrt{M_2}} \frac{1}{2\pi} \int_0^\infty \Big( \left(M_1 - \sqrt{M_1}\right) Q(\gamma_i) \mathrm{pdf}_\gamma^{LN}(\gamma) d\gamma$$
$$- \left(M_1 - 2\sqrt{M_1} + 1\right) \int_0^\infty Q(\gamma_i) \mathrm{pdf}_\gamma^{LN}(\gamma) d\gamma \int_0^\infty Q(\gamma_j) \mathrm{pdf}_\gamma^{LN}(\gamma) d\gamma \Big). \tag{17}$$

Next, detect the received signal of U2, subtract the demodulated signal of U1 from the received mixed signal, and perform M2-QAM demodulation to obtain the BER expression of U2 received signal as [26] [27, Eq. (5.1)]

$$BER_{U_2}^{LN} = \left(1 - BER_{U1}^{LN}\right) \times \left( 4\left(1 - \frac{1}{\sqrt{M_2}}\right) \int_0^\infty Q(\gamma) \mathrm{pdf}_\gamma^{LN}(\gamma) d\gamma \right.$$
$$\left. -4 \left( \left(1 - \frac{1}{\sqrt{M_2}}\right) \int_0^\infty Q(\gamma) \mathrm{pdf}_\gamma^{LN}(\gamma) d\gamma \right)^2 \right) + BER_{U_1}^{LN}. \tag{18}$$

In the Log-Normal channel, since there is no closed-form solution of $\mathrm{MGF}_\gamma^{LN}(s)$, we use the direct integration method to calculate the BER of each UE in the log normal distribution channel [27, Eq. (5.20)] as follows,

$$BER_{U_1}^{LN} = \frac{1}{\log 2(M_1)} \frac{4}{M_1 M_2} \sum_{i=1}^{\sqrt{M_2}} \sum_{j=1}^{\sqrt{M_2}} \frac{1}{\pi} \int_0^{\pi/2} \Big( \left(M_1 - \sqrt{M_1}\right)$$
$$\times \left[ \int_0^\infty \exp\left(\frac{x}{2\sin^2\theta}\right) \frac{10/\ln 10}{\sqrt{2\pi\sigma_L^2}\, x} \exp\left( -\frac{\left(10\lg(x/\gamma_i)\right)^2}{2\sigma_L^2} \right) dx \right] d\theta$$
$$- \left(M_1 - 2\sqrt{M_1} + 1\right) \frac{1}{\pi} \int_0^{\pi/2} \left[ \int_0^\infty \exp\left(\frac{x}{2\sin^2\theta}\right) \frac{10/\ln 10}{\sqrt{2\pi\sigma_L^2}\, x} \exp\left( -\frac{\left(10\lg(x/\gamma_i)\right)^2}{2\sigma_L^2} \right) dx \right] d\theta$$
$$\times \frac{1}{\pi} \int_0^{\pi/2} \left[ \int_0^\infty \exp\left(\frac{x}{2\sin^2\theta}\right) \frac{10/\ln 10}{\sqrt{2\pi\sigma_L^2}\, x} \exp\left( -\frac{\left(10\lg(x/\gamma_j)\right)^2}{2\sigma_L^2} \right) dx \right] d\theta \Big), \tag{19}$$

$$BER_{U_2}^{LN} = BER_{U_1}^{LN} + \left(1 - BER_{U_1}^{LN}\right) \times \left( 4(1 - \frac{1}{\sqrt{M_2}}) \right.$$
$$\times \frac{1}{\pi} \int_0^{\pi/2} \left[ \int_0^{\infty} \exp(\frac{x}{2\sin^2\theta}) \frac{10/\ln 10}{\sqrt{2\pi\sigma_L^2 x}} \exp\left( -\frac{\left(10\lg(x/\gamma)\right)^2}{2\sigma_L^2} \right) dx \right] d\theta$$
$$\left. -4(1 - \frac{1}{\sqrt{M_2}})^2 \left( \frac{1}{\pi} \int_0^{\pi/2} \left[ \int_0^{\infty} \exp(\frac{x}{2\sin^2\theta}) \frac{10/\ln 10}{\sqrt{2\pi\sigma_L^2 x}} \exp\left( -\frac{\left(10\lg(x/\gamma)\right)^2}{2\sigma_L^2} \right) dx \right] d\theta \right)^2 \right), \quad (20)$$

where $\gamma_i^{1/2} = \sqrt{2SNR}\left(\sqrt{P_1}/\sqrt{2(M_1-1)/3} + \left(2i-1-\sqrt{M_2}\right)\sqrt{P_2}/\sqrt{2(M_2-1)/3}\right)$,
$\gamma_j^{1/2} = \sqrt{2SNR}\left(\sqrt{P_1}/\sqrt{2(M_1-1)/3} + \left(2j-1-\sqrt{M_2}\right)\sqrt{P_2}/\sqrt{2(M_2-1)/3}\right)$,
$\gamma^{1/2} = \sqrt{2SNR}\sqrt{P_2}/\sqrt{2(M_2-1)/3}$, $Q(x) = \frac{1}{2\pi}\int_x^{\infty}\exp\left(-u^2/2\right)du$.

## 4. Simulation and Discussion

### *4.1. Parameters Settings*

The channel parameters and noise parameters are shown in Table 1, and their sources are based on references [12-15]. The details are shown in the following table.

Table 1. Simulation Parameters.

| Parameter | Value | Parameter | Value |
|---|---|---|---|
| UAV altitude $H_{\_UAV}$ | 30 m | Noise bandwidth $B$ | 50 MHz |
| Unit distance channel power $\beta$ | -60 dB | Ambient noise power $P_{am}$ | 16 dBm |
| RF noise variance $\sigma^2$ | -110 dBm | Noise bandwidth factor $I_2$ | 0.562 |
| UAV transmission signal power | 0.01 W | Noise bandwidth factor $I_3$ | 0.868 |
| LED and PD Plane height $H$ | 2.15 m | FET transconductance $g_m$ | 30 mS |
| PD receiving area $A_{rx}$ | 1 cm$^2$ | Open loop voltage gain $G$ | 10 |
| PD Responsiveness $\rho$ | 0.54 A/ W | PD capacitance per unit area $\eta$ | 1.12 $\mu$F/m$^2$ |
| PD field angle $\Psi$ | $60^o$ | FET channel noise figure $\Gamma$ | 1.5 |
| LED half power angle $\phi_{1/2}$ | $60^o$ | Absolute temperature of surrounding environment $T_A$ | $300^o$ K |
| Refractive index of condenser $n_c$ | 1.5 | | |
| Gain coefficient of filter $T_s$ | 1.0 | Distance of UEs $L$ | 100 m |

### *4.2. System Performance at Different UAV Positions*

Assuming that the height of the UAV is 30 m, the ground UE is fixed, the receiving radius r of the UE is set to 2.5 m, the transmission power of LED is set to 0.1W, the modulation orders M1 and M2 of U1 and U2 are 4, and the power distribution ratio is 0.85 and 0.15 respectively. The variation BER curve of UFVLC system at different UAV positions is obtained, which is shown in figure 3. It can be found that the BER of receiving U1 increases with the increasing horizontal distance X of UAV, and the BER of receiving U2 decreases with the increasing horizontal distance X of UAV. Meanwhile, the theoretical analysis results match well with the simulation results. In addition, The BER performance decreases significantly after considering the environmental noise.

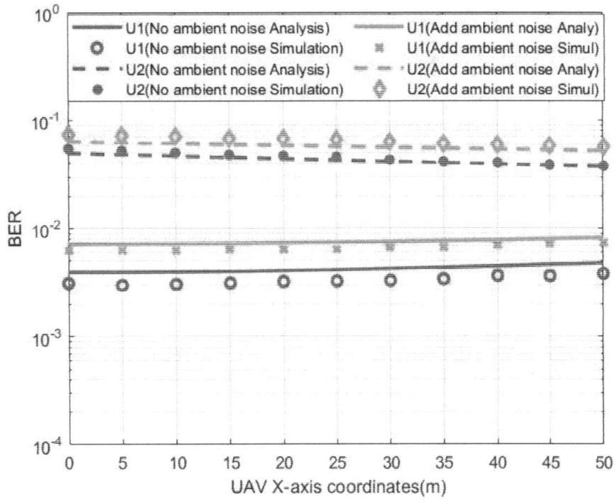


Figure 3. BER curve of UFVLC system at different UAV positions.

### 4.3. *Performance of Ground Users with Different Receiving Radius*

Assuming that the height of the UAV is 30 m, the ground UE is fixed, the position of the UAV is fixed as A (40,0,30), the transmission power of LED is set to 0.1W, the modulation orders M1 and M2 of U1 and U2 are 4, and the power distribution ratio is 0.85 and 0.15 respectively. The BER curve of the UFVLC system at different UE receiving radius is obtained, which is shown in figure 4. It can be found that the BER of receiving U1 and U2 increases with the increasing receiving radius r, and the theoretical analysis results match well with the simulation results. It is worth noting that the BER performance of the system decreases obviously after considering the environmental noise.

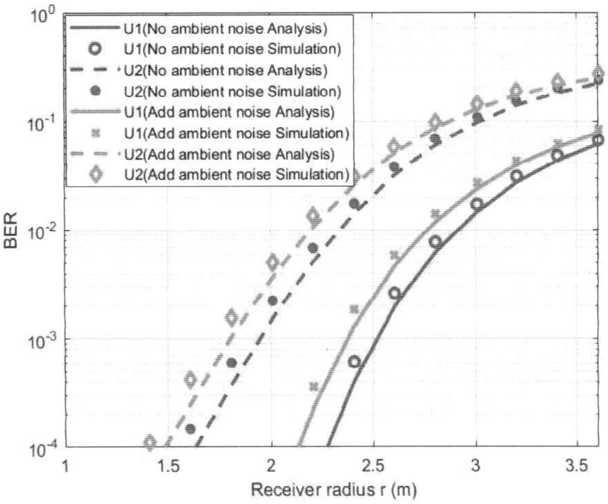


Figure 4. BER of ground users with different receiver radius.

## 5. Conclusions

We have studied the performance analysis of UAV-assisted outdoor fiber-based visible light communication system in this paper. The purpose of constructing this system is to provide high-speed and high-capacity communication for some users who are blocked by natural disasters or mountains. However, the research in this paper does not optimize the power and trajectory of UAVs, and only studies the performance considering the outdoor weak-turbulence channel environment. In the future, the research on multi-UAV assisted FVLC system for ground users and the auxiliary optimization of

UAV will further improve the flexibility of visible light communication, providing higher-speed and larger-capacity communication services.

**Acknowledgments**
This work was supported in part by the National Science Foundation of China (NSFC) under Grant 62161024, China Postdoctoral Science Foundation under Grant 2021TQ0136, the State Key Laboratory of Computer Architecture (ICT, CAS) Open Project under Grant CARCHB202019.

**References**

[1] Elgala, H., Mesleh, R., Haas, H. (2011) Indoor optical wireless communication: potential and state-of-the-art. IEEE Communications Magazine, 49(9): 56-62.
[2] Saadi, M., Wattisuttikulkij, L., Zhao, Y., et al. (1975) Visible light communication: opportunities, challenges and channel models. International Journal of Electronics &Informatics, 2(1): 1-11.
[3] Yu, L., Liu, Z., Wen, M., et al. (2021) Sparse code multiple access for 6G wireless communication networks: recent advances and future directions. IEEE Communications Standards Magazine, 5(2): 92-99.
[4] Ding, Z., Lei, X., Karagiannidis, G. K. (2017) A survey on non-orthogonal multiple access for 5G networks: Research challenges and future trends. IEEE J. Sel. Areas Commun., 35(10): 2181-2195.
[5] Saito, Y. et al. (2013) Non-orthogonal multiple access (NOMA) for cellular future radio access. In: IEEE VTC-Spring. Dresden. pp. 1–5.
[6] Yin, Popoola, L. Wu, W. O.X., Haas, H. (2016) Performance evaluation of non-orthogonal multiple access in visible light communication. IEEE Trans. Commun., 64(12): 5162-5175.
[7] Choi, D. H., Jung, B. H., Sung, D. K. (2014) Low-complexity maneuvering control of a UAV-based relay without location information of mobile ground nodes. In: IEEE Symposium on Computers and Communications (ISCC). Funchal. pp. 1-6.
[8] Huang, X., Zhang, J. A., Liu, R. P., et al. (2019) Airplane-aided integrated networking for 6G wireless: Will it work? IEEE Vehicular Technology Magazine, 14(3): 84-91.
[9] Yang, Y., Chen, M., Guo, C., et al. (2019) Power efficient visible light communication with unmanned aerial vehicles. IEEE Communications Letters, 23(7): 1272-1275.
[10] Pham, Q. V., Huynh-The, T., Alazab, M., et al. (2020) Sum-rate maximization for UAV-assisted visible light communications using NOMA: Swarm intelligence meets machine learning. IEEE Internet of Things Journal, 7(10): 10375-10387.
[11] Wang, Y., Yang, Y., Luo, T. (2020) Federated convolutional auto-encoder for optimal deployment of UAVs with visible light communications. In: IEEE International Conference on Communications Workshops (ICC Workshops). Dublin. pp. 1-6.
[12] Lee, S.-J., Jung, S.-Y. (2012) A SNR analysis of the visible light channel environment for visible light communication. In: 2012 18th AsiaPacific Conference on Communications (APCC). Jeju. pp. 709-712.
[13] Azizan, L.-H., Ab-Rahman, M.S., Jumiran, K. (2012) Analytical approach on SNR performance of visible light communication for modern lighting layout. In: International Conference on Innovation Management and Technology Research (ICIMTR). Malacca. pp. 332-336.
[14] Lu, H., Su, Z., Yuan, B. (2014) SNR and optical power distribution in an indoor visible light communication system. In: the 7th International Congress on Image and Signal Processing. Dalian. pp. 1063-1067.
[15] Arfaoui, M. A., et al. (2019) SNR statistics for indoor VLC mobile users with random orientation. In: IEEE ICC. Shanghai, pp. 1-6.
[16] Lee, I. E., Sim, M. L., Kung, F. W. L. (2009) Performance enhancement of outdoor visible-light communication system using selective combining receiver. IET Optoelectron., 3(1): 30-39.
[17] Komine, T., Nakagawa, M. (2004) Fundamental analysis for visible-light communication system

using led lights. IEEE Transactions on Consumer Electronics, 50(1): 100-107.
[18] Babatundi, O., Qian, L., Cheng, J. (2014) Downlink scheduling in visible light communications. In: the Sixth International Conference on Wireless Communications and Signal Processing (WCSP). Hefei. pp. 1-6.
[19] Ding, J., Wang, K., Xu, Z. (2014) Accuracy analysis of different modeling schemes in indoor visible light communications with distributed array sources. In: the 9th International Symposium on Communication Systems, Networks & Digital Sign (CSNDSP). Manchester. pp. 1005-1010.
[20] Prabu, K., Kumar, D. S. (2014) BER analysis of DPSK–SIM over MIMO free space optical links with misalignment. Optik-International Journal for Light and Electron Optics, 125(18): 5176-5180.
[21] Karp, S., et al. (2013) eds. Optical channels: fibers, clouds, water, and the atmosphere. Springer Science & Business Media, New York.
[22] Dios F., Rubio, J. A., Rodriguez, A., et al. (2004) Scintillation and beam-wander analysis in an optical ground station-satellite uplink. Applied Optics, 43(19):3866-3873.
[23] Jin, Y., Zhang, Y. D., Chalise, B. K. (2012) Joint optimization of relay position and power allocation in cooperative broadcast wireless networks. In: 2012 IEEE International Conference on Acoustics, Speech and Signal Processing (ICASSP). Kyoto. pp. 2493-2496.
[24] Zheng, K., Zheng, Q., Chatzimisios, P., et al. (2015) Heterogeneous vehicular networking: A survey on architecture, challenges, and solutions. IEEE communications surveys & tutorials, 17(4): 2377-2396.
[25] Zhu, Y.-J., Sun, Z.-G., Zhang, J.-K., Zhang, Y.-Y., Zhang, J. (2017) Training receivers for repetition-coded MISO outdoor visible light communications. IEEE Transactions on Vehicular Technology, 66(1): 529-540.
[26] Almohimmah, E. M., Alresheedi, M. T. (2020) Error analysis of NOMA based VLC systems with higher order modulation schemes. IEEE Access, 8: 2792-2803.
[27] Simon, M. K., Alouini, M.-S. (2000) Digital Communication Over Fading Channels. 1st ed., Wiley. Hoboken.

# Gyro Test Evaluation Based on Grey Clustering

**Wentao Guan[1*], Ronghui Jiao[1] and Chen Zhao[1]**

[1]China Academy of Space Technology, Beijing, 100094, China

xuyang_fu@mail.ccmu.edu.cn

**Abstract:** Based on the test and evaluation of satellite control sub-systems, this article applies the gray clustering evaluation method in the gray system theory to the lack of objective quantitative standards in the current test and evaluation, and establishes the evaluation index system and evaluation model of the two-float gyro. The model is verified by using telemetry data and the evaluation conclusion of the gyro is got. The establishment of the model provides an objective and quantitative standard for the evaluation, and reduces the subjectivity and ambiguity in the evaluation. The model and method can be applied to other components of the control sub-system, and it has promotion significance for the test and evaluation of the entire control sub-system.

## 1. Introduction

The model-based test and evaluation method is based on a series of test standards and judgment criteria transformed from interpretation knowledge to give a judgment on the advantages, value or meaning of the test target. The assessment is more comprehensive, with unified standards, and stronger objectivity. It has been used in the aerospace and other fields, and has achieved good results[1] [2] [3]. It is the development direction of spacecraft comprehensive test and evaluation technology. For example, literature [4] studies the application of analytic hierarchy and fuzzy theory models to the effectiveness evaluation of remote sensing satellites, literature [5] applies the same two models to the effectiveness evaluation of constellation satellite communication systems. At present, the satellite testing department of CAST has established system-level evaluation indicators and evaluation models. The applied methods include analytic hierarchy process, gray whitening weight function clustering method, Delphi survey method, fuzzy evaluation method and other evaluation methods. A analytic hierarchy process evaluation model was established and applied to the effectiveness evaluation of the Chang'e detector. A fuzzy evaluation model was also established to evaluate the test indicators of the Beidou-3 load subsystem.

At present, the electrical test and evaluation of the control system is mainly to extract the range of test data, and directly give the conclusion of qualified or unqualified based on the envelope criterion. This evaluation method is based on the subjective basis for the evaluation of some elements, especially the consistency of the data. The evaluation method is single and lacks the support of mathematical models. The evaluation conclusions are mostly qualitative judgments and lack quantitative standards. The entire system structure lacks a comprehensive, quantitative and objective judgment standard.

Starting from the characteristics of the control subsystem, this paper selects the key component of the two-floating gyro as the entry point for testing's evaluation. Combining the factors closely related to its comprehensive performance, using the gray clustering evaluation method in the gray system theory, the two-floating gyro evaluation index system and evaluation model are establishe. The model was verified by the model telemetry data, and the evaluation conclusion of the gyro was obtained. The establishment of the model provides an objective and quantitative standard for the evaluation, and reduces the subjectivity and ambiguity in the evaluation. The model and method can be applied to other components of the control sub-system, and it has promotion significance for the test and evaluation of the entire control sub-system.

## 2. Grey Cluster Evaluation Model

The performance evaluation of satellite components involves: the evaluated unit (component A, component B, component C); the evaluated item (power performance, temperature control performance, sensitivity, etc.); evaluation sample (digital performance of each item); evaluation gray Cluster (good effect, medium effect, poor effect, etc.).

The evaluation of each item is gray statistical evaluation, and the evaluation result is the pros and cons of each item (power performance, temperature control performance, sensitivity). The evaluation of each unit is gray cluster evaluation, the evaluation object is the unit, and the evaluation result is the pros and cons of each unit (part A, part B, and part C).

If $d_{i1}, d_{i2}, \ldots, d_{im}$ is the sample row of item $d_{i1}, d_{i2}, \ldots, d_{im}$ corresponding to unit $i$, the sample row $i$ is transformed into the evaluation value of the unit for each gray cluster $k=1, k=2, \ldots, k=m$ through transformation, which is called the gray cluster evaluation of the gray cluster to which the unit $i$ belongs[6].

$I$ : *evaluated unit*; $J$ : *evaluated item*; $K$ : *grey cluster*; $d$ : *sample matrix*

$\mathrm{POL}(j)$: *the polarity of item j in set J*

$f_{jk}$ : *whitening function of item for gray cluster*

$\sum_{j=1}^{m} c_{jk}$ : *global sum of the threshold value of the whitening function*

$\eta_{jk}$ : *the conversion coefficient* (*weight*) *of the whitening function* $f_{jk}$

$\sigma_{qk}$ : *the k gray evaluation value of q*, $q \in \{i, j\}$

$\sigma_{ik}$ : *The global sum of the converted whitening value*

$M_{\sigma}$ : *transformation from* $f_p(d_{ij})$ *to real number* $\sigma_{qk}$

Figure 1. formula symbol definition in this section

### 2.1. Grey evaluation elements

#### 2.1.1. Grey Evaluation System

The evaluated unit is recorded as $i, i \in I=\{1,2,\cdots, \omega\}$, record the evaluated item as $j, j \in J=\{1,2,\cdots, m\}$, record the evaluation grey cluster as $k, k \in K=\{1,2,\cdots, n\}$. And called $k=1$ as the up gray cluster, that is, "good", "more", "excellent", called $k=n$ as the low gray cluster, that is, "poor", "less" and "bad". The rest are middle gray, and the larger the serial number, the lower the cluster. Then $I$ is the set of evaluated units; $J$ is the set of evaluated items; $K$ is the set of evaluated grey cluster; $\{I, J, K\}$ is the grey evaluation system.

#### 2.1.2. Polarity of item set

Let $J$ be the set of evaluated items. $\mathrm{POL}(j)$ is the polarity of item $j$ in $J$.

$$\mathrm{POL}(j) \in \{\mathrm{POL}(\max), \mathrm{POL}(\min), \mathrm{POL}(\mathrm{mem})\} \tag{1}$$

Among them, $\mathrm{POL}(\max)$ is the maximum value polarity, $\mathrm{POL}(\min)$ is the minimum value polarity, and $\mathrm{POL}(\mathrm{mem})$ is the moderate value polarity. If $\mathrm{POL}(j)=\mathrm{POL}(\max)$, that means the items in the item set $J$ are unified to the maximum polarity, then $J$ is called regular, that is, the larger the item sample value, the better the item performance. The regularity of item set $J$ is the premise of grey evaluation.

### *2.1.3. Sample matrix*

Let $d_{ij}$ be the sample of unit $i$ for items $j$: $d_{ij}, i \in I, j \in J$, then call $\boldsymbol{d}$ the evaluated sample matrix

$$\boldsymbol{d} = \begin{array}{c} \begin{matrix} item1 & item2 & & itemm \end{matrix} \\ \begin{pmatrix} d_{11} & d_{12} & \cdots & d_{1m} \\ d_{21} & d_{22} & \cdots & d_{2m} \\ \vdots & \vdots & \vdots & \vdots \\ d_{\omega 1} & d_{\omega 2} & \cdots & d_{\omega m} \end{pmatrix} \begin{matrix} unit1 \\ unit2 \\ \\ unit\omega \end{matrix} \end{array} \tag{2}$$

Then $(d_{1j}, d_{2j}, \cdots, d_{\omega j}), j \in J$ is sample column of item $j$. $(d_{i1}, d_{i2}, \cdots, d_{im}), i \in I$ is sample row of unit $i$.

### *2.1.4. Whitening function*

The gray cluster $k$ is regarded as the morphological gray number. On the cognitive plane, the least information image with gray samples (whitening value) $d_{ij}$ as the abscissa and gray confirmation degree $f$ as the ordinate is called the whitening function of $k$ gray and denoted as $f_k$. The whitening function of item $j$ is $f_{jk}$.

### *2.2. Grey clustering formula*

**theorem 1**[6] Let $K$ be the set of gray cluster, $K = \{1, 2, \cdots, n\}$, $J$ be the set of items, $J = \{1, 2, \cdots, m\}$, $f_{jk}$ be the whitening function of item $j$ for gray cluster $k$, and $f_{j1}$ is the whitening function of the upper cluster, $f_{jn}$ is the whitening function of the last cluster, and the rest $f_{ji}, i \neq 1, i \neq n$ are the whitening functions of the middle cluster. Then the whitening function has the following formula, namely

1° whitening function of the upper cluster $f_{j1} = f_{j1}(c_{j1}, \infty) \Rightarrow f_{j1}(d_{ij}) = \begin{cases} d_{ij}/c_{j1}, & d_{ij} \in [0, c_{j1}] \\ 1, & d_{ij} \in [c_{j1}, \infty] \end{cases}$ (3)

2° whitening functions of the middle cluster $f_{ji} (i \in K, i \neq 1, i \neq n)$

$$f_{ji} = f_{ji}(-, c_{ji}, +) \Rightarrow f_{ji}(d_{ij}) = \begin{cases} d_{ij}/c_{ji}, & d_{ij} \in [0, c_{ji}] \\ -d_{ij}/c_{ji} + 2, & d_{ij} \in [c_{ji}, 2c_{ji}] \\ 0, & d_{ij} \in [2c_{ji}, \infty] \end{cases} \tag{4}$$

3° whitening function of the last cluster: $f_{jn} = f_{jn}(0, c_{jn}) \Rightarrow f_{jn}(d_{ij}) = \begin{cases} 1, & d_{ij} \in [0, c_{jn}] \\ -d_{ij}/c_{jn} + 2, & d_{ij} \in [c_{jn}, 2c_{jn}] \\ 0, & d_{ij} \in [2c_{jn}, \infty] \end{cases}$ (5)

4° $\sum_{j=1}^{m} c_{jk}$ is global sum of the threshold value of the whitening function

5° $\eta_{jk}$ is the conversion coefficient (weight) of the whitening function $f_{jk}$: $\eta_{jk} = c_{jk} / \sum_{j=1}^{m} c_{jk}$ (6)

6° The global sum of the converted whitening value is denoted as $\sigma_{ik}$, $\sigma_{ik}$ is the evaluation value of unit $i$ belonging to gray cluster $k$: $\sigma_{ik} = \sum_{j=1}^{m} f_{jk}(d'_{ij}) \eta_{jk}$ (7)

### 2.3. Grey Evaluation Rule

#### 2.3.1. General Grey Evaluation Rule

Let $\sigma_{qk}$ be the $k$ gray evaluation value of $q$ , $q \in \{i, j\}$ , $i \in I = \{1,2,\cdots,\omega\}$ , $j \in J = \{1,2,\cdots,m\}$ ; Let $M_\sigma$ be the transformation from $\{f_p(d_{ij})\}$ to real number $\sigma_{qk}$ .

$$M_\sigma: \ \{f_p(d_{ij})\} \to \sigma_{qk}, \ \ p \in \{k, jk\}, \ \ q \in \{i, j\}, \ \ K = \{1,2,\cdots, \ n\} \tag{8}$$

then 1° $\sigma_{qk}$ is $k$ gray cluster evaluation value of item( $q = j$ ) or unit( $q = i$ ); 2° $M_\sigma$ is the evaluation (transformation); 3° $\sigma_q = (\sigma_{q1}, \sigma_{q2}, \cdots, \sigma_{qn})$ is evaluation sequence of $q$ ; 4°If

$$\sigma_{qk^*} = \max \sigma_q = \max\{\sigma_{q1}, \sigma_{q2}, \cdots, \sigma_{qn}\} \tag{9}$$

Then $q$ belongs to gray cluster $k^*$ ; 5° If $k \in K, K = \{1,2,3\}$ , and $k = 1$ is up gray cluster, $k = 3$ is low gray cluster, $k = 2$ is middle gray cluster. It is said that the above gray evaluation is typical.

#### 2.3.2. Typical Grey Evaluation Rule

If $\sigma_q$ is typical gray evaluation sequence, $q \in \{j, i\}$ , $\sigma_q = (\overset{up}{\sigma_{q1}}, \overset{middle}{\sigma_{q2}}, \overset{low}{\sigma_{q3}})$ , then

- If $\sigma_{q1} > 0, \sigma_{q2} = 0, \sigma_{q3} = 0$ , $q$ strictly belongs to the up gray cluster; If $\sigma_{q1} = 0, \sigma_{q2} > 0, \sigma_{q3} = 0$ , $q$ strictly belongs to the middle gray cluster; If $\sigma_{q1} = 0, \sigma_{q2} = 0, \sigma_{q3} > 0$ , $q$ strictly belongs to the low gray cluster.
- If $\sigma_{q1} > \sigma_{q2} > \sigma_{q3}$ , $q$ is "middle in up"; if $\sigma_{q1} > \sigma_{q3} > \sigma_{q2}$ , $q$ is "lower in up";
- If $\sigma_{q2} > \sigma_{q1} > \sigma_{q3}$ , $q$ is "upper in middle"; if $\sigma_{q2} > \sigma_{q3} > \sigma_{q1}$ , $q$ is "lower in middle";
- If $\sigma_{q3} > \sigma_{q1} > \sigma_{q2}$ , $q$ is "upper in low"; if $\sigma_{q3} > \sigma_{q2} > \sigma_{q1}$ , $q$ is "middle in low";
- If "middle in up" or "lower in up", $q$ belongs to up cluster. If "upper in middle" or "lower in middle", $q$ belongs to middle cluster. If "upper in low" or "middle in low", $q$ belongs to low cluster.

## 3. Grey clustering evaluation of telemetry data from floating gyro

### 3.1. Evaluation process

Using the above gray clustering evaluation model and method, select the key component of the control system, the two floating gyro, and use its telemetry data to comprehensively and quantitatively evaluate the performance of the gyro.

A certain Beidou-3 satellite is equipped with a 3+1S two-floating top. This article intends to synthesize each gyro and obtain the quantitative evaluation result of each gyro. Using a typical gray clustering evaluate process, as shown in Figure 2.

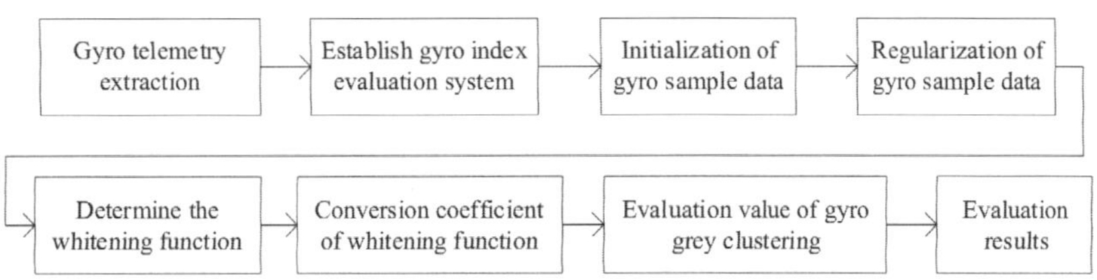


Figure 2. gray clustering evaluation process

### 3.2. Grey clustering evaluation index system

The statistics of the gyro-related telemetry data in the satellite test are shown in Table 1. The stability of the current telemetry of the gyro motor directly reflects the working state of the motor, which in turn reflects the stability of the angular momentum of the gyro rotor; and the temperature directly determines the damping coefficient of the floating liquid, which in turn affects the gyro sensitivity. Therefore, the power state, motor current, and temperature parameters of the gyro during steady-state operation are selected as evaluation indicators.

According to the grey evaluation definition, power supply state, motor current, temperature and other telemetry parameters are evaluated items, and GX, GY, GZ, GS gyro are evaluated units. The standard deviation of telemetry data was selected as the evaluation sample, as shown in Table 2. Four gyroscopes (clustering units) will be correlated and compared, and the characteristics of three parameters (clustering items) will be taken into account, and the evaluation will be carried out according to a unified scale.

Table 1. Gyro telemetry data

| Parameter | Minimum | Max | Average value | Standard deviation | Unit |
|---|---|---|---|---|---|
| GX power state | 4.04 | 4.48 | 4.147 | 0.03894 | V |
| GY power state | 4.02 | 4.32 | 4.117 | 0.03376 | V |
| GZ power state | 4.08 | 4.48 | 4.164 | 0.03494 | V |
| GS power state | 4.06 | 4.44 | 4.16 | 0.04576 | V |
| GX motor current | 0.058 | 0.076 | 0.067 | 0.002000 | A |
| GY motor current | 0.074 | 0.092 | 0.083 | 0.001732 | A |
| GZ motor current | 0.068 | 0.082 | 0.075 | 0.001732 | A |
| GS motor current | 0.090 | 0.112 | 0.101 | 0.003000 | A |
| GX temperature | 52.2 | 53.6 | 52.91 | 0.2210 | ℃ |
| GY temperature | 53.4 | 54.8 | 54.21 | 0.2340 | ℃ |
| GZ temperature | 53.4 | 54.4 | 54.00 | 0.1658 | ℃ |
| GS temperature | 52.8 | 53.6 | 53.26 | 0.1643 | ℃ |

Table 2. Gyro telemetry standard deviation

| Unit | Two floated gyro | Item 1<br>Power state U | Item 2<br>Motor current I | Item 2<br>Temperature T |
|---|---|---|---|---|
| 1 | GX | 0.03894 | 0.002000 | 0.2210 |
| 2 | GY | 0.03376 | 0.001732 | 0.2340 |
| 3 | GZ | 0.03494 | 0.001732 | 0.1658 |
| 4 | GS | 0.04576 | 0.003000 | 0.1643 |

According to the definition in Section 2.1, the evaluation index system has the following elements

- Uint: $i=1 \Rightarrow \text{GX}$ , $i=2 \Rightarrow \text{GY}$ , $i=3 \Rightarrow \text{GZ}$ , $i=4 \Rightarrow \text{GS}$
- Item: $j=1 \Rightarrow U$ power state , $j=2 \Rightarrow I$ motor current , $j=3 \Rightarrow T$ Temperature
- Clustering:
  $k=1 \Rightarrow$ grey cluster 1 "up"(good performance) ; $k=2 \Rightarrow$ grey cluster 2 "middle"(middle performance); $k=3 \Rightarrow$ grey class3 "low"(poor performance)

### *3.3. Original sample matrix*

Let $d_{ij}^0$ represent the sample of unit $i$ for item $j$. For example, $d_{11}^0 = 0.03894$ represents the sample of unit 1 (GX gyro) for item 1 (power state U standard deviation) as 0.03894. The following sample matrix $d_{ij}^0$ is based on Table 2.

$$\boldsymbol{d}^0 = \begin{array}{c} \begin{matrix} U & I & T \end{matrix} \\ \begin{pmatrix} d_{11}^0 & d_{12}^0 & d_{13}^0 \\ d_{21}^0 & d_{22}^0 & d_{23}^0 \\ d_{31}^0 & d_{32}^0 & d_{33}^0 \\ d_{41}^0 & d_{42}^0 & d_{43}^0 \end{pmatrix} \end{array} \begin{matrix} \mathrm{GX} \\ \mathrm{GY} \\ \mathrm{GZ} \\ \mathrm{GS} \end{matrix} = \begin{pmatrix} 0.03894 & 0.002 & 0.221 \\ 0.03376 & 0.001732 & 0.234 \\ 0.03494 & 0.001732 & 0.1658 \\ 0.04576 & 0.003 & 0.1643 \end{pmatrix} \tag{10}$$

*3.4. Initialization of sample matrix*

To unify the whitening function, the sample matrix $\boldsymbol{d}^0$ is initialized by column (i.e. by item). The sample matrix $\boldsymbol{d}^1$ of the initial value

$$\boldsymbol{d}^1 = \begin{pmatrix} 1.0000 & 1.0000 & 1.0000 \\ 0.8670 & 0.8660 & 1.0588 \\ 0.8973 & 0.8660 & 0.7502 \\ 1.1751 & 1.5000 & 0.7434 \end{pmatrix} \tag{11}$$

*3.5. Regularization of sample matrix*

According to the polarity definition of item sets in Section 2.1, the prerequisite for grey evaluation is that all evaluated item sets should meet maximum polarity (the larger the sample value, the better the performance). However, all items $j = 1,2,3,4$ are standard deviations of relevant telemetry. From the perspective of gyro working stability, the smaller the standard deviation is, the better the working stability is. The sample is of minimum polarity, which does not meet the conditions of grey evaluation. Reciprocal of each element $\boldsymbol{d}^1$, the sample matrix $\boldsymbol{d}'$ has maximum polarity.

$$\boldsymbol{d}' = \begin{pmatrix} 1.0000 & 1.0000 & 1.0000 \\ 1.1534 & 1.1547 & 0.9444 \\ 1.1145 & 1.1547 & 1.3329 \\ 0.8510 & 0.6667 & 1.3451 \end{pmatrix} \tag{12}$$

*3.6. Determine the whitening function*

According to the distribution of elements in each column of the sample, appropriate threshold value of whitening function was selected, as shown in Table 3

- Grey Class 1 (Upper Grey Class)

For $j = 1$, take $c_{11} = 1.15$ ; $j = 2$ , $c_{21} = 1.15$ ; $j = 3$ , $c_{31} = 1.3$ . The $j$ and $c_{j1}$ values are substituted into equation (3) to obtain the whitening function of the upper grey class

$$f_{11}(c_{11}, \infty) = f_{11}(1.15, \infty) , \quad f_{21}(c_{21}, \infty) = f_{21}(1.15, \infty) , \quad f_{31}(c_{31}, \infty) = f_{31}(1.3, \infty) \tag{13}$$

- Grey Class 2 (Middle Grey)

For $j = 1$ , take $c_{12} = 1$ ; $j = 2$ , $c_{22} = 1$ ; $j = 3$ , $c_{32} = 1$ . The $j$ and $c_{j2}$ values are substituted into equation (4) to obtain the whitening function of the middle grey class

$$f_{12}(-, c_{12}, +) = f_{12}(-, 1, +) , \quad f_{22}(-, c_{22}, +) = f_{22}(-, 1, +) , \quad f_{32}(-, c_{32}, +) = f_{32}(-, 1, +) \tag{14}$$

- Grey Class 3 (Low Grey)

For $j = 1$ , $c_{13} = 0.85$ ; $j = 2$ , $c_{23} = 0.67$ ; $j = 3$ , $c_{33} = 0.94$ . The $j$ and $c_{j3}$ values are substituted into equation (5) to obtain the whitening function of the low grey class

$$f_{13}(0, c_{13}) = f_{13}(0, 0.85) , \quad f_{23}(0, c_{23}) = f_{23}(0, 0.67) , \quad f_{33}(0, c_{33}) = f_{33}(0, 0.94) \tag{15}$$

*3.7. Calculate the conversion coefficients*

According to theorem 1, the global sum of the whitening function threshold values is $\sum_{j=1}^{m} c_{jk}$

$$\text{For } k = 1 , \quad j = 1,\ c_{jk} = c_{11} = 1.15; j = 2,\ c_{jk} = c_{21} = 1.15; j = 3,\ c_{jk} = c_{31} = 1.3 , \sum_{j=1}^{m} c_{j1} = \sum_{j=1}^{3} c_{j1} = c_{11} + c_{21} + c_{31} = 3.6 \tag{16}$$

In the same way, $k=2$, $\sum_{j=1}^{3} c_{j2}$ =3; $k=3$, $\sum_{j=1}^{3} c_{j3}$ =2.46

By substituting $c_{jk}$ and $\sum_{j=1}^{m} c_{jk}$ values into Equation (6), the conversion coefficient (weight) $\eta_{jk}$ of whitening function $f_{jk}$ can be obtained, as shown in Table 3

Table 3. Conversion coefficients of whitening functions

| | | $c_{jk}$ | $\sum_{j=1}^{m} c_{jk}$ | $\eta_{jk} = c_{jk} / \sum_{j=1}^{m} c_{jk}$ |
|---|---|---|---|---|
| $k=1$ | $j=1$ | 1.15 | 3.6 | 0.3194 |
| | $j=2$ | 1.15 | | 0.3194 |
| | $j=3$ | 1.3 | | 0.3611 |
| $k=2$ | $j=1$ | 1 | 3 | 0.3333 |
| | $j=2$ | 1 | | 0.3333 |
| | $j=3$ | 1 | | 0.3333 |
| $k=3$ | $j=1$ | 0.85 | 2.46 | 0.3455 |
| | $j=2$ | 0.67 | | 0.2724 |
| | $j=3$ | 0.94 | | 0.3821 |

*3.8. The evaluation value of grey clustering*

Substitute $\sigma_{ik} = \sum_{j=1}^{m} f_{jk}(d_{ij}^{'})\eta_{jk}$ into equation (7), get

$$i=1, k=1: \sigma_{ik} = \sigma_{11} = \sum_{j=1}^{m} f_{j1}(d_{1j}^{'})\eta_{j1} = f_{11}(1)\eta_{11} + f_{21}(1)\eta_{21} + f_{31}(1)\eta_{31} = 0.8333 \quad (17)$$

$$i=1, k=2: \sigma_{ik} = \sigma_{12} = \sum_{j=1}^{m} f_{j2}(d_{1j}^{'})\eta_{j2} = f_{12}(1)\eta_{12} + f_{22}(1)\eta_{22} + f_{32}(1)\eta_{32} = 0.9999 \quad (18)$$

Similarly, the following table data can be obtained

Table 4. Grey clustering evaluation value

| | | $\sigma_{ik} = \sum_{j=1}^{m} f_{jk}(d_{ij}^{'})\eta_{jk}$ |
|---|---|---|
| $i=1$ | $k=1$ | 0.8333 |
| | $k=2$ | 0.9999 |
| | $k=3$ | 0.7784 |
| $i=2$ | $k=1$ | 0.9011 |
| | $k=2$ | 0.8788 |
| | $k=3$ | 0.6707 |
| $i=3$ | $k=1$ | 0.99 |
| | $k=2$ | 0.7993 |
| | $k=3$ | 0.5281 |
| $i=4$ | $k=1$ | 0.7826 |
| | $k=2$ | 0.7242 |
| | $k=3$ | 0.8252 |

### 3.9. Grey clustering evaluation

The grey clustering evaluation is carried out from the grey evaluation value of each unit according to Formula (9)

① $i=1$, GX gyro $\sigma_1$: $\sigma_1 = \begin{pmatrix} \sigma_{11} & \sigma_{12} & \sigma_{13} \\ 0.8333 & 0.9999 & 0.7784 \end{pmatrix}$ (grey cluster 1, 2, 3), $\sigma_{1k^*} = \max\{\sigma_{11}, \sigma_{12}, \sigma_{13}\} = 0.9999 = \sigma_{12}$

Indicates $k^* = 2$, $(i=1) \in \{k^* = 2\}$, the working stability of GX gyro is gray class 2: "middle". As $\sigma_{12} > \sigma_{11} > \sigma_{13}$, GX is "upper in middle".

② $i = 2$, GY gyro $\sigma_2$ : $\sigma_{2k^*} = \max\{\sigma_{21}, \sigma_{22}, \sigma_{23}\} = 0.9011 = \sigma_{21}$ . Indicates $k^* = 1$ , $(i=2) \in \{k^* = 1\}$, the working stability of GY gyro is gray class 1: "up". As $\sigma_{21} > \sigma_{22} > \sigma_{23}$, GY is "middle in up".

③ $i = 3$, GZ gyro $\sigma_3$: $\sigma_{3k^*} = \max\{\sigma_{31}, \sigma_{32}, \sigma_{33}\} = 0.99 = \sigma_{31}$. Indicates $k^* = 1$, $(i=3) \in \{k^* = 1\}$, the working stability of GZ gyro is gray class 1: "up". As $\sigma_{31} > \sigma_{32} > \sigma_{33}$, GZ is "middle in up".

④ $i = 4$, GS gyro $\sigma_4$ : $\sigma_{4k^*} = \max\{\sigma_{41}, \sigma_{42}, \sigma_{43}\} = 0.8252 = \sigma_{43}$ . Indicates $k^* = 3$ , $(i=4) \in \{k^* = 3\}$,the working stability of GS gyro is gray class 3: "low". As $\sigma_{43} > \sigma_{41} > \sigma_{42}$, GZ is "upper in low".

### 3.10. Evaluation results of gyro grey clustering

By grey clustering evaluation, the stability of each gyro is: $i=1$, GX "upper in middle"; $i=2$, GY "middle in up"; $i=3$, GZ "middle in up"; $i=4$ ,GS "upper in low".

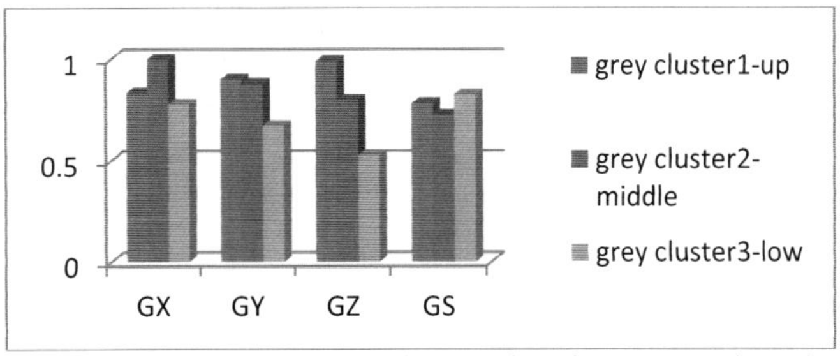


Figure 3. Histogram of evaluation results of gyro grey clustering

As shown in Figure 3, the gray clustering evaluation results of GX, GY and GZ gyros are relatively concentrated and have good consistency. The evaluation values of gray cluster 1 and 2 are significantly higher than that of gray cluster 3. However, the evaluation conclusion of GS gyro deviates from that of GX, GY and GZ gyro to a certain extent. The evaluation value of gray cluster 1 and 2 is low, and that of gray type 3 is high. In fact, GS gyro belongs to different models and its comprehensive state is inconsistent with the other three gyros. Therefore, the evaluation conclusion reflects the validity of the evaluation model from the side.

## 4. Conclusion

Based on control subsystem test evaluation as the background, this paper selected the gyro as evaluation object, application of grey clustering evaluation method of the grey system theory, set up two floating gyro evaluation index system and evaluation model, and the model is verified by using telemetry data, finally got the evaluation conclusion of gyro, show the effectiveness of the model from the side, A useful attempt has been made on the control system test and evaluation technology.

**Reference**

[1] Zhan Xianglin, Meng Yanjun, Zhang Weiye. Effectiveness Evaluation of Aircraft Hydraulic Energy System Based on Improved Grey Clustering Method[J]. Computer Integrated Manufacturing Systems,2021,38(7),45-49.

[2] Huang Kui, Zhang Siyu, Wang Zhen. An Aviation Equipment Reliability Evaluation Model Based on Grey Clustering[J]. Ship Electronic Engineering,2021,41(1),128-131.

[3] Li Qing, Yang Xiaohui, Liu Zhensheng. Graded warning scheme of transmission line galloping based on grey clustering analysis[J]. Electrical Measurement and Instrumentation.2020,57(17),45-51.

[4] Xiang Lei, Yang Xin, Zhang Yang. Effectiveness Evaluation for Satellite System Based on Analytic Hierarchy Process and Fuzzy Theory[J]. Computer Integrated Manufacturing Systems, 2013,30(2): 55-61.

[5] Yang Haitao. Research on Methods of Constellation Satellite Communication System Effectiveness Evaluation[D]. Master dissertation of National University of Defense Technology, 2007.

[6] Deng Julong. Basis of Grey theory[M]. Wuhan: Huazhong University of Science and Technology Press, 2002,397-421.

# State assessment of 750kV electric energy metering device based on grey theory and fuzzy analytic hierarchy process

**Kun Wang**[1]

[1] State Grid Gansu Electric Power Company Marketing Service Center, Lanzhou, Gansu, 730050, China

*Corresponding author's e-mail: wangkun@gs.sgcc.com.cn

**Abstract.** It is necessary to evaluate the operation state of 750kV electric energy metering device to predict stop loss. On the basis of previous studies, the state indexes and evaluation methods of electric energy metering devices are improved: the quantitative state indexes are described by nonlinear mathematical scoring model, the qualitative state indexes are clustered by grey theory, and the state weight of each index is determined based on the improved analytic hierarchy process, Finally, the state evaluation results are obtained with the help of weighted average model and fuzzy mathematics theory. The improved state evaluation strategy is more perfect and practical, and realizes the all-round supervision of the operation state of the electric energy metering device.

## 1. Introduction

The 750kV electric energy metering device (EEMD) is an important tool for accurate measurement and accurate fair trade settlement between power generation and transmission[1]. The connection unit of 750kV EEMD is complex, and there are many indicators affecting the operation state. Properly selecting and classifying the evaluation indicators has become the primary problem to be solved in the state evaluation. A feasible scheme is to divide each state index into three parts: basic state, operation state and auxiliary decision-making[2]. It seems more reasonable to divide the status indicators from four aspects: historical operation records, real-time operation performance, component configuration and working condition environment[3].

In this paper, the hierarchical model of state index of 750kV EEMD is introduced, and the quantitative state index is described by nonlinear mathematical scoring model; The fuzzy theory and grey theory are used to cluster the state indexes and determine the state evaluation vector; The state weight vector is determined based on the improved analytic hierarchy process, and the state evaluation results are obtained with the help of weighted average model and fuzzy mathematics theory; Finally, an application example shows the rationality and effectiveness of the method.

## 2. Construction of state index hierarchical model

According to the operation characteristics of EEMDs, the status indicators can be divided into two categories: technical indicators and management indicators. Among them, the technical and management indicators can be further divided into a number of indicator subdivisions. In order to cover all levels of indicators of the energy metering device and simplify the complexity of the hierarchical model, this paper improves the indicator framework to obtain the improved hierarchical model of state indicators as shown in Table 1.

Table 1. State index level model of EEMD

| | Technical indicators | | Management indicators | |
|---|---|---|---|---|
| | Performance Indicators | Reliability | Component Configuration | Work environment |
| PT | Field error calibration | Family defect | Model selection | Voltage、Frequency fluctuation |
| | Laboratory calibration | Historical faults | Auxiliary configuration | Load properties |
| CT | Field error calibration | Family defect | Model selection | Current、Frequency fluctuation |
| | Laboratory calibration | Historical faults | Auxiliary configuration | Load properties |
| Energy meter | Field error calibration | Family defect | Model selection | Voltage、Current、Frequency fluctuation |
| | Laboratory calibration | Historical faults | Auxiliary configuration | Power factor、Harmonic |
| Secondary circuit | PT secondary pressure drop error | Family defect | Secondary wire cross-sectional area | Switching load transient |
| | CT/PT secondary load | Historical faults | | |

Status indicators can be divided into two categories: quantitative status indicators and qualitative status indicators. For the scoring of each status index of the electric energy metering device, the deduction value can be calculated according to the degree of status quantity deterioration. To address the problem of large bias in evaluation results due to the use of expert scoring evaluation method in the existing literature, this paper uses a nonlinear evaluation model to score the quantitative status indicators, as shown in Table 2. Analysing the historical data of expert ratings, the quantitative indicators can be divided into three evaluation models for calculation: (1) for the error measurement value, the closer to the zero value the higher the rating of the indicator, so the full trapezoidal scoring model is used for calculation; (2) for the frequency deviation value, the smaller the deviation the higher the rating of the indicator, so the descending half trapezoidal scoring model is used for calculation; (3) for the discrete indicators, the stepped scoring model is used for calculation.

Table 2. Quantitative status index system

| Scoring Model | Function expression |
|---|---|
| Full trapezoidal scoring model | $f(x)=\begin{cases}0, x\leq x_1 \cup x\geq x_4\\ \frac{1}{x_2-x_1}(x-x_1), x_1<x<x_2\\ 1-\frac{1}{x_4-x_3}(x-x_3), x_3<x<x_4\\ 1, x_2<x<x_3\end{cases}$ |
| Descending half trapezoidal scoring model | $f(x)=\begin{cases}1, x\leq x_1\\ 1-\frac{1}{x_2-x_3}(x-x_3), x_1\leq x\leq x_2\\ 1, x>x_2\end{cases}$ |
| Stepped scoring model | $f(x)=\begin{cases}1, x\leq x_1\\ 0.5, x_1<x<x_2\\ 0, x\geq x_2\end{cases}$ |

## 3. Cluster analysis of state indicators

After obtaining the score of each state index according to the scoring model, it is necessary to determine the state category to which the score value of the state index belongs, that is, the cluster analysis of the state index. According to the different index characteristics, this paper adopts fuzzy clustering method for quantitative state indexes and grey clustering method for qualitative state indexes, and the possibility of affiliation with each state category is used as the final evaluation result.

### *3.1. Fuzzy cluster analysis of quantitative state index*

Using the characteristics of fuzzy relation synthesis, fuzzy clustering divides the change interval of the evaluated index, and analyses the correlation degree of the index belonging to each level through the mapping of membership function. The specific steps are as follows:

(1) Establish a comment set of evaluation indicators $V$; Example of comment sets is $V = (\text{Excellent}, \text{Good}, \text{Medium}, \text{Poor})$;

(2) Construct the membership function of evaluation index $F$;

(3) Fuzzy evaluation index;

(4) Maximum membership principle.

### *3.2. Grey clustering analysis of qualitative state indexes*

Grey clustering is to classify the whitening number of clustering objects for different clustering indicators according to several grey categories, and judge the type of clustering objects by calculating the comprehensive effect of all indicators, that is, the process from grey to white. The specific mathematical expression is as follows:

Set $N$ clustering objects, $M$ clustering indexes and $S$ Grey classes, $d_{ij}$ the sample of the i-th clustering object to the j-th clustering index, the sample matrix $D = (d_{ij})_{N<>M}$. $F$ is mapping, $Opf_{jk}(d_{ij})$ is the sample of $d_{ij}$, $f_{jk}(d_{ij})$ is the whitening weight function of the j-th index k grey class, $\sigma_{ik}$ is the weight of mapped grey clustering:

$$F: Opf_{jk}(d_{ij}) \rightarrow \sigma_{ik} \in [0,1] \tag{1}$$

Defined $\lambda_j^k$ as the critical value of j index k grey class and the weight of j index k grey class, then:

$$\eta_{jk} = \frac{\lambda_j^k}{\sum_{j=i}^{M} \lambda_j^k} \tag{2}$$

Grey clustering coefficient of object i belongs to K grey class:

$$\sigma_{ik} = \sum_{j=1}^{M} f_{jk}(d_{ij}) \times \eta_{jk} \tag{3}$$

The set of grey clustering coefficients $\sigma_{ik}$ for each grey class to which object i belongs is the grey clustering set $R$. The greater the value of clustering coefficient in the set, the greater the degree that the index belongs to a certain grey class.

## 4. Weight analysis of status indicators

Since the degree of importance of each state assessment index varies, the relative importance of each index, i.e., weights, needs to be analysed. Hierarchical analysis has been widely used to determine the weights of multi-level, multivariate, structurally complex systems, and is an effective weight calculation method that combines quantitative and qualitative analysis. The state index of 750kV EEMD is affected by many factors, which is suitable for the situation where it is difficult to evaluate the decision results directly and accurately. The calculation steps of the improved hierarchical analysis method are as follows:

(1) Establish the hierarchy as shown in Table 1;

(2) Determine the importance matrix $A = (a_{ij})_{m\times m}$, where $a_{ij}$ is defined as follows:

$$a_{ij} = \begin{cases} 2, a_i \text{ is important than } a_j \\ 0, a_j \text{ is important than } a_i \\ 1 \end{cases} \tag{4}$$

(3) Construct the judgment matrix $B = (b_{ij})_{m\times m}$, where $b_{ij}$ is defined as follows:

$$b_{ij} = \begin{cases} \frac{r_i - r_j}{r_{max} - r_{min}} \times (d_m - 1) + 1, r_i \geq r_j \\ [\frac{|r_i - r_j|}{r_{max} - r_{min}} \times (d_m - 1) + 1]^{-1}, r_i \leq r_j \end{cases} \tag{5}$$

where $r_i = \sum_{j=1}^{m} a_{ij}$, $r_{max} = \max(r_i)$, $r_{min} = \min(r_i)$, $d_m = r_{max}/r_{min}$.

(4) Calculate the optimal transfer matrix $L = (l_{ij})_{m\times m}$, where $l_{ij}$ is defined as follows:

$$l_{ij} = \frac{1}{m}\sum_{k=1}^{m} lg\frac{b_{ik}}{b_{jk}} \tag{6}$$

(5) Construct $B^* = (b_{ij}^*)_{m\times m}$, which is the fitted consistent matrix of $L$, and $b_{ij}^*$ is defined as follows:

$$b_{ij}^* = 10^{l_{ij}} \tag{7}$$

(6) Find the vector of state indicator weights $\Omega$. The eigenvector a corresponding to the maximum eigenvalue of $B^*$ is calculated, and after normalization, the improved weight vector $\Omega = [\omega_1, \omega_2, \ldots, \omega_m]$, which indicates the relative importance of each condition indicator, is obtained. According to the initial state judgment matrix $R$ in the cluster analysis in Section 3.2 and the weight vector $\Omega$, and principle of harmony law, integrated status judgment vector $P$ for status assessment indicator categories can be calculated as follows:

$$P = \Omega \times R \tag{8}$$

From the principle of maximum affiliation, it is known that the greater the value of affiliation in the vector set P, the greater the possibility that the indicator category belongs to the corresponding level, i.e., the status evaluation result of the indicator category is the status level corresponding to the maximum affiliation value. For different status assessment results, operation and maintenance managers can make management decision plans. If the assessment result is excellent, the maintenance cycle of the equipment can be extended appropriately to save manpower and material resources; when the assessment result is poor, the maintenance cycle of the equipment should be shortened or the equipment should be overhauled immediately to reduce or avoid the production loss caused by the equipment problem.

## 5. Case analysis of condition assessment

Take the state evaluation of secondary circuit as an example. The status indexes of the secondary circuit of a 750kV gateway EEMD are scored, and the results are shown in Table 3.

Table 3. Scores of secondary loop status indicators

| Indicator type | Indicator name - Number | Index score |
|---|---|---|
| Quantitative index | PT secondary pressure drop error-1 | 0.825 |
| | PT secondary load-2 | 0.4 |
| | CT secondary load-3 | 0.75 |
| | Historical fault-4 | 0.8 |
| | PT secondary wire cross-sectional area-5 | 0.8 |
| | CT secondary wire cross-sectional area-6 | 0.8 |
| | Switching load transient-7 | 0.6 |
| Qualitative index | Secondary circuit closeness-8 | 8.9 |
| | Historical faults -9 | 9.5 |

For the quantitative state indicators, the measured values of PT secondary pressure drop error, PT secondary load and CT secondary load are applied to the intermediate scoring model according to Section 2; historical faults and switching load transient are suitable for the step-down scoring model; PT secondary conductor cross-sectional area and CT secondary conductor cross-sectional area are suitable for the larger, the better scoring model. After determining the parameters of each scoring model according to the method described in Section 2, the scores of the measured values of each state index are obtained; the average of multiple experts' scores was used as the final score for all qualitative indicators.

After determining the parameters of each scoring model according to the method, the scores of the measured values of each state index can be obtained; For quantitative status indicators, the score value of each indicator corresponds to the membership function model of "excellent, good, medium, low and poor" of each comment, as shown in Fig 1.

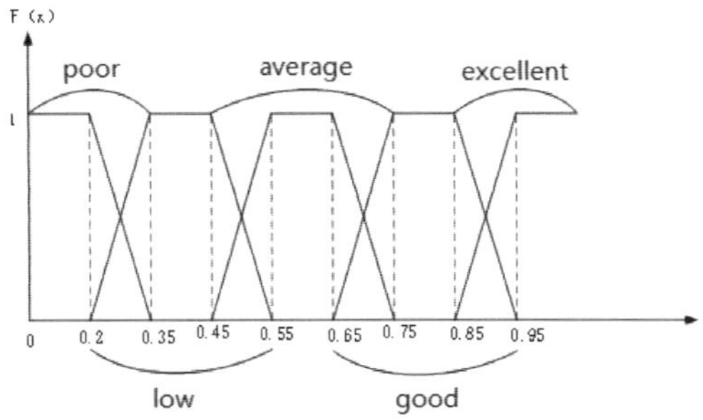


Figure 1. Membership function of quantitative state index

For qualitative status indicators, the scoring value of each indicator corresponds to the whitening weight function model of "excellent, good, medium, low and poor" of each grey category, as shown in Fig 2.

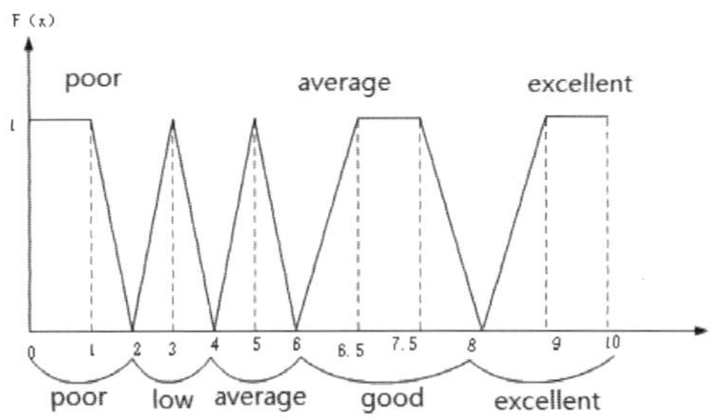


Figure 2. Whiten weight function of qualitative index

The scoring values of each qualitative state index in Table 1 range from 0 to 10. In Figure 2, odd discrete partition values of 1, 3, 5, 7 and 9 are used as the critical values of the whitening weight function of the lower limit measure of difference class, the moderate measure of low class, the moderate measure of medium class, the typical side of good class and the upper limit measure of superior class. Taking the critical value as the center, the numerical expansion is carried out, and the discrete domain is transformed into a continuous domain, which is in line with the clustering definition of the score value of qualitative state index.

Rank the status indicators covered by performance indicators, reliability, component configuration and working condition environment to obtain the comprehensive clustering matrix to which the score value of each status indicator. The weight $\Omega_1 = [0.56,0.22,0.22]$ of the state index covered by the performance index, then the combined clustering of the performance index calculated by the weighted average model the composite clustering phase quantity $P_1 = \Omega_1 \times R_1 = [0,0.22,0,0.78,0]$. Similarly, the integrated clustering phase $P_2 = [0,0,0,0.44,0.56]$ for reliability index, $P_3 = [0,0,0,0.72,0.28]$ for component configuration, and $P_4 = [0,0.17,0.61,0.22,0]$ for operating environment can be obtained.

According to the comprehensive clustering phasors $P_1$, $P_2$, $P_3$, $P_4$ and the maximum membership principle, it is likely that the evaluation results of performance index, reliability index, component configuration index and working condition environment index are "good", "excellent", "good" and "medium" respectively. Similarly, the technical and management indicators and the evaluation results of the whole secondary circuit are likely to be "good", indicating that the secondary circuit is in good operation.

The clustering results of the performance indicators are analysed and the initial clustering matrix of PT secondary pressure drop error, PT secondary load, and CT secondary load is :

$$R_1 = \begin{bmatrix} r_1 \\ r_2 \\ r_3 \end{bmatrix} = \begin{bmatrix} 0 & 0 & 0 & 1 & 0 \\ 0 & 1 & 0 & 0 & 0 \\ 0 & 0 & 0 & 1 & 0 \end{bmatrix} \quad (9)$$

The assessment presented as $r_2$ is low due to a low PT secondary load score. From the weights $\Omega_1$ of the three sub-indicators, we can see that the weight of PT secondary load is low, which makes the combined clustering result of the performance indicators good more likely.

Thus, when the difference in affiliation between the categories is not significant, information on similar affiliation categories should be appended to the maximum affiliation principle to help evaluators perform further screening and testing.

## 6. Conclusion

This paper improves the state index and evaluation method of 750kV EEMD. The example calculation results and analysis show that the improved state evaluation strategy can realize the all-round supervision of the operation state of 750kV EEMD.

## Acknowledgement

This work is support by Science and Technology Project of State Grid Gansu Electric Power Company(No. 52273119100B).

## References

[1] Huang Liping, Wang Hui, Du Weihua. Operation evaluation methods for energy metering devices[J]. East China Electric Power, 2014, 42 (11): 2322-2326.

[2] Du Weihua, Cao Yi, Li Da. Application of state evaluation technology in management of gateway electric energy metering device[J]. East China Electric Power, 2013, 41 (10): 2107-2110.

[3] Tang Farong. The application of the fuzzy synthetic evaluation method based on the specialist grading system in the load forecasting of power system[J]. Journal of Electric Power, 2011, 32( 5) : 380-382.

# Study on Processing Technology of Aluminium-based silicon carbide material for Shafting Installation

**Lin Li[1], Hao Li[1], Bin Zhang[1], Tianming Wang[1], Junyu Yu[1], Lei Mao[1] ,Jialin Wei[1], Ning Huang[1*], and Lei Chen[1]**

[1] Beijing Spacecrafts Co., Ltd., Beijing 100094, China

*Corresponding author's e-mail: huangning161@126.com

**Abstract:** Aluminium-based silicon carbide (SiCp/Al) composites are widely used in spacecraft structures due to their excellent properties. Aiming at the high-precision machining requirements of thin-walled shell components installed in SiCp/Al shafting in aerospace field. The structural characteristics of thin-walled shell components are analysed. Combined with material characteristics and mechanical properties, a single clamping vertical horizontal conversion machining scheme is formulated. Match the right machine tools and PCD tools. Then the metal cutting process is simulated by AdvantEdge, and the tool structure parameters and cutting process parameters are optimized. Finally, the process route suitable for installing housing components is formulated, with dimension accuracy of H6, cylindricity of 0.003mm and coaxiality of 0.006mm, which solves the difficult problem of machining high-precision shaft mounting housing.

## 1. Introduction

Aluminium-based silicon carbide (SiCp/Al) composite is a new material with excellent mechanical and physical properties such as high specific strength and stiffness, low coefficient of thermal expansion, high yield strength, low density, good thermal conductivity, wear resistance and fatigue resistance [1-3].Due to the lightweight, high strength and high stability of SiCp/Al composites and the secondary processing with traditional equipment, SiCp/Al composites are widely used in satellite supports, satellite bearings, satellite antennas and other devices, and have become indispensable materials in the aerospace field [4-6].

SiCp/Al mainly achieves dimensional accuracy and surface roughness through machining. SiCp/Al contains SiC particles with high hardness, high strength and strong corrosivity, and the hardness is higher than that of common tools. Therefore, SiCp/Al is a difficult machining material. During cutting, the tool wear is serious, which shortens the service life of the tool. With the increase of SiC, the machinability of SiCp/Al becomes worse and worse, which limits the improvement of cutting speed, machining efficiency and accuracy [7-9]. In order to improve the processing efficiency and quality of SiCp/Al, scholars at home and abroad have studied the processing technology. Ding et al. [10] carried out the wear test of machining SiCp/Al with PCB and PCD tools. The results show that PCD tools have better wear resistance and fracture resistance than PCBN tools. Dabade et al. [11] used CBN tools to study the machining surface quality of SiCp/Al under different process parameters. The results show that the feed rate is an important factor affecting the cutting force and surface roughness. In addition, Paulo Davim and Muthu Krishnan [12-13] found that the increase of milling rate will also accelerate tool wear, and the decrease of feed rate will reduce machining efficiency. The cutting test of SiCp/Al composites conducted by Zhang Wenfeng et al. [14] pointed out that good comprehensive benefits can be obtained when the cutting speed is 30~40m/min. Wu Zhenyu et al. [15] found that when the cutting depth is 0.5mm and the feed rate is 0.02mm/z, the cutting speed is in the range of 10~40m/min, and the milling surface quality is better. Wang Zhida et al. [16] studied the influence of laser assisted cutting technology on SiCp/Al processing quality. The results show that the cutting stress of SiCp/Al decreases after heating, and the laser assisted cutting technology can improve the surface roughness. Wang Dazhen et al. [17] studied the precision machining process of SiCp/Al with a volume fraction of 45%. The results show that with the increase of cutting speed and the decrease of feed rate, the roughness will decrease, and there will be a very thin machining layer on the ultra-precision machined surface. Ge Yingfei et al. [18] conducted experimental research on ultra precision machining of 15% SiCp/Al, the results show that the feed rate has a significant effect on the surface roughness, followed by the cutting speed and the cutting depth. Cao Dewei et al. [19] controlled the deformation of motor shell by using YG6X and PCD cutting tools to process 30% volume fraction SiCp/Al, optimizing rough turning semi finish turning finishing machining route, arranging heat treatment to eliminate residual stress and other process methods. To sum up, SiCp/Al processing presents problems such as low cutting speed, small feed rate and serious tool wear, The SiCp particles in the cutting process are broken, fallen off and scratched, which will cause poor surface quality. Therefore, in order to obtain good machining efficiency and quality, we need to reasonably select cutting tools and cutting parameters, and also need to be combined with the method of planning reasonable process route.

The shaft mounting shell component of a spacecraft is made of SiCp/Al, and the machining accuracy directly determines the transmission performance of the moving device. Due to the difficult machinability of SiCp/Al, there are some problems such as serious tool wear, out of tolerance of machining size and low machining efficiency.

Therefore, this paper focuses on the characteristics of high shape and position accuracy, complex structure and difficult machining of shafting installation shell components, mainly from the selection of machining equipment, tool selection, cutting parameters and process flow design. Formulate a scheme to solve the deformation and accuracy out of tolerance in the processing of SiCp/Al shafting installation shell components, and overcome the processing technology problems of shafting installation shell components.

## 2. Housing material and structure analysis

### *2.1 Performance analysis of Aluminium-based silicon carbide materials*

Due to the different content of silicon carbide, the mechanical and physical properties of the prepared SiCp composites are also changed.For the shafting installation housing component in this study, the volume fraction of silicon carbide of Aluminium-based silicon carbide composite material selected is 17%, which is manufactured by powder metallurgy method. It is made by vacuum hot pressing after mixing aluminum matrix powder and silicon carbide reinforced particles. The heat treatment state is T6. The measured mechanical and physical properties are shown in Table 1.

Table 1. Mechanical and physical properties of Aluminium-based silicon carbide composites

| Name | Tensile strength M Pa | Yield Strength M Pa | Elongation after breaking % | Hardness stiffness HBS | Coefficient of thermal conductivity W/m·K | Young's modulus G Pa | Coefficient of thermal expansion $10^{-6}$/K a |
|---|---|---|---|---|---|---|---|
| 17%Si $C_p$/Al | ≥490 | ≥400 | ≥3.5 | 180 | ≥155 | 105±5 | 17±1 |

It can be seen from table 1 that the hardness of 17% Aluminium-based silicon carbide material is very high and it is difficult to process. If the traditional processing method is adopted, it is easy to produce chip buildup, serious tool wear, and the phenomenon of tool tip collapse in the cutting process. The tool life is greatly reduced, it is difficult to obtain high-quality machining surface, and the corresponding shape and position, dimensional tolerance and roughness are difficult to ensure.

*2.2 Structural analysis of housing parts*

The shafting installation housing adopts split housing structure, as shown in Figure 1. The installation interfaces of three gears and related gear accessories shall be provided inside the shafting mounting housing components, and the structural stiffness of the installation of the transmission pair shall be ensured. The transmission accuracy of the gear directly determines whether the device can realize transmission, extraction, shear and other movements, and the whole movement process requires high reliability.Bearing mounting surface and bearing mounting holes are distributed on two parts, which need to be assembled into components for combined machining to ensure the dimensional and position accuracy of bearing holes.The shaft installation housing components adopts split housing structure, and the overall dimension is 219.5mm × 90mm × 158mm.

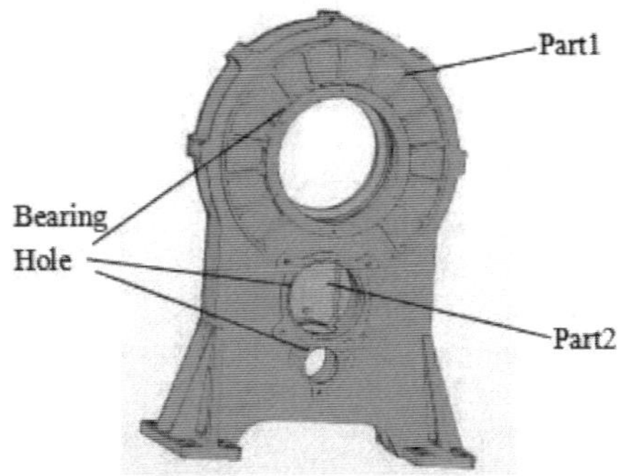


Figure 1. shafting mounting housing components model

The housing components has the structural features of complex structure, easy deformation of thin-wall and high accuracy of shape and position and difficult processing:

(1) The housing component is a thin-walled part with complex structure, with poor rigidity and minimum wall thickness of 1mm. It is easy to cause tool tremor during processing. Many bearings are installed, which is easy to deform and exceed the tolerance during processing. Therefore, it is necessary to study and select the processing tools.

(2) The design accuracy of parts is high, the dimensional accuracy of bearing hole is 0.006mm, and the cylindricity is 0.003mm,and the upper and lower coaxiality of shafting installation hole is 0.006mm. It is necessary to reasonably select the processing mode and processing machine tool;

The component device has high requirements for shape and position accuracy, dimensional accuracy and surface roughness of parts. The maximum bearing hole is Φ62mm,and the maximum dimensional tolerance of bearing mounting hole is 0.008mm, and the roughness of bearing hole is required to be Ra0.8. The parallelism between each bearing hole and the bottom surface of the shafting mounting housing components is better than 0.012mm.The coaxiality between the upper and lower bearing holes is better than 0.006mm.The cylindricity is better than 0.003mm.The perpendicularity between the middle bearing installation end face and the right bearing hole is better than 0.015mm.The parallelism between the left and right bearing holes is better than 0.005mm. The bearing installation surface and bearing installation holes are distributed on the upper part of the shafting housing and the lower part of the shafting installation housing.

## 3. Study on processing technology

Due to the high hardness silicon carbide particles contained in the part material, the peeling of silicon carbide particles is easy to cause product edge collapse, serious tool wear and other problems in the machining process, and it is difficult to ensure the machining consistency, surface quality and machining accuracy. Therefore, the part processing technology needs to reasonably select the processing equipment, processing tools and processing parameters, Formulate reasonable processing technology.

*3.1 Selection of machine tool*

In view of the complex structure and high-precision geometric tolerance of the housing parts, it is necessary to select suitable machine tools and design reasonable tooling for one-time clamping. According to the characteristics of the shaft system installation housing components, it can be seen that simple milling and traditional turning can not meet the requirements of dimensional accuracy and shape and position accuracy of parts, but can only be realized by vertical boring. The machining machine with high machining accuracy must be selected, and the requirements of high stiffness, high stability, high retention and high automation should be met. Therefore, the Swiss weiliming 528jig five axis machine tool is selected in this study. Its positioning accuracy is 2 microns and the repeated positioning accuracy is 1.5 microns. It can realize the vertical and horizontal conversion, ensure that all housing components are processed at one time, and avoid the loss of accuracy due to repeated clamping.

*3.2 Tool selection*

The Mohs hardness of silicon carbide particles dispersed in Aluminium-based silicon carbide composites is 9.5, which is second only to the hardest diamond. The hardness is higher than that of common tools (such as high-speed steel tools and cemented carbide tools), which will cause severe tool wear in the process of machining. Therefore, high-speed steel tools are not suitable for the processing of Aluminium-based silicon carbide materials. PCD tool has high hardness and toughness of diamond, which is higher than that of single crystal diamond, and the blade matrix has high bending strength. It is the best choice for processing this material, but the manufacturing of PCD tool is difficult and the cost is high [20-21]. Therefore, in this paper, four kinds of cutting tools made of PCD, PVD, CBN and YG6X were selected to carry out processing comparative tests on SiCp/Al composites, and the wear morphology of the machined cutting tools was analyzed and studied by using a variety of analysis and testing means such as scanning confocal microscope, scanning electron microscope (SEM), PGI1240 surface measuring instrument and X3000 stress tester.

Fig. 2 is a photograph of tool wear morphology taken by scanning confocal microscope(1:80)and scanning electron microscope(1:300) respectively. It can be seen that the tools of four different materials have different degrees of wear, and the scratches parallel to the cutting direction are clearly visible in the wear morphology of the tool surface. Through the comparison of the wear amount of the four tools, it can be seen that YG6X tool > CBN tool > PVD tool > PCD tool, and YG6X tool has the phenomenon of tool tip collapse in the cutting process, while PCD tool has almost no wear.This is because the Aluminium-based silicon carbide material is easy to produce chip accumulation tumor in the cutting process. The exposed silicon carbide particles continuously scratch the cutting edge, and the particles on the transition surface are continuously pressed into or pulled out of the matrix, resulting in bulges or pits on the material surface, resulting in the scratch of the tool cutting edge. YG6X tool is a cemented carbide tool with high density caused by fine grain structure. Generated debris tumoris larger than the other three tools.It is easier to form bulges or pits to scratch the tool, and its wear is the largest. The high hardness of PCD tool makes it almost free of chip accumulation tumor on the tool face, so the wear is the smallest.

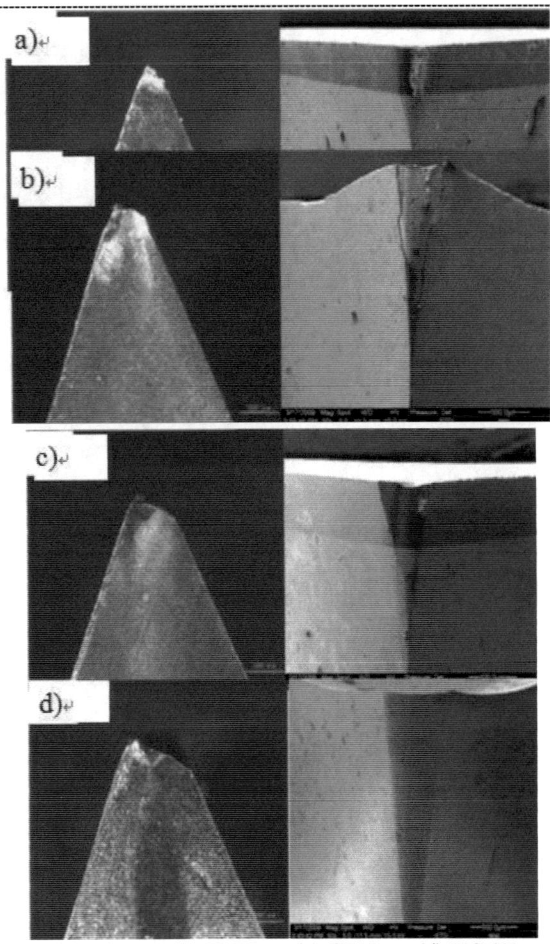


(a) PCD (b) PVD (c) CBN (d) YG6X

Figure 2. Wear morphology of four cutting tools after cutting

Therefore, combined with the practical experience of parts processing, when processing 17% SiCp/Al materials, taking into account the processing cost and processing quality, cemented carbide YG6X tools with relatively low price and good wear resistance can be selected in the rough processing process, and PCD tools can be selected in semi finishing and finishing.

*3.3 Optimization of machining program and tool structure parameters*

When PCD tool is used to mill the bearing bore of parts, the rotating speed of the machine tool is greater than 1000 rpm and the cutting amount is less than 0.1mm. Based on the physical simulation of metal cutting process AdvantEdge, the physical simulation of the milling process of bearing hole of SiCp / Al shell assembly is carried out (as shown in Figure 3). With the goal of balancing the cutting force, the axial force is reduced during finish machining (using the residual body stiffness, large cutting depth / small cutting width milling mode), the feed speed in the NC machining program is improved (as shown in Figure 4), and the deformation of the assembly is controlled in the machining process.

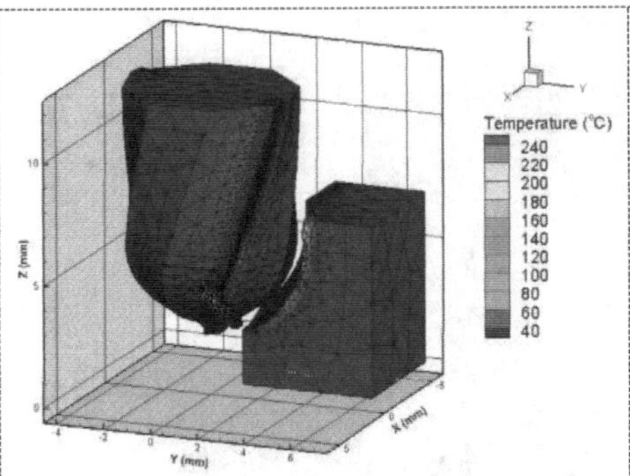


Figure 3(a). Finite element simulation

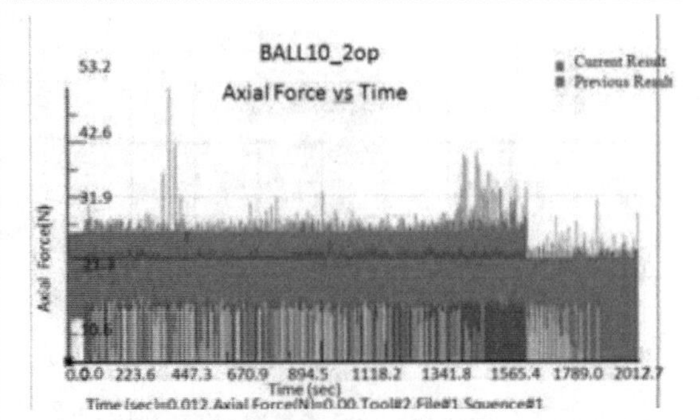


Figure 3(b). Metal cutting process simulation of milling process

```
N191 G01 X122.
N192 Y.018
N193 X133.
N194 Y63.
N195 X33.93
N196 G02 X52.209 Y48.933 I-33.93 J-63.
N197 X49.479 Y33.048 I-7.372 J-6.91
N198 G01 X46.018 Y31.821
N199 G02 X40.225 Y31.204 I-8.05 J48.057
N200 X30.802 Y31. I-9.26 J210.108
N201 G01 X-30.815
N202 G02 X-39.972 Y31.193 I-.147 J210.677
N203 X-46.034 Y31.821 I1.803 J46.947
N204 X-54.114 Y38.004 I1.74 J10.644
N205 X-52.356 Y48.776 I8.954 J4.069
N206 X-33.93 Y63. I52.356 J-48.776
......
```

```
N191 G01 X122. F3719.71
N192 Y.018 F5063.78
N193 X133. F3718.23
N194 Y63. F3719.71
N195 X33.93F3719.71
N196 G02 X52.209 Y48.933 I-33.93 J-63. F4488.02
N197 X49.479 Y33.048 I-7.372 J-6.91 F5511.83
N198 G01 X46.018 Y31.821 F5344.42
N199 G02 X40.225 Y31.204 I-8.05 J48.057 F5258.95
N200 X30.802 Y31. I-9.26 J210.108 F5222.85
N201 G01 X-30.815 F5195.92
N202 G02 X-39.972 Y31.193 I-.147 J210.677 F5259.9
N203 X-46.034 Y31.821 I1.803 J46.947 F5334.92
N204 X-54.114 Y38.004 I1.74 J10.644 F5709.18
N205 X-52.356 Y48.776 I8.954 J4.069 F3717.86
N206 X-33.93 Y63. I52.356 J-48.776 F3717.93
......
```

Figure 4. NC program before and after optimization

According to the above analysis, when turning, the rotation speed of the machine tool shall be more than 800 rpm to obtain better surface finish and reduce tool wear. Due to the poor impact resistance of PCD tools, the cutting amount is generally less than 0.2mm, and the cutting amount during finishing is less than 0.1mm.

At the same time, the typical turning parameters of PCD tool machining are selected: v = 80m / min, f = 0.002mm/r, AP = 0.02mm. The cutting simulation research of aluminum based silicon carbide machining with three PCD turning tools with different tip shapes (as shown in Figure 5) is carried out, and the shell semi finishing and finishing tools are selected.

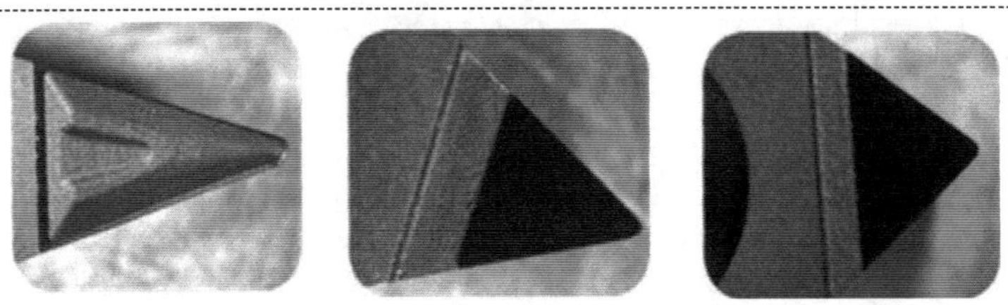

(a)35° (b)55° (c)80°

Figure 5. PCD turning tool

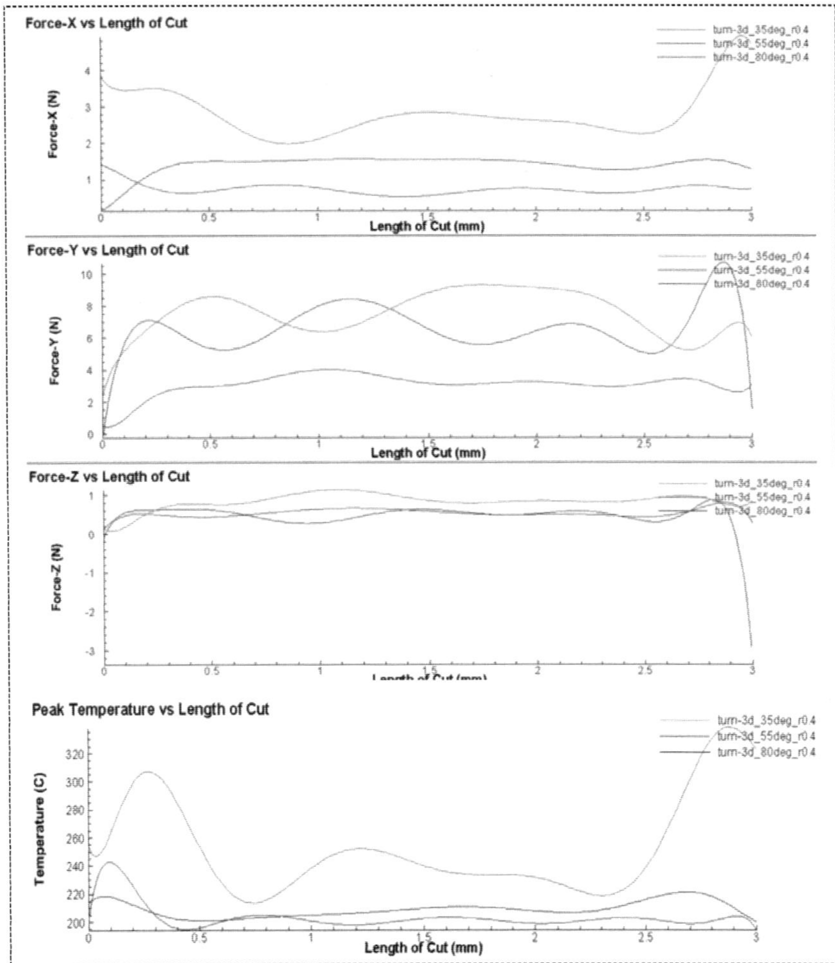


Figure 6. Simulation diagram of Aluminium-based silicon carbide cutting

As shown in Figure 6, the cutting simulation results of machining Aluminium-based silicon carbide with three different tools of 35°, 55°and 80°are shown respectively. The simulation results show that the cutting forces of 35°, 55°and 80°are 8N, 3N and 7n respectively, and the cutting temperatures are 260°C, 200°Cand 220°Crespectively. The above cutting parameters and three kinds of cutting tools are used for turning test to detect the roughness of the test bar after turning. The roughness of the test bar after machining with 35°, 55°and 80°turning tools is 0.21um, 0.09 um and 0.31um respectively. 55°/r0. 4 turning tools are better than the other two groups.

According to the cutting force, cutting thermal simulation results and test results, 55°turning tool is more suitable for semi finishing and finishing of Aluminium-based silicon carbide.

### *3.4 Process design*

In order to ensure the surface quality of housing parts and meet the tolerance accuracy requirements such as coaxiality and parallelism, the process methods of rough machining finish machining, separate machining and secondary machining after components are adopted. First, roughen the two parts of the shafting installation housing respectively, and leave 3mm machining allowance (single side) at the position with high accuracy of geometric tolerance and dimensional tolerance. Then assemble the two parts of the shaft system installation housing, and carry out combined finish machining on the bearing hole and the position with high geometric tolerance accuracy. The main processing process of housing components is as follows:

(1) Wire cutting: before rough machining, install the shafting on the upper and lower parts of the housing for wire cutting, remove most of the allowance, reduce the amount of cutting tools and reduce the processing cost;

(2) Rough machining + heat treatment: rough machining the upper and lower surfaces of the shafting installation housing and heat treatment strengthening. Allowance shall be reserved at the position with high accuracy of geometric tolerance and dimensional tolerance, allowance shall be reserved for the thickness of the upper and lower surfaces of the shafting installation housing, and other dimensions shall be processed in place;

(3) Semi-finishing after components: assemble the upper and lower parts of the shafting housing, process the pin hole, install the pin, and then semi finish the parts. Take the installation surface of the driving mechanism as the datum plane, and the processing bottom surface is the vertical processing surface;

(4) Finish machining after components: clamp and align the parts on the machine tool, finish boring each hole and surface, disassemble the parts after machining, and process the bearing mounting hole at the lower part of the shafting mounting housing. PCD tool with tip shape of 55 ° shall be selected during finishing;

(5) Monitoring + error compensation: strictly control the temperature of the processing site during the processing process to avoid excessive temperature change, resulting in out of tolerance processing. After processing, the accuracy measurement is realized using CMM. According to the temperature of the measurement site, the temperature difference of the measurement results is converted into the part processing program in the compensation NC machine tool.

Among them, due to the influence of measuring tools, ambient temperature and other factors, there will be errors between on-line measurement and precision measurement. In the process of finish machining, it is necessary to compare and detect them, so as to provide machining basis for machine tool operators. At the same time, the finishing and precision detection process need to be repeated many times, that is, the online precision measurement is carried out when machining the allowance of 0.1 ~ 0.2mm each time, and the temperature compensation is carried out for the machining program according to the on-site temperature, so as to judge the measurement error caused by factors such as environment and personnel, and make corresponding compensation in the last precision machining, so as to meet the requirements of machining accuracy.

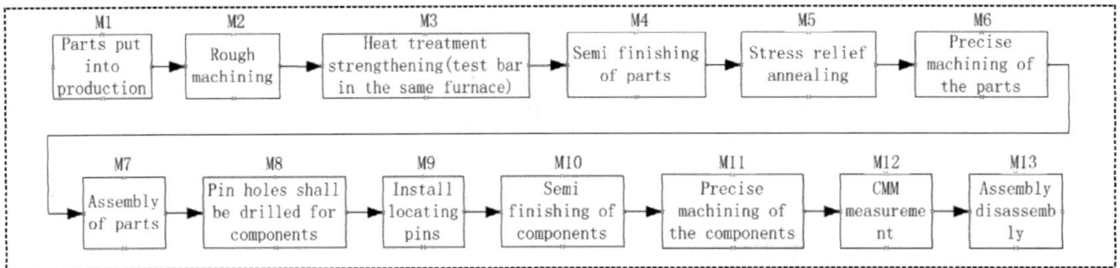


Figure 7. Technological process

## 4. Machining accuracy results

Based on the above comparative tests and process methods, two parts of the shafting mounting housing components made of aluminium-based silicon carbide are machined. Dimensional and position tolerances were tested on CMM and the results were shown in Table 2. It can be seen that the accuracy of hole size, coaxiality, cylindricity and parallelism of components Part 1 and Part 2 have met the design requirements, and the final components has been completed as shown in Figure 8.

Table 2.Shape and position tolerance of shafting mounting housing components

| Product name | Project | Tolerance requirements（mm） | Measured value（mm） | |
|---|---|---|---|---|
| | | | Parts1 | Parts2 |
| shafting mounting housing components | Coaxiality | Φ0.006 | Φ0.005 | Φ0.004 |
| | Cylindricity | Φ0.003 | Φ0.002 | Φ0.002 |
| | Parallelism | 0.005 | 0.003 | 0.003 |
| | Inner hole size | 0~ +0.006 | +0.003 | +0.003 |
| | Bearing hole size | 0~ +0.008 | +0.005 | +0.003 |

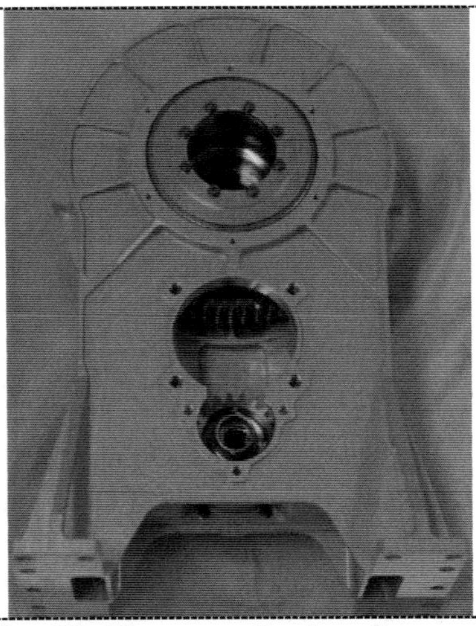

Figure 8. Shafting mounting housing components

## 5. Conclusion

According to the structural characteristics of shafting mounting housing components, by specifying reasonable process flow and selecting reasonable processing equipment and cutting tools, the problem of difficult processing of Aluminium-based silicon carbide composites is overcomed, and the processing accuracy of high-precision thin-walled Aluminium-based silicon carbide parts is realized. The following conclusions are obtained:

(1) When machining Aluminium-based silicon carbide composites, the wear of YG6X tool is the largest and that of PCD tool is the smallest.

(2) Considering the factors of processing cost and quality, when processing 17% SiCp/Al materials, low-cost and good wear-resistant cemented carbide cutters can be used during roughing, and PCD cutters can be used during semi-finishing and finishing.

(3)The high-precision requirements of shape and position tolerance of housing parts can be realized by rough-finish machining, separate machining and combination machining after components.

**Acknowledgments**

This work is supported by Beijing Nova Program (Z201100006820101).

**References**

[1] Lu J., Wu Q. Y., Zhang L., et al. Study of 2024/SiCp aluminium maxtrix composites sheets prepare method. World nonferrous Metals, 2019(16): 39-41.

[2] Miracle D B. Metal matrix composites–From science to technological significance. Composites Science and Technology, 2005, 65(15): 2526-2540.

[3] ANDREWES C J E, FENG H, LAU W M. Machining of an aluminum/SiC composite using diamond inserts. Journal of Materials Processing Technology, 2000, 102(1-3): 25-29.

[4] Zhou J. L., Huang S. T., Xu L. F., et al. Research Status of High Speed Cutting for SiCp/Al Composite Materials. Tools Technology, 2008(07): 3-7.
[5] Cui Y., Li L. F., Li J. L., et al. High volume fraction SiC/Al composites forspace-based optomechanical structure. Optics and Precision Engineering, 2007(08): 1175-1180.
[6] Zhang J. H., Zhang Y. M., Han J. C., et al. Design, fabrication and testing of space-borne SiC mirror. Optics and Precision Engineering, 2006(02): 179-184.
[7] Lv D. J. Research on Al/SiC Composite Machining Technology. Electro — Mechanical Engineering, 2011, 27(05): 29-32.
[8] Li X, Seah W. Tool wear acceleration in relation to workpiece reinforcement percentage in cutting of metal matrix composites. Wear, 2001, 247(2): 161-171.
[9] Tan H. L., Wang C. R. Study on the application of NUMERICAL control milling in sic particle reinforced Aluminum matrix composites. Internal Combustion Engine & Parts, 2019(21): 93-94.
[10] [1] Ding, X. , W. Y. Liew , and X. D. Liu . Evaluation of machining performance of MMC with PCBN and PCD tools. Wear -Lausanne (2005).
[11] [1] Dabade, U. A. , et al. Surface finish and integrity of machined surfaces on Al/SiCp composites. Journal of Materials Processing Technology 192.none(2007):166-174.
[12] Davim J. P. , Baptista A. M. Relationship between cutting force and PCD cutting tool wear in machining silicon carbide reinforced aluminium. Journal of Materials Processing Tech, 2000, 103(3):417-423.
[13] [1] Muthu krishnan N. ,Murugan M., Rao K. P. Machinability issues in turning of Al-SiC (10p) metal matrix composites. International Journal of Advanced Manufacturing Technology, 2008, 39(3-4):211-218.
[14] Zhang W. F., Pan X. N. The Optimum Velocity of Cutting SiCp Reinforced Alum inium Matrix Composite with PCD Tool. Materials Engineer, 1999(01): 14-16.
[15] Wu Z. Y., Wang X. G., Sun F. H., et al. Research on High-speed Milling Processes of SiC Particle Reinforced Aluminium Matrix Composites. Tools Technology, 2004(03): 15-18.
[16] Wang Z. D., Zhai C. T., Yu P., et al. Simulation and Experimental Study on Laser Assisted Micro Cutting of Aluminium-based Silicon Carbide. Modular Machine Tool & Automatic Manufacturing Technique, 2020(10): 59-63.
[17] Wang D. Z., Feng P. L., Zhao Q. L., et al. Study on the Ultraprecision Machining Technology of Aluminum-Matrix Composite Astronautic Parts. JOURNAL OF ASTRONAUTICS, 2006, 27: 1341-1346.
[18] Ge Y. F. , Xu J. H. , Yang H., , et al. Experiments of ultra-precision turning of SiCp / Al composites. Optics and Precision Engineering, 2009, 17(07): 1621-1629.
[19] Cao D. W., Kong L. G., Yang J. Aluminium Matrix Silicon Carbide Motor Housing Parts Processing. WeiTe Motors, 2021, 49(08): 51-53.
[20] Liu H. Wear Property of Al-based/SiC Composites Cutting Tools During Precision Milling. Foundry Technology, 2014, 35(12): 2830-2832.
[21] Wang H, Yang J, Liu Z. Experimental research on the tool wear during precision milling process of SiCp/Al composite materials. Materials Science & Technology, 2012, 20(05): 12-15.

# Elastic Distortion Calculation of a Controlled-Clearance Piston-Cylinder Assembly up to 1500 MPa Using Finite Element Method

**Chunguang Liu[1, 2], Zhongli Zhang[1], Bian Hong[1], Can Wang[1*], Zhenghao Lin[1], Miaolin Feng[2], Jinming Zhang[1]**

[1] Shanghai Institute of Measurement and Testing Technology, 1500 Zhangheng Road, Shanghai, 201203, China

[2] Shanghai Jiao Tong University, 800 Dongchuan Road, Shanghai, 200240, China

* Corresponding author's e-mail: wangc@simt.com.cn

**Abstract.** A controlled-clearance piston-cylinder assembly is modelled using finite element method (FEM). Elastic distortion of the piston-cylinder engagement gap is calculated by the FEM model and then is input to a fluid dynamics model to solving the pressure distribution which works as load in the FEM model. Accurate distortion-pressure solutions are obtained for measurement pressure of (100~1500) MPa by iterative calculations until a convergence of interest is reached. Results are generated for various jacket pressures maintaining balance. Dramatic rapid decent are found near the gap exit after a steady profile for both distortion and pressure distribution curves. Gap exit width can be two orders of magnitude smaller than the nominal gap clearance, and comparable to the irregularity of the piston and cylinder, introducing risks of inaccuracy or potential damage.

## 1. Introduction

Piston gauge has been used as a standard pressure measuring instrument with high accuracy, high reproducibility and high reliability. Based on Pascal's law, pressure is measured by the weight applied to the piston which falls slowly in the cylinder to maintain hydrostatic balance. For high pressure or ultra-high pressure (no strict definition, usually above 150 MPa), additional jacket pressure surrounding the outer surface of cylinder is needed to reduce the clearance between piston and cylinder avoiding measurement failure due to excessive piston fall speed. Such controlled-clearance (CC) type of piston gauge was first proposed by Johnson and Newhall [1] and has been widely accepted and used by national metrology institutes [2-4]. It has been intensively studied in the past using various techniques [4-15]. Among these studies, finite element method (FEM) has been proven available and efficient to determine the distortion of piston and cylinder, especially in the piston-cylinder engagement gap [11-15]. This study develops a numerical model using FEM and fluid mechanics to calculate the elastic distortion of a CC piston-cylinder assembly (PCA). The PCA is used in a CC piston gauge developed by Gu et al. [8] for high precision pressure measurement based on Heydemann-Welch (H-W) model [6], which is more an experiment-based method and ignores the pressure distribution in the engagement gap between piston and cylinder. The pressure distribution and PCA deformation in the engagement gap are calculated iteratively by fluid mechanics and FEM, respectively. Results are generated and analyzed for measurement pressure from 100 MPa to 1500 MPa with various jacket pressures maintaining balance.

## 2. Modelling

### *2.1. PCA description*

The schematic diagram of the PCA of the CC piston gauge is shown in figure 1. The measurement pressure $p_m$ is generated and injected by a tube sealed at the bottom of the cylinder. The mass is applied on the top of the piston to create weight balancing $p_m$. Jacket pressure $p_j$ is injected and sealed in the space between the outer surface of the cylinder and a surrounding sleeve (not shown) made of steel. The pressure $p_j$ is individually generated so that its magnitude can be freely adjusted during operation. The pressure transmitting fluid is sealed inside of the cylinder with a tiny leak flow through the piston-cylinder engagement gap with a length $L$ = 15mm from the cylinder top exit. The piston falls down to fill the space of the outflow fluid. The clearance of the engagement gap is small so that the piston falls slowly and steadily without breaking the balance. During operation, the balance is guaranteed by adjusting $p_j$ to neutralize the expansion caused by the fluid pressure load in the gap. Under zero pressure, the engagement gap is uniform with a clearance $h_0$ = 2.15 $\mu$m.

The piston is made of tungsten carbide (WC) with 10% cobalt (Co), with total length of 47 mm, and diameter $r_{p0}$ = 2.063 mm. The cylinder is made of steel, with inner diameter in the engagement section $r_{c0}$ = 2.0673 mm and outer diameter of 26 mm. The Young's modulus, and Poisson ratio of the piston are $E_p$ = 630 GPa and $\mu_p$ = 0.22. These quantities for the cylinder are $E_c$ = 206 GPa, $\mu_c$ = 0.28.

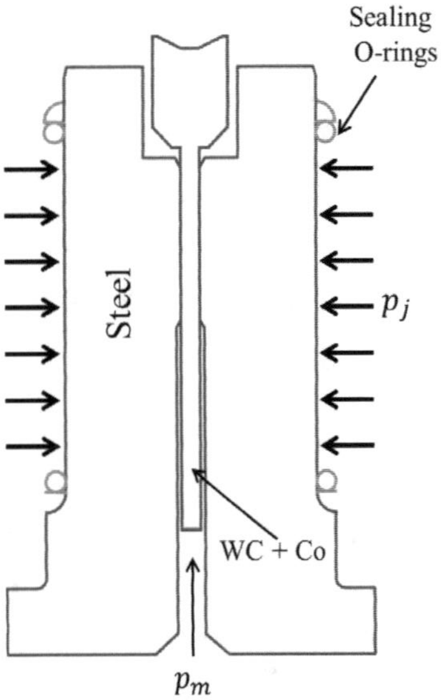


Figure 1. Schematic diagram of PCA of the CC piston gauge.

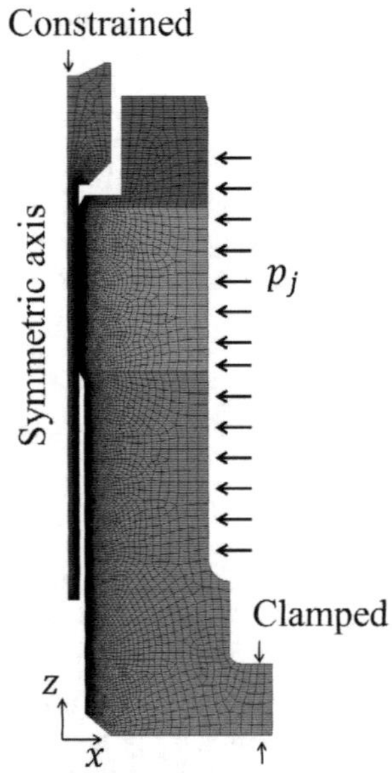


Figure 2. 2-D FEM model of the PCA

A mixture of glycerol and ethylene glycol is used as the pressure transmitting fluid. The density $\rho(p)$ and dynamic viscosity $\eta(p)$ of the fluid vary with pressure as shown in figure 3 and figure 4.

### *2.2. FEM Modelling*

The structural elastic deformation of the PCA is solved in a static mechanics model using FEM as shown in figure 2. A two-dimensional model is used because the PCA is symmetric about its vertical centre axis. To capture precise deformation of the PCA, element size of $5\times 10^{-3}$ mm is assigned for the surfaces for both the piston and the cylinder. Near the exit of the engagement gap, where the deformation is found changing dramatically, the elements are further refined to $1\times 10^{-3}$ mm. The lower skirt of the cylinder is clamped. A uniform jacket pressure is applied on the outer surface of the cylinder. The piston's vertical motion is constrained.

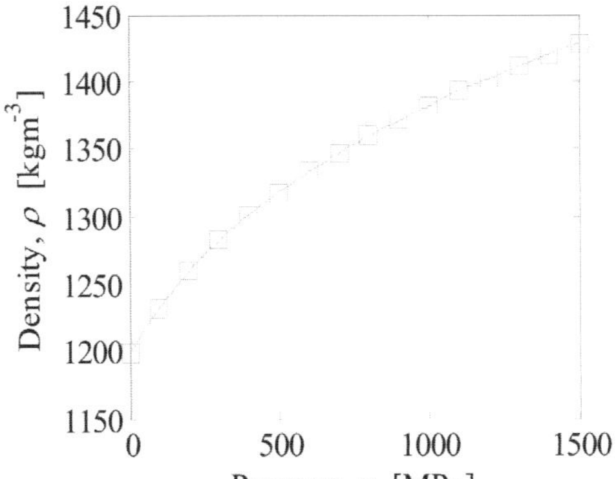


Figure 3. Density versus pressure of the pressure transmitting fluid of a mixture of glycerol and ethylene glycol.

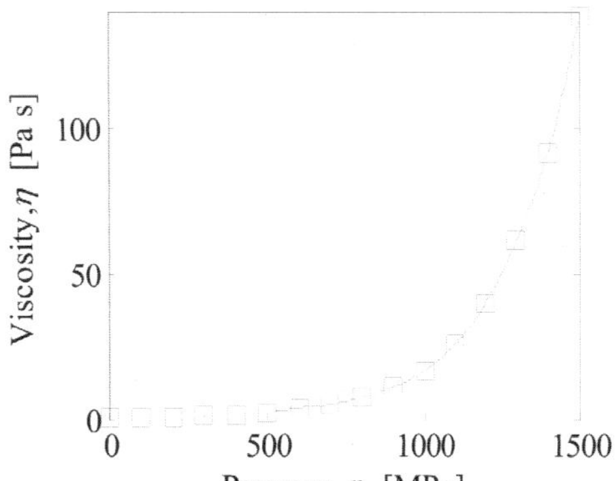


Figure 4. Dynamic viscosity versus pressure of the pressure transmitting fluid of a mixture of glycerol and ethylene glycol.

*2.3. Calculation of pressure distribution*

The fluid in the piston-cylinder engagement gap is considered to be Newtonian, permanent, single-phased flow. Although the piston is rotating and falling at the time during operation, the structure is axisymmetric, and the coupling between the vertical and circumferential flows are weak and neglectable for low operational rotating speed. For this reason, the flow is assumed to be one-dimensional, viscos, and laminar. The pressure distribution $p_z(z)$ along the engagement gap has the entrance pressure $p_z(0) = p_m$ and exit pressure $p_z(L) = 0$, where $z$ is the vertical coordinate with $z = 0$ at the engagement gap entrance. $p_z$ can be derived from the Navier-Stokes equations as [13]:

$$p_z = p_m - \frac{p_m}{D}\int_0^z \frac{\eta(p_z)}{\rho(p_z)h_z^3} d\tau , \tag{1}$$

$$D = \int_0^L \frac{\eta(p_z)}{\rho(p_z)h_z^3} dz , \tag{2}$$

where $h_z(z)$ is the distorted clearance profile of the engagement gap.

Solutions of $p_z$ and $h_z$ are determined iteratively until a convergence of interest is reached. $h_z$ is calculated by the elastic deformation results obtained from the FEM model in Section 2.2. For iteration step $i$, $p_z^{(i)}$ is calculated from equation (1) with $h_z^{(i-1)}$ calculated from the previous iteration step $i - 1$. Then $p_z^{(i)}$ is input into the FEM model as the load in the engagement gap to determine the elastic distortion profile $h_z^{(i)}$ which is used to calculate $p_z^{(i+1)}$ in the next iteration $i + 1$. A program written in C language is developed to load results from models, to do iterative calculations, and to determine convergence. The convergence criteria are

$$\max\left(\left|h_z^{(i+1)} - h_z^{(i)}\right|\right) < 10^{-5}\ \mu m \tag{3}$$

and

$$\max\left(\left|p_z^{(i+1)} - p_z^{(i)},\right|\right) < 10^{-5}\ \text{MPa}. \tag{4}$$

The results are considered to be converged when both equation (3) and equation (4) are satisfied.

## 3. Results and Discussion

The deformed PCA of the FEM model for $p_m$ = 1500 MPa and $p_j$ = 650 MPa is shown in figure 5. Radial displacements are plotted as smooth contours. The cylinder is squeezed by $p_m$ and $p_z$ on the inner surface and $p_j$ on the outer surface such that negative and positive displacements are found on the inner and outer surfaces of the cylinder, respectively. Most part of the piston below the engagement gap exit has negative displacement due to $p_m$ and $p_z$. Engagement gap has tapered shape near the exit due $p_j$ and zero pressure at the exit. In order to capture accurate gap deformation, nodal displacements of the surfaces of piston and cylinder within the engagement gap are exported for pressure calculation.

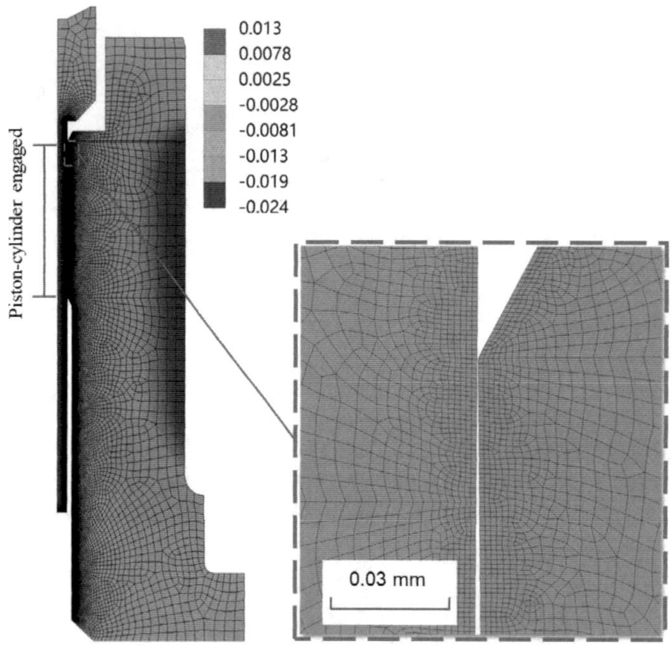


Figure 5. Deformed PCA of the FEM model. Radial displacements (unit: mm) are shown in contour plot. Results near the exit of the piston-cylinder engagement gap are zoomed. $p_m$ = 1500 MPa, $p_j$ = 650 MPa.

Converged distortion results are shown in the form of radii of piston and cylinder versus normalized engagement gap length in figure 6, including $p_m$= (0, 300, 900, 1500) MPa with $p_j$ = (0, 200, 420, 650) MPa, respectively. ($p_m$, $p_j$) = (0, 0) MPa means nominal radii under zero pressure. In general, cylinder has larger deformation than the piston due to smaller Young's modulus. The piston has a nearly uniform distortion except locations close to the exit. Rapid decent at the exit is also found for the cylinder which explain the tapered shape observed in figure 5. The changes of radii around the exit are stiffer for larger $p_m$. For $p_m$ = 300 MPa, the distorted radius of the cylinder is smaller than the nominal value meaning the expansion caused by $p_z$ is overwhelmed by the squeeze effect of $p_j$. For larger measurement pressure ($p_m$ = 900 MPa, 1500 MPa), expansion is dominant.

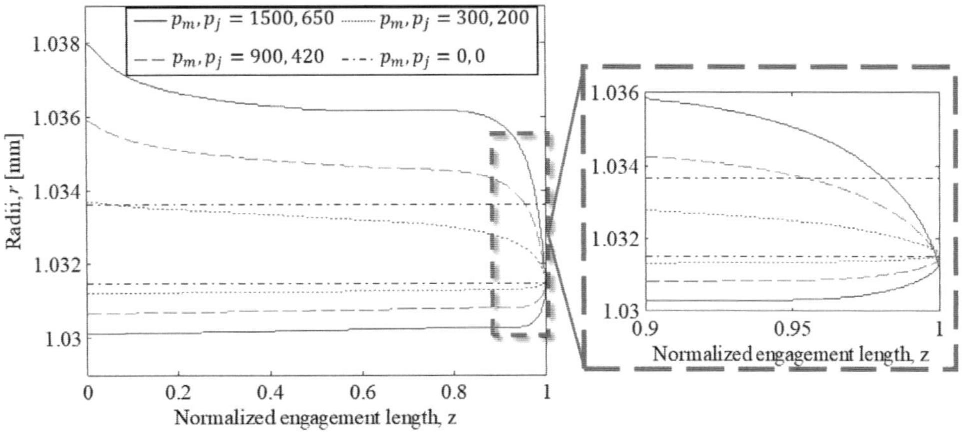


Figure 6. Distorted radii profiles versus the piston-cylinder engagement length $z$. The results are from converged $p_z$ - $h_z$ iterative calculation. Unit for $p_m$ and $p_j$ is MPa.

Converged engagement gap clearances at the exit $h_z(L)$ are shown in figure 7 for various $p_m$ and $p_j$. $h_z(L)$ decreases as $p_j$ increases as expected and more sensitive to $p_j$ for lower $p_m$ scenarios. For larger $p_m$, the magnitude of $h_z(L)$ can be very small to maintain pressure balance. For $p_m$ = 1500 MPa, $h_z(L)$ is about (0.02~0.04) $\mu$m which is two orders of magnitude smaller than the nominal clearance, and is comparable to irregularities of the piston and the cylinder. This can bring risks of inaccuracy due to contact or potential scratch on the surfaces.

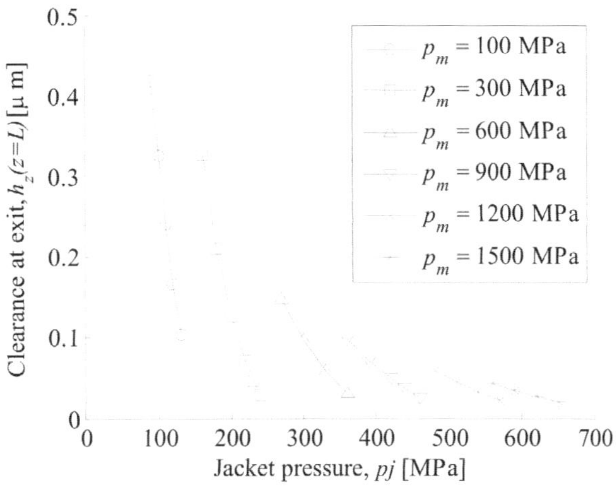


Figure 7. Deformed PCA of the FEM model. Radial displacements (unit: mm) are shown in contour plot. Results near the exit of the piston-cylinder engagement gap are zoomed. $p_m$=1500 MPa, $p_j$=650 MPa.

Rapid decent near the piston-cylinder engagement exit is more dramatic for pressure distribution $p_z$ as shown in figure 7. Pressure and length are plotted in normalization form. Pressure drop is relatively slow and steady until 59%, 81%, and 88% of $p_m$ decrease in the last 1% length of the engagement gap for $p_m$ = 300 MPa, 900 MPa, and 1500 MPa, respectively. These results, working as load, are responsible for the corresponding rapid decent of gap clearance shown in figure 6. Compared to other models in the literature [12-14], this PCA has a relatively large nominal gap clearance and a "soft" cylinder such that the fluid flow has to be controlled at the exit where $p_z = 0$ and squeeze effect of $p_j$ is able to overwhelm the expansion caused by $p_z$.

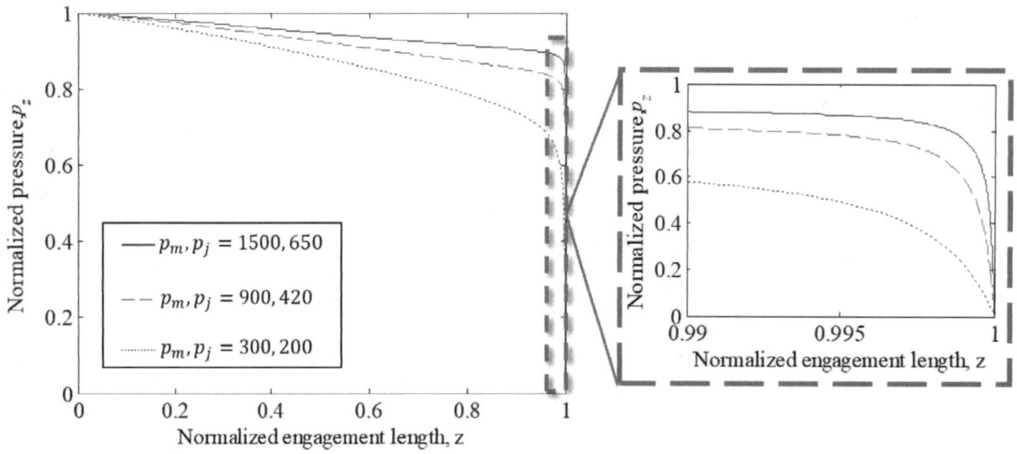


Figure 8. Distorted Normalized pressure distribution $p_z$ versus the piston-cylinder engagement length $z$. The results are from converged $p_z$ - $h_z$ iterative calculation. Unit for $p_m$ and $p_j$ is MPa.

## 4. Conclusions

This study established a finite element model of piston-cylinder assembly used in a controlled-clearance piston gauge operated as a measurement instrument for ultra-high pressure. Elastic distortion of the piston and the cylinder is calculated from the finite element model. The pressure distribution is analytically determined. Pressure and distortion are calculated iteratively until a convergence of interest is reached. Results are generated for measurement pressure from 100 MPa to 1500 MPa, with various jacket pressures that can maintain valid measurement operation with slow fluid flow. The results show rapid decent for both clearance and pressure near the exit of the piston-cylinder engagement section due to large nominal clearance and material property of the cylinder. The minimum clearance can be two orders smaller than the nominal clearance and comparable to the irregularities of the piston and the cylinder, introducing risks of inaccuracy or damage due to contact.

**Acknowledgements**
This work was funded by the Science and Technology Research Planning Project of the State Administration for Market Regulation (Project No.: 2021MK031).

**References**

[1] Johnson, D.P., Newhall, D.H. (1953) The piston gage as a precise pressure-measuring instrument. Trans. Am. Soc. Mech. Engrs., 75: 301-310.
[2] Nishibata, K., Yamamoto, S., Kaneda, R. (1980) A dead-weight piston manometer for use up to 2 GPa. Jpn. J. Appl. Phys., 19: 2245-2256.
[3] Borovkov, V.M., Kuznetsov, D.I., Sekoyan, S.S., Shchipunov, A.N., Aslanyan, A.E., Gavrilkin, S.M. (2015) Creation of a national primary standard for the unit of pressure in the range of 10-1600 MPa. Measurement Techniques, 57: 1233-1237.
[4] Kajikawa, H., Kobata, T., Ooiwa, A. (2007) Development of a new controlled-clearance pressure balance for hydraulic high-pressure standard. In: SICE Annual Conference. Takamatsu. pp. 384-387.
[5] Dadson, R., Greig, R., Horner, A. (1965) Developments in the accurate measurement of high pressures. Metrologia, 1: 55-67.

[6] Heydemann, P.L.M., Welch, B. (1975) Piston gages. In: Le Neindre, B., Vodar, B. (Eds.), Experimental Thermodynamics Volume II: Experimental Thermodynamics of Non-reacting Fluids. Butterworths, London. pp.147-202.

[7] Dadson, R.S., Lewis, S.L., Peggs, G.N. (1982) The Pressure Balance: Theory and Practice. HMSO, London.

[8] Gu, S., Wang, L., Liu, J., Dai, X., Pan, Z., Huang, Z., Zhu, J., Ma, Z., Lin, D. (2008) A development of dead weight piston gauge for pressure to 1500 MPa. Shanghai Measurement and Testing, 29: 10-14.

[9] Olson, D.A., Gelany, S.A., Eltawil, A.A. (2012) Use of a higher order Heydemann–Welch model to characterize a controlled clearance piston gauge. Measurement, 45: 2469-2471.

[10] Hong, B., Pan, Z., Gu, S., Zhang, Z. (2018) An Investigation of the high pressure controlled-clearance deadweight piston gauge. In: The 2018 5th International Conference on Systems and Informatics (ICSAI 2018). Nanjing. pp. 734-738.

[11] Molinar, G., Sabuga, W., Robinson, G., Legras, J. (1998) Comparison of methods for calculating distortion in pressure balances up to 400 MPa –EUROMET Project #256. Metrologia, 35: 739-759.

[12] Sabuga, W., Molinar, G., Buonanno, G., Esward, T., Rabault, T., Yagmur, L. (2005) Calculation of the distortion coefficient and associated uncertainty of PTB and LNE 1 GPa pressure balances using finite element analysis—EUROMET project 463. Metrologia, 42: S202-S206.

[13] Sabuga, W., Molinar, G., Buonanno, G., Esward, T., Legras, J., Yagmur, L. (2006) Finite element method used for calculation of the distortion coefficient and associated uncertainty of a PTB 1 GPa pressure balance—EUROMET project 463. Metrologia, 43: 311-325.

[14] Dogra, S., Yadav, S., Bandyopadhyay, A.K. (2010) Computer simulation of a 1.0 GPa piston-cylinder assembly using finite element analysis (FEA). Measurement, 43: 1345-1354.

[15] Sabuga, W., Prazak, D., Rabault, T. (2014) Recent progress in high pressure metrology in Europe. EPJ Web of Conferences, 77: 00006.

Journal of Physics: Conference Series **2264** (2022) 012014 doi:10.1088/1742-6596/2264/1/012014

# Implementation of White Rabbit Time Synchronization System in State Acquisition System of High-energy Physics Experimental Device

**Zhanfei Yang[1,a], Yichao Ma[1,b], Wenhui Yang[1,c], Yiyuan Zhang[1,d]**

[1]School of electrical and control engineering,Shaanxi University of science and technology,xian,shanxi,710021,china

[a]email:1173210512@qq.com, [b]email:mayichao@sust.edu.cn

[c]email:739162666@qq.com, [d]email:1978341783@sust.edu.cn

**Abstract.**The state acquisition system of high energy physics experimental device has very high requirements for the clock synchronization accuracy of each node. The White Rabbit (WR) distributed synchronous timing technology can realize multi node sub nanosecond clock distribution within several kilometers, and meet the requirements of clock synchronization accuracy of the system. This paper presents an implementation of WR timing system with ZYNQ 7045 as the core in the state acquisition system. This paper introduces the hardware design and implementation scheme of collector node, the implementation scheme of WR node, and the statistical data of WR-PTP time synchronization experiment results. The test results show that this method can realize the time synchronization between the nodes of the state collector, and the accuracy is better than 1 nanosecond. It can provide high-precision time synchronization function for the multi node signal acquisition system of high-energy physics experiment.

## 1. Introduction

Modern high-energy particle physics experimental devices are numerous and distributed in a wide space [1-2]. In order to ensure the normal operation of each device and the reliability of experimental data, its working state must be collected and fed back. The realization of multi node and high-precision condition monitoring has very high accuracy and stability requirements for the clock of each node of the condition monitoring system.

WR-PTP protocol is a sub nanosecond time synchronization technology based on optical fiber ethernet [3], which can realize time synchronization between tens of thousands of nodes in a wide range (< 10km) [4], and is widely used in distributed network measurement and control, 5G base station network time synchronization[5]. In order to meet the requirements of high-precision time synchronization between each node of the state acquisition system of high-energy physics experimental device, WR-PTP protocol is selected to realize sub nanosecond time synchronization of each node.

## 2. Materials and Methods

*2.1.collector electronics design scheme*

*2.1.1. Overall system architecture*

The overall design of the system is shown in Figure 1. Relying on WR time synchronization technology, a high-precision global time service network is established to realize sub nanosecond time synchronization of each monitor node. In terms of specific modules, the architecture is based on white rabbit (WR) protocol, including WR clock reference source, WR switching network, WR node, front-end electronics fee, DAQ system, etc. Among them, WR clock reference source is composed of a GPS timing machine and a WR top-level interactive machine; WR switching network is composed of multiple WR switches; WR node consists of monitor front-end electronics.

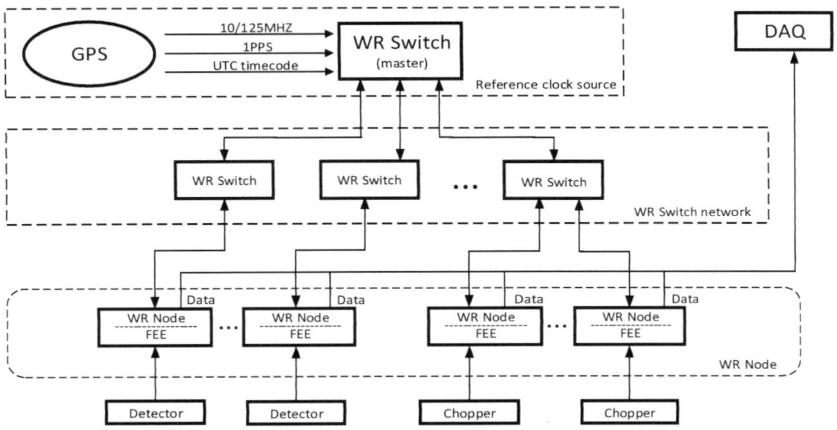


Figure 1. Electronic layout scheme of state acquisition system

*2.1.2 Hardware design and implementation of signal collector based on ZYNQ + FPGA*

On the premise of meeting the operation requirements of WR time synchronization system and the signal acquisition quality requirements of spectrometer supporting equipment, the hardware of signal collector based on ZYNQ + FPGA is designed and developed, and its principle diagram is shown in Figure 2. The acquisition of status signal is realized based on low-cost FPGA chip 10m08sae144c8g, and WR-PTP time synchronization technology and data processing are realized based on zynq7000 chip xc7z045ffg900. After the FPGA chip completes the data acquisition, it sends the data to the ZYNQ chip through serial communication. The two combine with each other to realize high-precision spectrometer equipment status signal acquisition and data transmission. In addition, the level selection function is added at the signal receiving end to provide 5V or 24V signal level selection to meet the application requirements of different equipment.

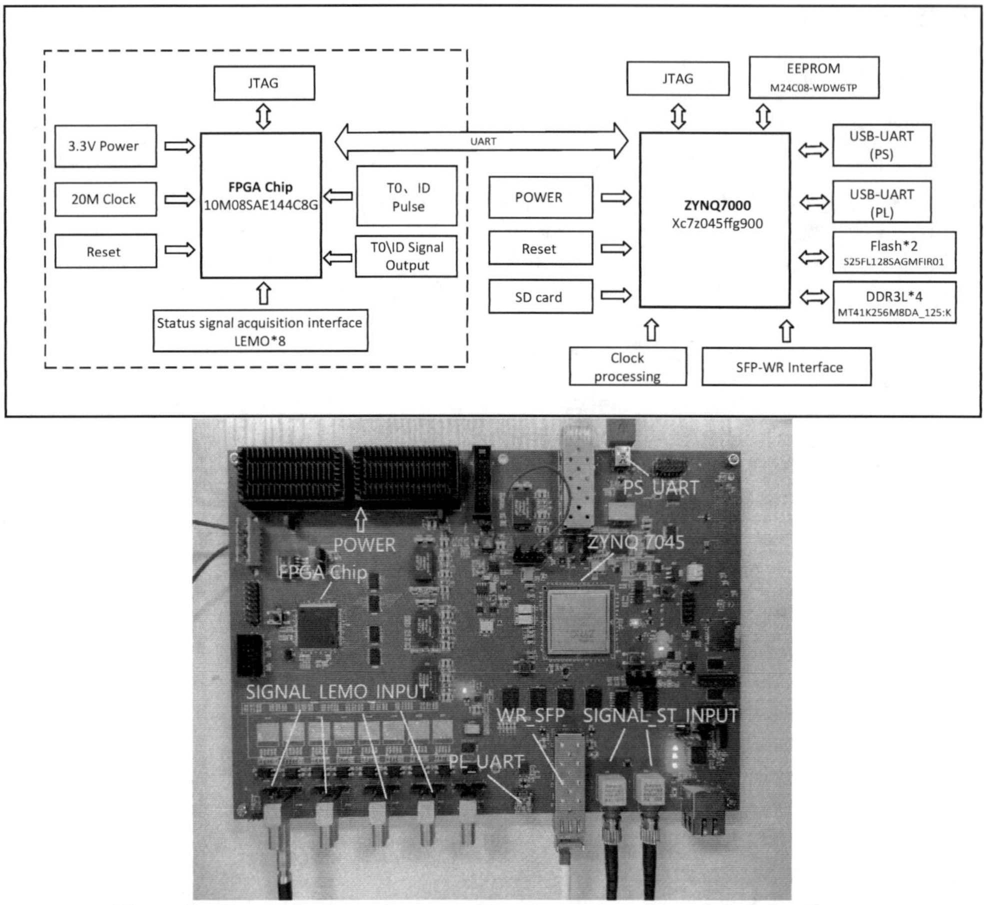


Figure 2. Circuit design scheme and hardware diagram of signal collector

*2.2.Principle of WR time synchronization technology and Realization of WR-PTP node*

White rabbit protocol is a high-precision distributed timing protocol based on optical fiber Ethernet [6]. As shown in Figure 3 (a), it is a chain time synchronization model of white rabbit protocol. Firstly, the White Rabbit protocol needs to build the clock loop between the master and slave nodes through the optical fiber transmission link, and then complete the high-precision time synchronization of the master and slave nodes based on PTPv2 protocol and clock data fusion technology. During synchronization, master is the time service node and slave is the time service node. The master terminal and slave terminal are connected through optical fiber. The synchronization includes time synchronization and clock phase synchronization.

*2.2.1 Time synchronization*

Time synchronization is based on PTPv2 synchronization principle, which was introduced by IEEE in 2008 [7]. In the process of synchronization, the time difference between the master and slave is corrected mainly by measuring the round-trip delay [6]. The synchronization process is shown in Figure 3 (b).

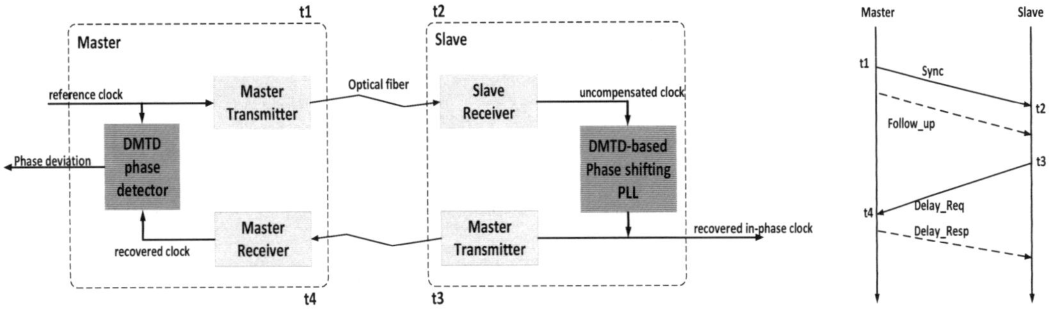


(a) (b)

Figure 3.White Rabbit protocol chain time synchronization model and PTP synchronization mechanism

The specific process of PTPv2 synchronization message exchange is as follows: (1) the master node sends a sync message to the slave node, records the sending time as t1, and passes this t1 timestamp through follow_up message is sent to the slave node; (2) The slave node receives the sync message and records the receiving time as t2. The slave node receives follow_up message, obtain t1 timestamp; (3) Slave node sends delay_Req message and record the local sending time t3; (4) Master node receives delay_Req message, record the receiving time t4 and send it through delay_Resp message is sent to slave node; (5) Slave node receives delay_Resp message, obtain the t4 timestamp. So far, the slave node obtains the t1, t2, t3 and t4 timestamp, calculates the time difference between the master and slave using the timestamp,and adjusts the local time.

According to the PTP protocol, the link delay of the master and slave can be calculated: $\text{delay}_{ms} = \frac{(t_4 - t_1) - (t_3 - t_2)}{2}$ ; The time difference between master and slave is $\text{offset}_{ms} = t_2 - t_1 - delay_{ms} = \frac{t_2 - t_1 + t_3 - t_4}{2}$.

*2.2.2 clock phase synchronization*

On the basis of full compatibility with PTPv2 protocol, White Rabbit protocol adopts synchronous Ethernet, Dual mixer time difference technology(DMTD) and WR synchronous link model to complete clock phase synchronization and improve the synchronization accuracy to sub nanosecond level [8].

(1)Synchronous Ethernet

WR technology adopts synchronous Ethernet to realize the same clock frequency of each node. As shown in Figure 4, in standard Ethernet,the clocks of each node operate independently. In synchronous Ethernet, all nodes form a clock network topology. Sub nodes or sub switches use clock and data recovery (CDR) technology to recover the clock from the data link, and eliminate the jitter caused by the clock recovery circuit from the PLL inside the node. The recovered clock is not only the system clock of the slave node, but also the reference clock of the next node. In this way, the clock frequency of all nodes of the whole network is consistent with that of the root node [9].

 2264 (2022) 012014 doi:10.1088/1742-6596/2264/1/012014 

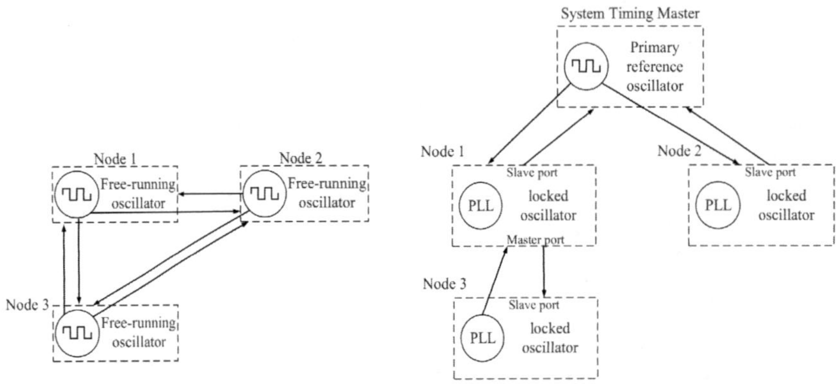


(a) Standard Ethernet (b) Synchronous Ethernet

Figure 4. Comparison between standard Ethernet and synchronous Ethernet

(2)Dual mixer time difference technology(DMTD)

In order to further improve the time synchronization accuracy, the white Rabbit protocol uses the phase detector to accurately measure the phase difference between the data recovery clock and the local clock and correct the time stamp, so as to further improve the time synchronization accuracy of PTPv2 technology and bring it into the sub nanosecond level.

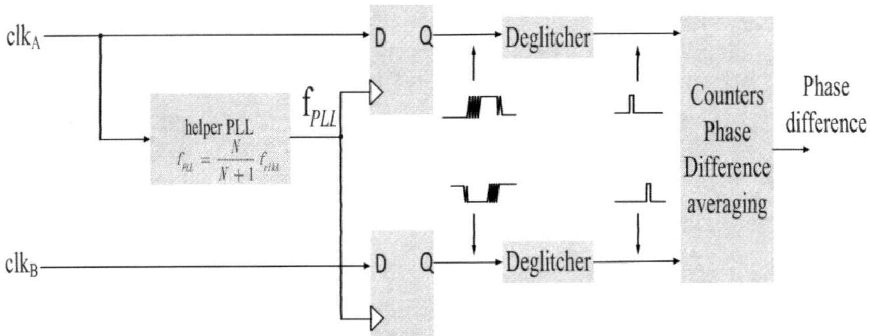


Figure 5. Principle of DMTD phase detection

The operation principle of DMTD is shown in Figure 5. ClkA and clkB represent the local clock and the recovery clock in the data stream respectively. They have the same frequency but different phase. An auxiliary clock signal is generated by an external phase-locked loop, and a small difference ($f_{PLL} = \frac{N}{N+1} f_{clkA}$)exists between the frequency of the signal and the frequency of the measured signals ($clk_A$ and $clk_B$ ). Therefore, after sampling, D flip-flop will output a very low-frequency signal [6]. By measuring the phase difference of the output signal of D flip-flop, the phase difference of the measured signal can be calculated. The phase detector, filter control circuit and voltage controlled oscillator at the node constitute a phase-locked loop circuit to realize the phase locking and phase adjustment of the slave node clock.

(3)WR synchronous link model

The synchronization link of WR master-slave node is shown in Figure 6. The total delay of the round-trip link can be regarded as composed of three parts [11]: (1) Hardware delay of transceiver circuit of master-slave node( $\Delta_{TXM}$ , $\Delta_{TXS}$ , $\Delta_{RXM}$ , $\Delta_{RXS}$ ); (2) Bit sliding delay caused by word alignment operation in serial parallel conversion circuit( $\varepsilon_M$ , $\varepsilon_S$ ); (3) Optical fiber link transmission delay( $\delta_{MS}$ , $\delta_{SM}$ ).

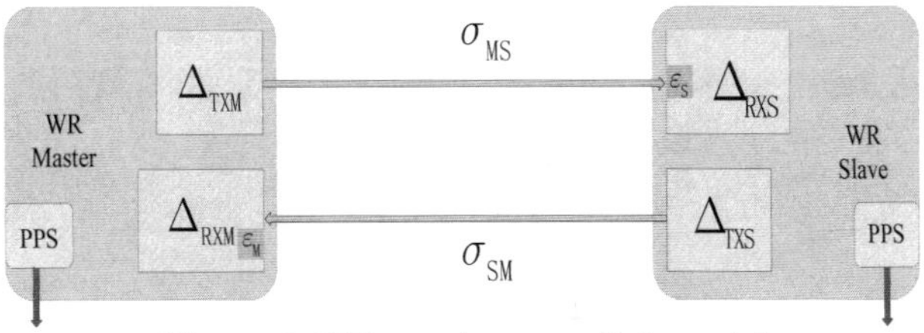


Figure 6. WR synchronous link model

The refractive index of optical fiber to different wavelength light is different, which leads to the difference of transmission delay of optical fiber link( $\delta_{MS}$ and $\delta_{SM}$ ).WR defines the fiber asymmetry coefficient $\alpha$ describe their relationship as $\alpha = \frac{\delta_{MS}}{\delta_{SM}} - 1$. so as to achieve automatic calibration.

*2.2.3 WR PTP node implementation*

This paper is based on WRPC-v4.2 (White Rabbit PTP core) kernel[8] realizes WR time synchronization function on ZYNQ 7045 chip, and customizes peripheral hardware system, FPGA system and embedded system according to actual needs. As shown in Figure 7, it is the implementation block diagram of WR-PTP node, and its specific composition is as follows:

The peripheral hardware system mainly includes SFP optical transceiver, USB-UART and other interface circuits, clock processing circuit, power supply system and external memory (flash, EEPROM, DDR3L). The FPGA system is designed in VHDL language,mainly including latticemicro32 soft core processor (LM32), Minic network interface controller, MAC layer implementation (WR Endpoint), Wishbone bus, 1 PPS generator, UART interface, IRIG-B time encoder, etc. The design of embedded software system is based on LM32 soft core, mainly including WR-PTP protocol software, soft PLL, network protocol.

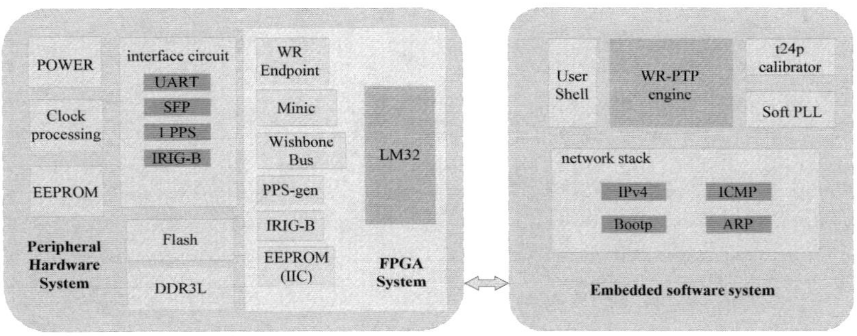


Figure 7. Implementation block diagram of WR-PTP node

## 3. Results & Discussion

*3.1. WR-PTP time synchronization test of equipment*

The test environment uses a WR switch and a signal collector node, and uses a 1909m long single-mode optical fiber to form a master-slave WR network structure. The two devices use their internal clock to regularly output a second pulse(PPS), and the WR-PTP time synchronization accuracy is given by measuring their deviation value. By connecting the PPS output of WR switch and the PPS signal of signal collector node to the oscilloscope at the same time, the synchronization results of two second pulses can be observed. The test results are shown in Figure 8 below. The Yellow waveform in the figure is the PPS signal of signal collector node, and the green waveform is the PPS signal generated by WR switch.

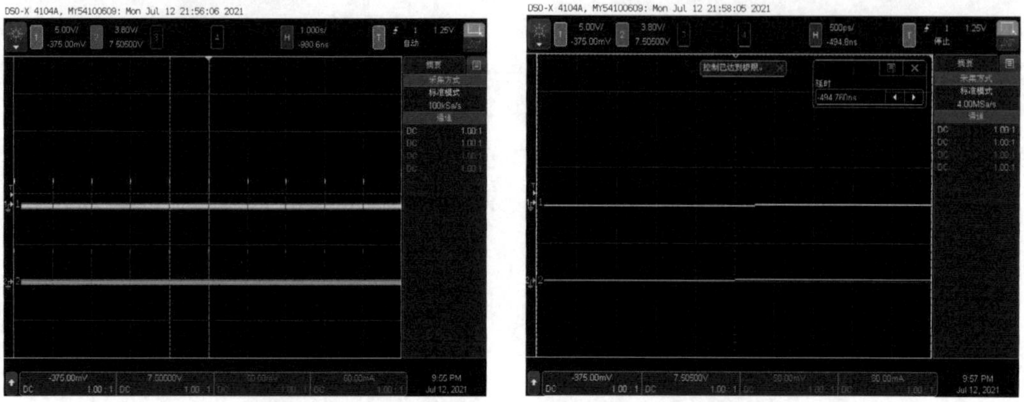

Figure 8. synchronization results of two PPS signals

In the test process, the delay of two PPS signals is continuously measured for 2000 times, and then the measured data are statistically processed. After Gaussian fitting, the probability density distribution of delayed data is shown in Figure 9 below, which conforms to the normal distribution. The measurement results show that the average delay of the two PPS signals is 250ps and the Sigma value is 23.12ps. The PPS signals of WR master-slave nodes can achieve sub nanosecond synchronization.

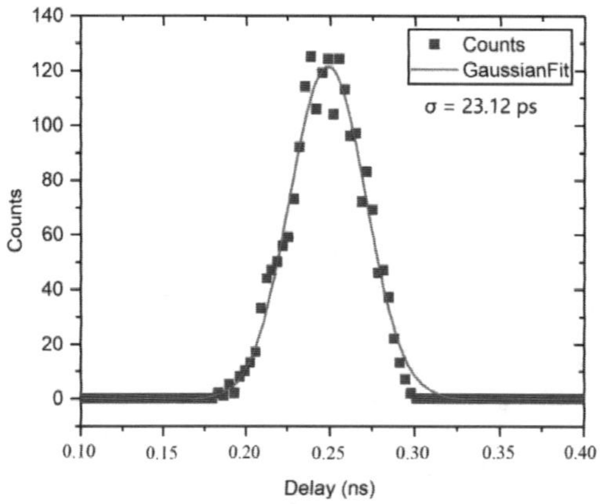


Figure 9. statistics of WR synchronization accuracy test results

## 4. Conclusions

In this paper, WR time synchronization technology is used to realize high-precision time synchronization between the state acquisition nodes of the state acquisition system of high-energy physics experimental device. WR is a new improved time synchronization method based on PTP protocol and synchronous Ethernet.It improves the synchronization accuracy to sub nanosecond level through DMTD and WR synchronization link model, and is suitable for multi node and wide range. The test results show that the WR time synchronization network realized in this paper can realize the time synchronization accuracy between the nodes of the state collector, which is better than 1ns. It solves the time service requirements of long-distance multi node and high-precision of the state acquisition system of the high-energy physics experimental device, and makes the system run accurately, reliable and easy to maintain, It can be used as the key supporting technology of the state acquisition system of high-energy physics experimental device.

**References**

[1]. LI,J.H.,et al.(2020)Discussion on key Equipment Installation Technology of Spallation Neutron Source in China. Electromechanical Engineering Technology. 49(02):179-182.
[2]. Cheng,H.,ZHANG,W.,WANG,F.W.,et al.(2019)Multidisciplinary Application of Spallation Neutron Source in China.Physics.48(11):701-707.
[3]. PENG,CH.J.,YANG,J.,et al.(2018)Development and Application of WR Technology.Journal of Astronautic Metrology and Measurement.38(05):7-15.
[4]. LI,P.J.,LI,W.,ZHU,W.X.,et al.(2019)Overview of Network Time Synchronization Protocols. Computer engineering and Application.55(3):30-38.
[5]. LI,Y.Y.,WU,J.F.,WANG,K.(2020)Test and Analysis of Precise Time Synchronization Protocol for 5G Network Application.Global positioning system.45(05):90-96.
[6]. GONG,G.H.,LI,H.M.,(2017)High Precision Distributed Timing Technology Based on Optical Fiber Ethernet.Navigation, Positioning and Timing, 4(06):68-74.
[7]. LI,EN,SONG,W.ZH.,LUO,Q.S.,et al.(2019)Research on Construction Strategy of High Precision Time Service System in Optical Network.Optical communication technology.43(9):58-62.
[8]. DERVISKADIC,ASJA.,RAZZAGHI,REZA.,WALGER,QUENTIN.,et al. (2020)The White Rabbit Time Synchronization Protocol for Synchrophasor Networks. IEEE transactions on smart grid.11(1):726-738.
[9]. MATTIA RIZZI.,MACIEJ LIPINSKI.,PAOLO FERRARI.,et al.(2018)White Rabbit Clock Synchronization: Ultimate Limits on Close-In Phase Noise and Short-Term Stability Due to FPGA Implementation.IEEE Transactions on Ultrasonics, Ferroelectrics, and Frequency Control.65(9):1726-1737.
[10]. CUI,G.L.,XIANG,Y.,HUA,Y.,(2021)Research on Metastable Suppression Method of Digital Dual Mixer Time Difference Technology.Journal of time and frequency.44(1):10-16.
[11]. ZHU,W.CH.,KANG,B.,et al.(2020)Research and Implementation of Sub Nanosecond Time Synchronization Technology Based on WR-PTP.Optical communication technology.44(12):6-10.

# Research and Design of Scalable Advanced Application Based on Dual-Core Smart Meter

**Xinglin Guo[1], Bensong Zhang[2], Xiaofei Fan[1], Shenchen Pan[1], and Haowen Wu[3*]**

[1] Shenzhen Power Supply Bureau Co., Ltd. , Shenzhen, Guangdong, 518001, China

[2] CSG Digital Grid Research Institute Co., Ltd. , Guangzhou, Guangdong, 510700, China

[3] Digital Grid Branch of CSG Digital Grid Research Institute Co., Ltd. , Guangzhou, Guangdong, 510700, China

[*]Corresponding author's e-mail: aiziyuanbao@163.com

**Abstract.** The dual-core smart meter based on the embedded operating system adopts a new software and hardware architecture design, and is equipped with software and hardware resources suitable for scalable advanced applications, which can support application software to form a unified large-scale market. Based on the software and hardware platform of the dual-core smart meter, this paper designs and implements application management functions suitable for advanced applications, and the design realizes advanced functions such as load identification and contract power purchase. It verifies the powerful software and hardware platform expansion capabilities of the dual-core smart meter. It has laid a certain technical foundation for supporting the construction of next generation of smart meter application ecological market.

## 1. Introduction

IR46 standard is the recommended smart meter technical recommendations of the International Organization of Legal Metrology. China as a member of international legal metrology organization, has the obligation to actively promote the implementation of IR46 standard in China [1]. Therefore, China officially issued and implemented the JJF1245 series of technical specifications, which laid a technical and legal foundation for the development and promotion of the next generation of smart meters. According to the provisions of JJF1245 series technical specifications, the basic functions of next generation of smart meter software are divided as shown in Table 1. The non-legal software part encapsulates the time-of-use rate function, fee control function, display function, power freezing function and event recording function of the previous generation of meters, while reserving technical application space for expandable advanced functions.

In the process of research and development of next generation of smart meter technology, three technical solutions have been formed in the industry: dual-core smart meter solution, single-core smart meter solution, split energy meter solution, but they are ultimately unified in the dual-core smart meter solution [2]. The dual-core smart meter scheme consists of two "MCU", thus forming a metering unit and a management unit. Legal software runs independently on the metering unit, and non-legal software runs on the management unit. The separation of the two through serial ports is realized on the physical level, which is conducive to certification testing and more conducive to the installation and operation of advanced applications. In addition, through reasonable resource allocation, the management core supports in the form of an independent APP, which is used to meet the differentiated functional requirements such as non-interventional load identification, orderly charging, power quality monitoring, etc., to support digital grid construction and power market reform, and create a more ambitious imagination space for the power market [3-4].

Table 1. Functional division of legal software and illegal software for next generation smart meters.

| Function item | Legal software part | Illegal software part |
|---|---|---|
| 1 | Measurement of electrical parameters | Calculation, storage and freezing of tariff electric energy |
| 2 | Electric energy measurement | Calculation and storage of the maximum demand (including the maximum demand for each rate) |
| 3 | Clock or clock maintenance | Display method and display of illegal content |
| 4 | Protection of equipment-specific parameters | Event records except for legal system related functions |
| 5 | Freezing of electrical parameters and electrical energy data | Implementation of external data communication interface (such as infrared, RS485, carrier, wireless, etc.) |
| 6 | Event records of reset, upgrade, time calibration, and modification of device-specific parameters | Control function |
| 7 | Software interface of legal system related software and illegal system related software | Alarm function |
| 8 | Methods to ensure the safety of legal functions | Fee control function |
| 9 | Identification of legal system related software | Other unmentioned functions |
| 10 | Display of legal content | --- |

## 2. Dual core smart meter and management unit

To meet the above needs and the requirements of domestic and international technical standards, a typical application scheme for next generation of smart meters is proposed in this paper as shown in Figure 1. The management core software architecture divides the software for managing MCUs into driver layer, kernel layer, middle layer and application layer. The kernel layer, driver layer and middle layer constitute the software platform to meet the stability and versatility requirements of the software platform; the application layer realizes the basic non-legal metering functions of the meter and flexibly implements the advanced application functions. The application layer software completely isolates the implementation details of the hardware platform, realizes system calls through a standard API interface, and realizes data distribution and processing based on the meter communication protocol.

The software platform has powerful task scheduling, resource allocation and business support capabilities. This meets the requirements of supporting multi-process address space isolation, supporting online upgrade of extended applications, supporting independent upgrade of individual applications, supporting Increment OTA mechanism, relieving the pressure of massive data transmission in communication channels, supporting perfect application management functions, and guaranteeing stable and reliable upgrade and operation of advanced application software. The dual-core smart meter management unit is configured with software and hardware resources that can expand advanced applications, supporting flexible expansion of advanced applications, facilitating the formation of a unified scale market for application software and reducing software and hardware development and management costs.

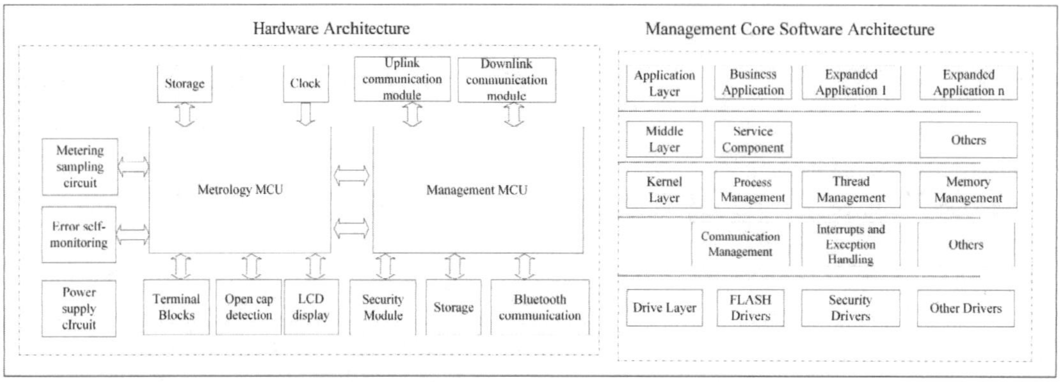


Figure 1. Typical application of next generation smart meters.

The management unit MCU chip uses high-speed embedded memory and has a Cortex-M4 core. The hardware schematic of the management unit is shown in Figure 2.

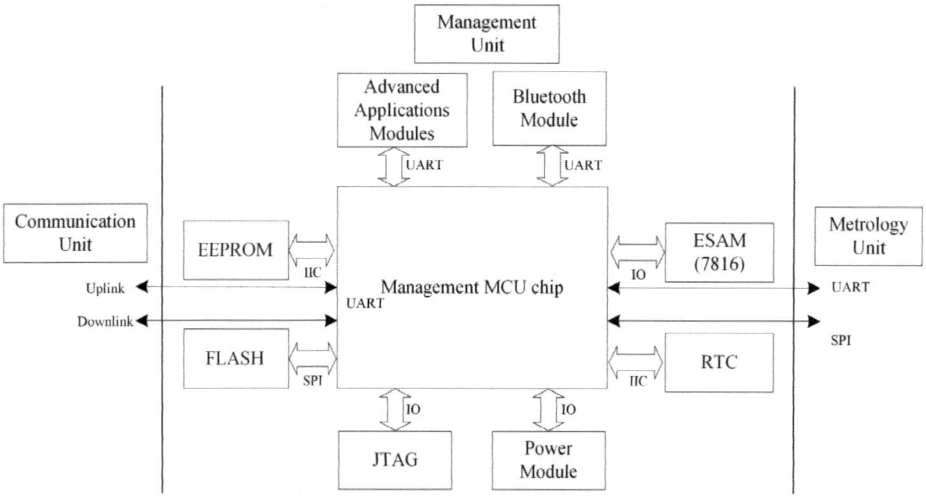


Figure 2. Management unit hardware schematic diagram.

## 3. Advanced application management design

### *3.1. Application store*

In order to support the construction of the application ecological market based on next generation of smart meters and to realize the functions of application uploading and application distribution, the overall business design of the application store is shown in Figure 3 below with reference to the model of smartphone application store front and back design and considering the business logic of the existing metering automation system and electricity energy data platform.

The application store contains four main modules, namely, the general backend module, the user center module, the frontend display module, and the communication interface module [5].

(1) General backstage module: it can manage all the functions and services in the system, and can set the system functions flexibly and have the highest authority of the system.

(2) User center module: the operation console for application developers and application users, according to the user role corresponding to the rights, application developers can realize the application upload, application current status view and other functions, application users can query the application application distribution details.

(3) Foreground display module: it shows all the applications that have been put on the shelves, and according to modern Web site design thinking, it implements regional division of all application display pages, and provides application users with functions such as searching applications, viewing application details, and applying for application delivery.

(4) Communication interface module: it provides an interface for the caller to obtain task information issued by the application. According to the agreed request protocol and the parameters required by the request, the interface caller can obtain the task data set issued by the application and the corresponding information of the application.

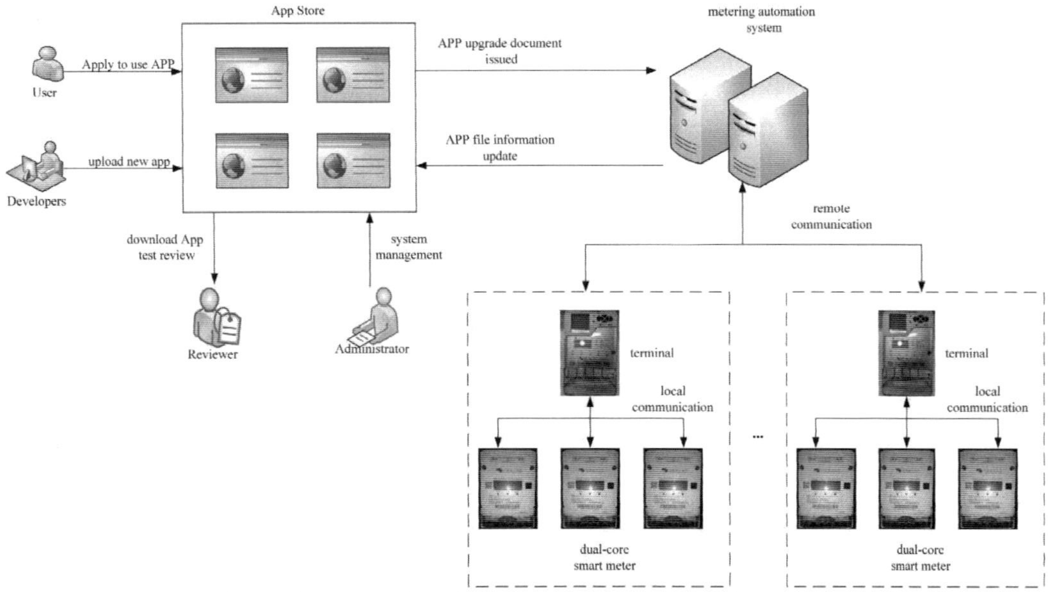


Figure 3. Simplified diagram of application store APP upgrade information interaction.

### *3.2. Increment OTA*

There are two types of online upgrades for the dual core smart meter management unit software, which are Increment OTA, Full OTA. The two upgrade types mainly differ in creating the upgrade package and are suitable for different application scenarios. The batch upgrade of the meter software may lead to a long time high load occupation of the communication channel from the master station to the meter. Considering that the existing carrier communication technology is still not popularly applied on a large scale, this paper adopts Increment OTA technology, as shown in Figure 4 below.

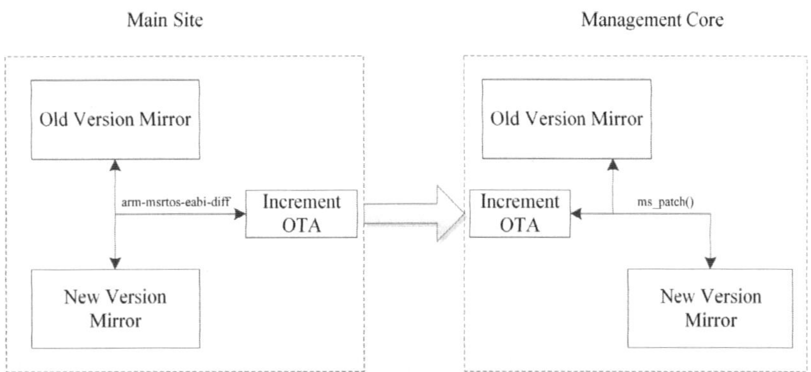


Figure 4. Increment OTA upgrade process.

Full OTA upgrade does not require the old version image file, it is generally used to upgrade the entire firmware, and the upgrade file is issued for the new version image upgrade. Increment OTA needs to create a differential package by comparing the old version of the image file with the new version of the image file through the differential tool. Generally speaking, Increment OTA is much smaller than Full OTA, and after the management unit receives the differential package, it is merged with the old backup image to create the new version of the image file.

### *3.3. Software backup and rollback design*

In theory, it is impossible for software testing to completely find all defects in software. It is difficult to completely fix all defects in the application software to be upgraded by the management unit. Due to the software's own defects, the impact of the operating hardware and software environment, and external electromagnetic interference, the management unit software may fail to operate normally after the upgrade is completed, and the normal operation of the software has to be guaranteed by rolling back to the previous stable version.

The hardware platform selected for the dual-cell smart meter management unit has two non-volatile memories, the MCU's internal Nor Flash and its external Nor Flash. The internal Nor Flash is used to store BootLoader, OS image, installed service and application images and boot scripts. After the system is powered on and booted, the BootLoader is started from the internal Nor Flash, then the OS is loaded, and finally the services and applications are loaded and run. The external Nor Flash is used to store the meter business data and the service and application images to be upgraded. Due to the limited space in the internal Nor Flash, only an area is created in the internal Nor Flash to store the backup software version of the last stable management unit.

## 4. Typical scalable application research and design

Next generation of smart meters is designed to last for more than 10 years, and will most likely require upgraded and updated software features to meet the development needs of new businesses. Facing the development needs of electric internet of things, next generation of electricity meters will undertake increasingly important edge computing tasks. Based on the powerful data collection function of next generation of electricity meters, it is expected that the demand for advanced applications such as non-intrusive load identification, orderly charging, and power quality monitoring will become stronger [6-7]. Based on the above requirements, this paper briefly introduces the design and implementation of two typical advanced application functions of contract power purchase and load identification based on the hardware and software platform of dual-core smart electricity meter.

### *4.1. Application store*

#### *4.1.1. Overview of trading power clearing*

The current power clearing process of power market trading starts from reading meter power in the metering center of provincial companies and ends with obtaining daily power clearing results in the trading center. According to the needs of the existing metering center and customer service center, the data categories read by the smart meters in operation are as follows.

The first category is the reading of power data, which is collected once every hour, including the total positive active power. The second category is to read the load curve data, including power, voltage and current, which is collected once every 15min.

Each meter reports 339KB of data per day at 15min collection interval. If the collection interval is increased to 5min according to future demand, the reported data will be about 1M. The future participation in electricity market trading may reach millions of meters. At present, the above-mentioned amount of data that needs to be read has put greater pressure on the communication network for uploading data, and it cannot detect power outages and other situations that cause power abnormalities in a timely manner, which has a greater impact on the timeliness of daily power clearing. In addition, after completing the data reading, the metering center needs to perform time-sharing power calculation according to the rate periods determined by the power trading contract. After the processing is completed, it can be pushed to the customer service center for power accounting, and the power clearing process of the transaction is shown in Figure 5 below. The huge amount of time-sharing and rate-sharing power calculation will cause huge pressure on the metering center, and once the centralized data processing "centralized" system of the metering center encounters a failure, it will have a huge impact and make it difficult to verify the power of electricity market transactions retroactively.

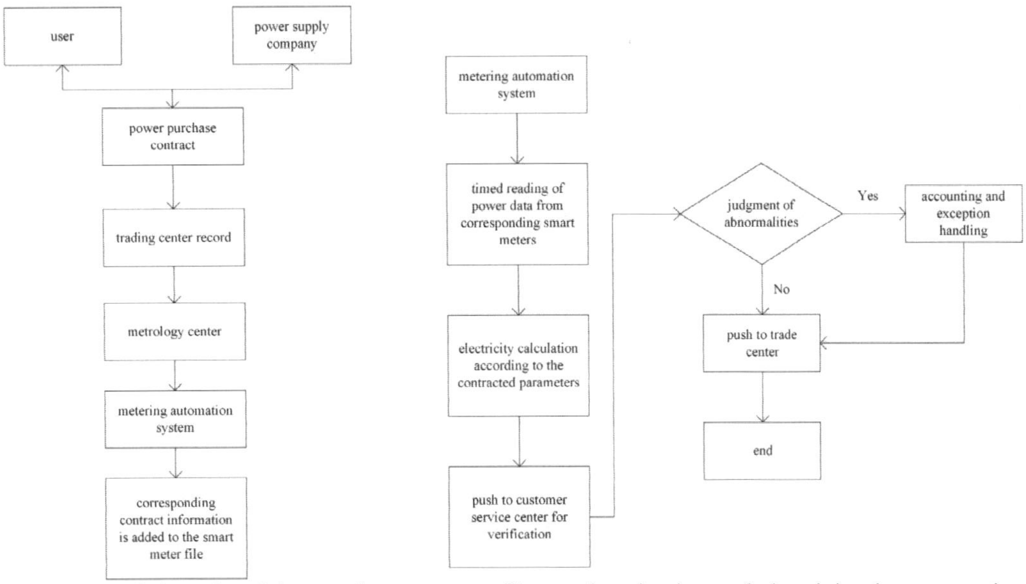


Figure 5. Existing electricity purchase contract file synchronization and electricity data processing flow.

### *4.1.2. Design of transaction power clearing function based on dual-core smart meters*

In order to meet the functional requirements of power market trading for power clearing, a power clearing APP for power market trading is designed based on next generation of smart meter software platform, as shown in Figure 6 below.

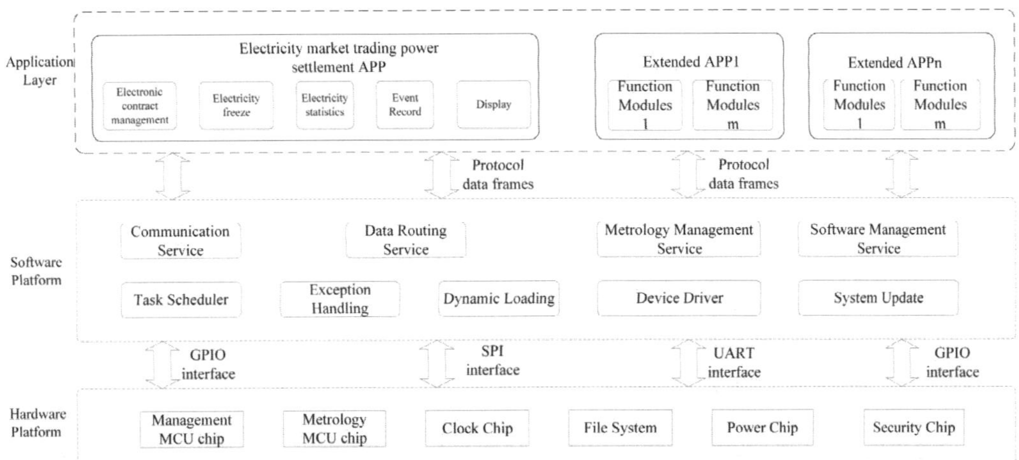


Figure 6. Software architecture design of dual-core smart meter management unit with power clearing function.

The power market trading power clearing APP receives and parses the electronic contracts sent to meters by the metering automation system, and obtains parameters such as rate periods and power freeze. At the meter side, it realizes time-sharing power freeze with minimum 5min interval, forms daily statistical reports and monthly statistical reports of time-sharing and rate-sharing power, and marks abnormal events to facilitate the investigation and processing of abnormal situations, and stores the process data of time-sharing power calculation for a long time at the meter end to facilitate the verification of process data. By reporting daily and monthly electricity processing results and supporting process data verification, it reduces the pressure of data processing verification on communication network, metering center and customer service center by uploading large amount of process data. When uploading the daily clearing power processing results, each meter reports only 11KB of data per day according to the 5min collection interval (reducing the data volume by about 90%), which greatly reduces the amount of uploaded data. In addition, the software platform based on next generation of meter operating system can flexibly expand and upgrade the functions of the electricity market trading power clearing APP to meet the expansion needs of future electricity market trading for data processing at the meter side. The flow of transaction power clearing APP file synchronization and power data processing is shown in Figure 7 below.

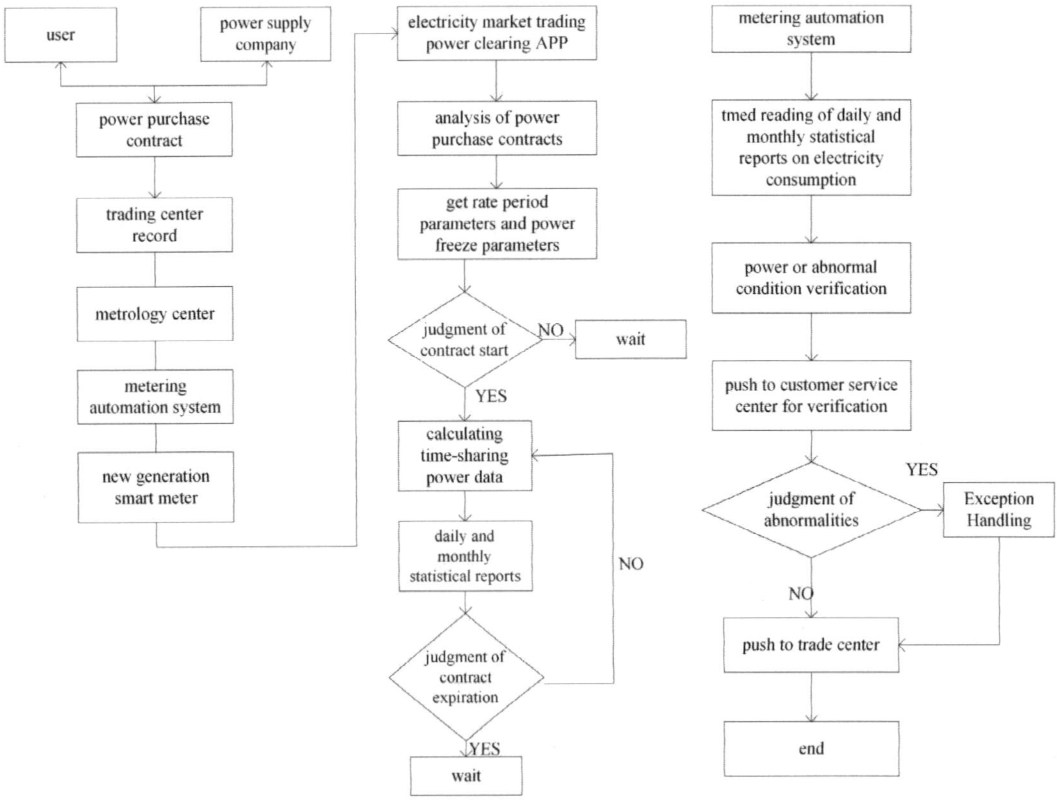


Figure 7. File synchronization and power data processing flow of the transaction power clearing APP.

### *4.2. Load identification function design and implementation*

#### *4.2.1. Overview of load identification technology*

Non-intrusive load identification technology refers to the installation of monitoring equipment at the power entrance and obtaining the type of individual loads in the load cluster, their operating conditions, accumulated power consumption and other power consumption parameters by means of voltage and current signal analysis at that location [8-9]. The essence of non-intrusive load identification is load decomposition. In other words, the total user load information is decomposed into the information of each power-using device, and then the power consumption information such as the energy consumption of the power-using device and the power consumption pattern of the user is obtained [10]. The technology implementation path is shown in Figure 8 below.

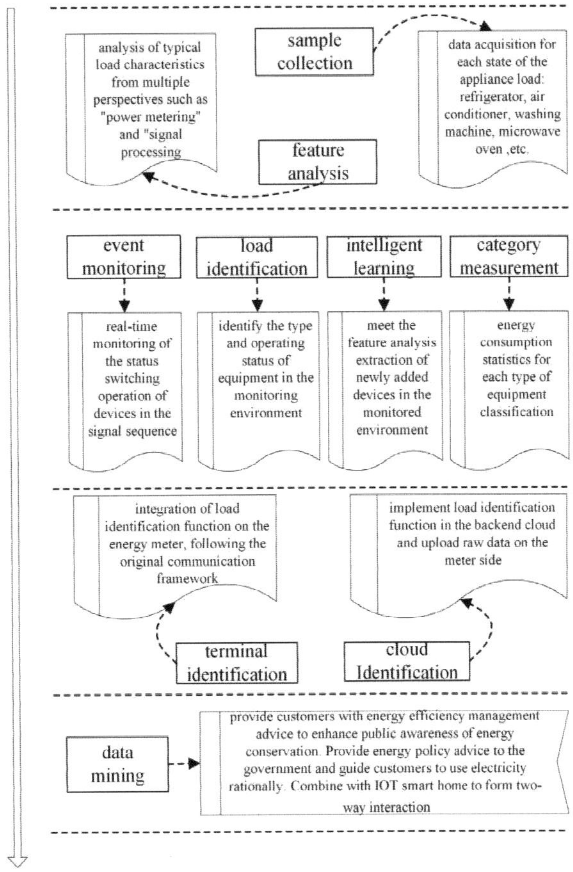


Figure 8. Implementation path of load identification technology.

### *4.2.2. Design of load identification function based on dual core smart meter*

The load identification APP runs on top of the software platform of the dual-core smart meter and completes the load identification function together with the load identification module. The load recognition APP is mainly responsible for realizing the functions of handshaking and initialization of the load recognition module, reading the data of the load recognition module, and upgrading the algorithm library of the load recognition module. The design principle of load identification APP is to be able to complete the functions related to meter load identification without coupling other business applications or services as much as possible. The load identification APP interacts externally mainly with the file interface and virtual bus service components provided by the software platform. The design of the hardware and software architecture of the management unit with load recognition is shown in Figure 9.

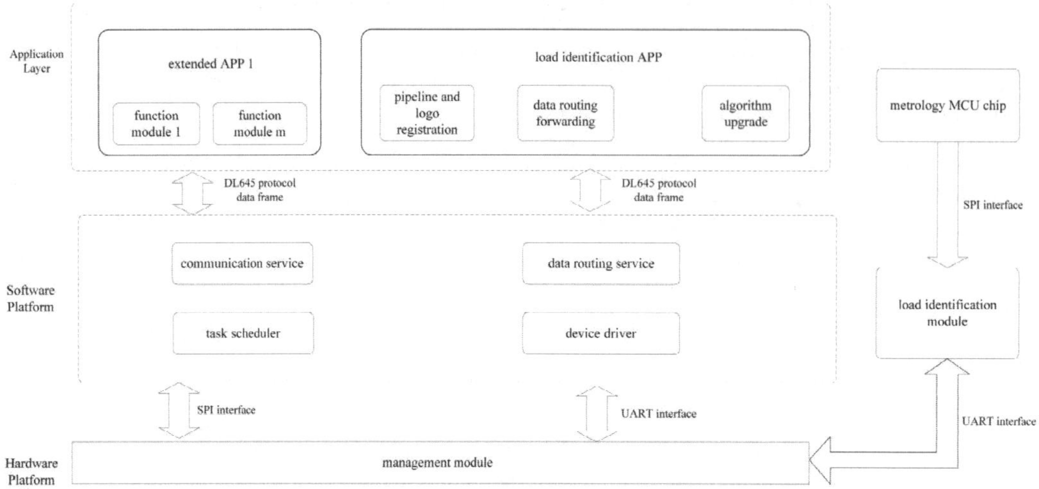


Figure 9. Software and hardware architecture design of management unit with load identification function.

The load identification APP business functions are divided into the following parts, and the functional operation flow is shown in Figure 10 below.

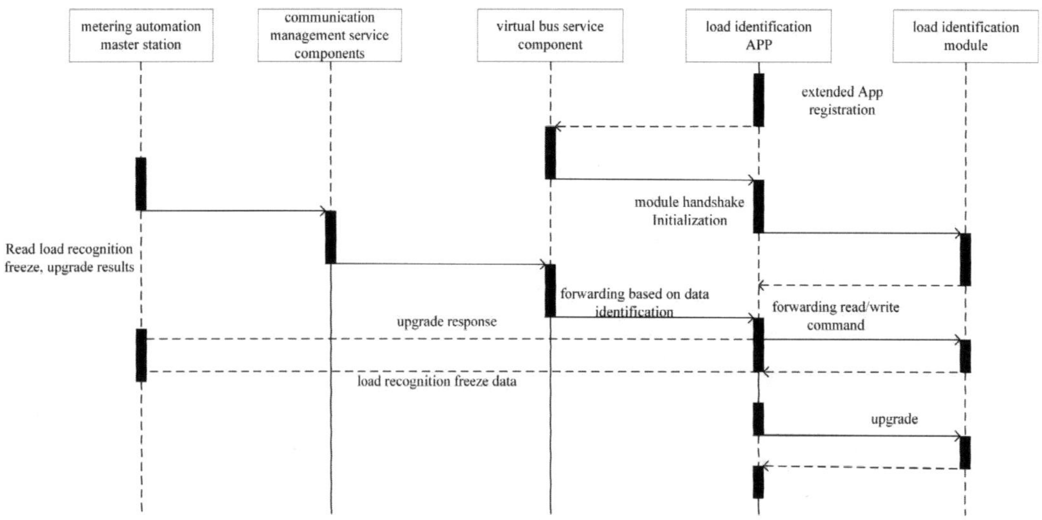


Figure 10. Operation process of load identification function.

(1) Registering with the virtual bus service component regarding the pipeline number and data identification to be used by the load identification module.

(2) After the dual-core smart meter is powered on, the load identification APP and the load identification module perform handshaking and parameter initialization.

(3) The load identification app is responsible for data routing and forwarding when the main metering automation station reads the data of the designated dual-core smart meter load identification module.

(4) After receiving the complete upgrade package, the load identification app needs to perform protocol conversion, and after completing the protocol conversion of the upgrade package, it will forward it to the load identification module to upgrade the load identification module.

The load identification module is mounted on the back of the dual-core smart meter management unit, and the algorithm is embedded inside the module, as shown in Figure 11 below.

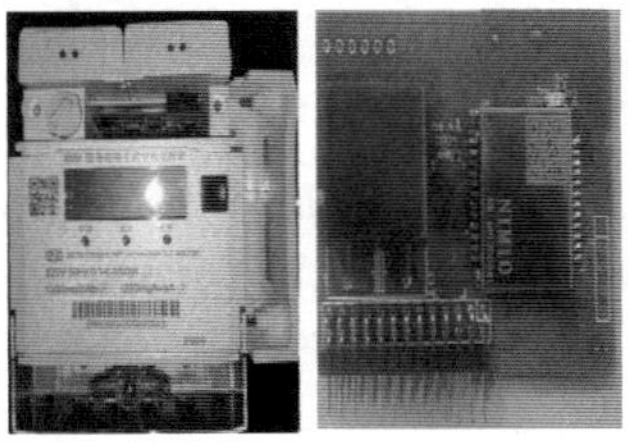

Figure 11. Load identification module for power meter prototype and management unit.

## 5. Conclusion

This paper introduces the software and hardware technical solutions of next generation smart meter management unit. Based on next generation smart meter dual-core hardware architecture and software platform architecture, the application management functions to meet next generation smart meter application ecological market are designed, and the flexible expansion and upgrade of advanced application APP is supported. Load identification function and contract power purchase function are designed and implemented, which verifies the powerful ability of next generation smart meter software and hardware platform to support advanced application functions. Based on the above function design and implementation, it has laid a certain technical foundation to meet the differentiated business needs of next generation of smart meters during their 15-year life span and build a grand application ecological market in the future.

**Acknowledgments**

This research was supported by the research project on Research on the core unit technology of management unit and measurement terminal for new smart meters of China Southern Power Grid Co., Ltd. (JY-O-JL-20-002)

## References

[1] Liu Mingjie, Zhou Lin, Miao Changsheng, et al. (2017) Study on the development discussion of new generation smart meter. Electrical Measurement & Instrumentation, 54: 94-100.

[2] Zhou Zhimei, Zhao Dongyan, Zhang Haifeng, et al. (2017) Research on the key technology of the dual MCUs on the energymeter for the IR46 standard. Application of Electronic Technique, 43: 7-11+19.

[3] Hou Xingzhe, Liu Xingzhi, Zheng Ke, et al. (2020) Technical Prospect of A New Generation of Smart Energy Meter in the Ubiquitous Power Internet of Things environment. Electrical Measurement & Instrumentation, 57: 128-131.

[4] Zhang Leping, Hu Shanshan, Li Xiangfeng, et al. (2018) The Preliminary Study on Interactive E-Meter Design under Electricity Transaction Reform. Electrical Measurement & Instrumentation, 55: 146-152.

[5] Xiao Yile, Yang Lixin, Bai Zhihua, et al. (2019) Technical scheme of on-line program upgrading in next generation of multi-core smart meter. Micro controllers and Embedded System Applications, 12: 54-57.

[6] Liuying, Liuyan, Yankai, et al. (2021) Peak Load Probability Estimation of Smart City Based on Smart Meter. Electrical Measurement & Instrumentation, 58: 166-171.

[7] Li Jingli, Yang Xuchen, Zhang Linjuan, et al. (2021) Research on hierarchical control strategy of orderly charging of large-scale electric vehicles. Electrical Measurement & Instrumentation, 58: 168-175.

[8] Chen Chen, Gao Pinghang, Jiang Jiange, et al. (2021) A deep learning based non-intrusive household load identification for smart grid in China. Computer Communications, 177: 176-184.
[9] Li Li, Huang Youjin, Xiong Wei, et al. (2021) Non-intrusive load identification based on inproved convolutional neural network. Electrical Measurement & Instrumentation, 4: 1-8.
[10] Wang Huinan, Xue Jianli, Liu Jiayi. (2021) Non-intrusive Load Monitoring Based on Smart Meter Data. Electrical Measurement & Instrumentation, 58: 178-182.

# Application of Joint Filter Based on EMD and IIR in SINS

**Jinlong Xing**[1,a], **Tijing Cai** [1,b] **and Gongliu Yang**[1,2,c*]

[1] School of Instrument Science and Engineering, Southeast University, Nanjing, Jiangsu, 210018, China

[2] School of Instrumentation and Optoelectronic Engineering, Beihang University, Beijing, 100191, China

[a]email: 230198299@seu.edu.cn, [b]email: caitij@seu.edu.cn (C.T.),

[*]Corresponding author's e-mail: [c]yanggongliu@buaa.edu.cn

**Abstract.** To solve the problem that the interference of vibration noise in the strapdown inertial navigation system (SINS) affects navigation accuracy, a joint noise reduction method based on Empirical Mode Decomposition (EMD) decomposition and Infifinite Impulse Response (IIR) low-pass digital filter was proposed. On the basis of EMD decomposition, Finally, SINS vehicle test is carried out. The results show that the combined noise reduction method based on joint noise reduction method based on EMD decomposition and IIR low-pass digital filter has an obvious effect on the noise reduction of accelerometer signal. After filtering, the horizontal positioning error of INS is better than 0.29 nautical miles, and the navigation accuracy is improved by 31%. Therefore, the joint noise reduction method based on EMD decomposition and IIR low-pass digital filter proposed in this paper can effectively improve the navigation accuracy of SINS.

## 1. Introduction

In the SINS is installed directly on the carrier of aircraft or vehicle. The engine vibration and aerodynamic disturbance will cause carrier vibration, which will pollute the real measurement signal of the inertial device accelerometer. Therefore, the measurement signal of SINS accelerometer contains not only the motion information of the carrier to be measured, but also many other interferences and coupling signals. Therefore, it is important to study an appropriate algorithm to preprocess the output data of accelerometer, remove noise and retain the true information of carrier motion before inertial navigation solution[1].

Wavelet decomposi tion and EMD are often used to reduce noise for non-stationary, non-linear signals. Empirical mode decomposition (EMD) decomposes signals according to their own characteristics. It has good adaptability, simple and efficient decomposition process. However, in its high frequency IMF component, there will be a phenomenon that noise signals overlap with useful signal modes. In order to solve this problem, Liu Haijiang et al. [2] proposed a method of combining EMD with wavelet threshold, and verified the validity of the method. However, wavelet analysis has a large amount of computation, complex decomposition process and unstable effect. In the paper [3-4], the noise-assisted signal analysis method was introduced into the EMD theory, and the ensemble empirical mode decomposition (EEMD) theory was proposed. EEMD suppressed the mode aliasing to a great extent. However, if the white noise added in the EEMD method is not suitable, it will cause problems such as reconstruction errors and IMF pseudo-components.

In this article, according to the corresponding characteristics of the accelerometer signal and the signal processing method of EMD, we propose a joint noise reduction method based on EMD decomposition and IIR low-pass filtering. The signal-to-noise aliasing phenomenon in the high-frequency IMF component can improve the noise reduction effect of the signal. And it is verified by the inertial navigation algorithm. The results show that the accuracy of SINS is effectively improved after pre-filtering.

## 2. EMD Filter Design

EMD noise reduction is a new signal analysis method proposed by Huang et al. [5], which can decompose the signal into a series of eigenmode functions IMF and residual term res with frequencies arranged from high to low. The residual term is extracted from the signal to be decomposed and has good adaptability [6], which enables it to decompose different signals quickly and efficiently. The eigenmode function IMF needs to satisfy:

(1) In the entire data segment, the number of extreme points and the number of zero-crossing points must be equal or differ by at most one.

(2) At any time, the mean value of the upper envelope formed by the local maximum points and the lower envelope formed by the minimum points is zero.

The EMD filtering process is as follows: set the signal as $X(t)$ , $M_1$ as the average value of the upper and lower envelopes, then the function $H_{10}=X(t)-M_1$ , and $H_1$ as the eigenmode function to be selected. When $H_1$ satisfies certain screening conditions, it is an inherent modal component. In the second screening, $H_{10}=X(t)-M_1$ obtained from the first screening is used as the original signal. then there are

$$\begin{aligned} H_{11} &= H_{10} - M_{11} \\ H_{12} &= H_{11} - M_{12} \\ &\vdots \\ H_{1i} &= H_{1(i-1)} - M_{1i} \end{aligned} \tag{1}$$

$H_{1i}$ is obtained by decomposing formula (1), and $i$ is the number of sievings until the $i$-th $H_{1i}$ is an eigenfunction. At this time, let $C_1=H_{1i}$, and $C_1$ be the first eigenmode function of the signal. The remaining signal is

$$R_1=X(t)-C_1 \tag{2}$$

Repeat the above steps for $R_1$ to obtain $C_2$ . Similarly, $n$ eigenmode functions that meet the conditions can be obtained. From high frequency to low frequency, there are

$$\begin{aligned} R_1 &= X(t) - C_1 \\ R_2 &= R_1 - C_2 \\ &\vdots \\ R_n &= R_{n-1} - C_n \end{aligned} \tag{3}$$

Until $R_n$ does not satisfy the condition of decomposition, that is, monotonic function, EMD decomposition ends. The commonly used sieving standard function is

$$SD = \sum_{i=0}^{T} \frac{\left|H_k(t) - H_{k-1}(t)\right|^2}{H_{(k-1)}^2(t)} \tag{4}$$

Among them, the value of $SD$ is between 0.2 and 0.3.

After EMD decomposition of the original signal, several EMD components and residuals (res) are obtained, and the original signal $X(t)$ is expressed as:

$$X(t) = \sum_{i=1}^{n} IMF_i(t) + r(t) \tag{5}$$

In formula (5): $IMF_i(t)$ is each IMF component; $n$ is the number of IMF components; $r(t)$ is the residual term.

For the reconstruction after signal decomposition, the correlation coefficient threshold method is generally used. The correlation coefficient between each IMF component and the original signal is calculated by formula (6), and the first minimum value is the distinguishing threshold. The IMF component corresponding to this point is defined as the transition IMF, and the IMF components before and after the critical point are respectively Defined as noise IMF and useful IMF, the transition IMF and useful IMF are reconstructed and the denoised signal after EMD decomposition is obtained.

$$r_i\left(IMF_i, X(t)\right) = \frac{Cov\left(IMF_i, X(t)\right)}{\sqrt{Var\left[IMF_i\right] Var[X(t)]}} \tag{6}$$

In formula (6): $IMF_i$ represents the $i$-th IMF.

## 3. Design of IIR Digital Filter

The digital filter processes the input digital signal through related algorithms and programs, so as to achieve the function of the analog filter to filter the specific frequency of the signal, and can achieve high-pass, low-pass, band-pass and other filtering functions. Digital filters can be divided into Finite Impulse Response (FIR) and IIR . Compared with the FIR filter, the IIR filter has the advantages of simple structure, less parameters and higher efficiency, and is suitable for computer-aided design. IIR only needs to use a few orders to complete the amplitude frequency characteristics of fir with multiple orders. If there are no special requirements for linearity, IIR can obtain better out of band attenuation characteristics under the same gate scale and the same frequency speed [7]. In this paper, the IIR filter is designed. The transfer function expression and corresponding difference equation of IIR digital filter are [8] respectively:

$$H(z) = \frac{\sum_{t=0}^{M} b_r z^{-r}}{1 + \sum_{k=1}^{N} a_k z^{-k}} = \frac{Y(z)}{X(z)} \tag{7}$$

$$y(n) = \sum_{k=1}^{N} a_k y(n-k) + \sum_{k=0}^{M} a_k x(n-k) \tag{8}$$

Where, $a_k$ and $b_k$ are filter coefficients respectively.

In this article, the Butterworth low-pass filtering digital filter is selected, and its amplitude square function is:

$$A\left(\Omega^2\right) = \frac{1}{1+\left(\frac{j\Omega}{j\Omega}\right)^{2N}} \tag{9}$$

Where $N$ is the filter order.

## 4. Noise reduction scheme based on EMD and IIR digital filtering

Due to factors such as engine vibration, road surface, and the measurement error of the accelerometer itself, there is usually noise in the collected data, and the error diffusion through the inertial navigation algorithm seriously affects the SINS navigation accuracy. The frequency band of the acceleration signal itself is relatively low, mainly included in the low frequency part, and the high frequency part is mainly noise. The traditional EMD noise reduction method is to directly remove the decomposed first two-order high-frequency IMF components, but there may still be some useful information in the high-frequency components, such as acceleration spikes or sudden changes.

Therefore, the traditional EMD decomposition will make the modes appearing in the high-frequency IMF components aliased and mixed with useful signals. Using MATLAB software to build IIR low-pass digital filter to filter high-frequency IMF components, and then reconstruct with low-frequency IMF components to achieve the purpose of improving the noise reduction effect. Because the driving behavior under the acceleration condition is simple, the working condition is relatively stable, and the frequency band of the acceleration spike or sudden change information is low, the use of low-pass filtering will not lose the useful information in the high-frequency IMF, which can achieve the purpose of improving the noise reduction effect. The noise reduction steps are:

(1) Perform EMD decomposition on the original signal to obtain several IMF components and a res remainder;

(2) Divide IMFs into noise IMFs, transition IMFs and useful IMFs based on the correlation coefficient method;

(3) Perform IIR low-pass filtering on the noise IMF and transition IMF to obtain the filtered signal;

(4) Perform EMD reconstruction on the filtered signal and the useful IMF component to obtain a noise reduction signal.

The noise reduction process is shown in Figure 1.

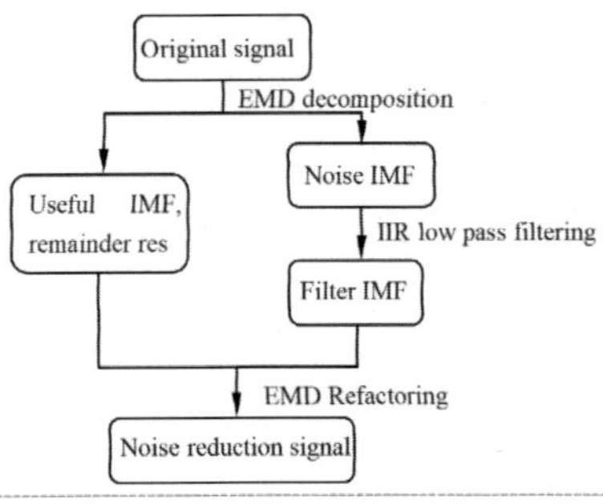


Figure 1. Process of improve noise reduction.

## 5. Test and Analysis

In order to verify the effectiveness of the initial joint filtering based on EMD and IIR, the self-developed high-precision optical fiber strapdown inertial navigation system (accuracy 0.002°/h) was mounted on the navigation car for sports car experiments, and the three-axis inertial navigation system was tested. The original data of the accelerometer and the fiber optic gyroscope, the sampling frequency is 200Hz, and the test equipment is shown in Figure 2.

Figure 2. Vehicle-mounted SINS test equipment.

Taking the y-axis(forward) accelerometer signal as an example, the output data collected from the accelerometer is as follows:

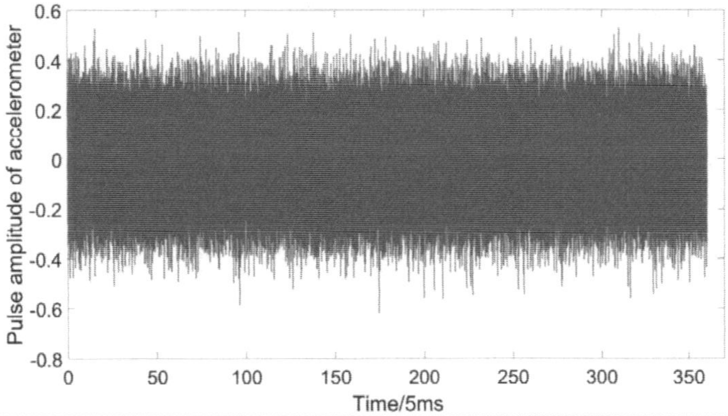


Figure 3. Y-axis (forward) accelerometer of SINS raw data curve.

According to the EMD algorithm introduced in the first chapter, the signal processing program is compiled using matlab, and the EMD decomposition is performed on the original signal of the y-axis acceleration. The number of decomposition layers is set to 12, and the result is shown in Figure 5. After the original accelerometer signal is decomposed, 12 IMF components and a residual term res are obtained. The IMF components are arranged in order of frequency, and there is obvious signal-to-noise aliasing in the first four IMF components.

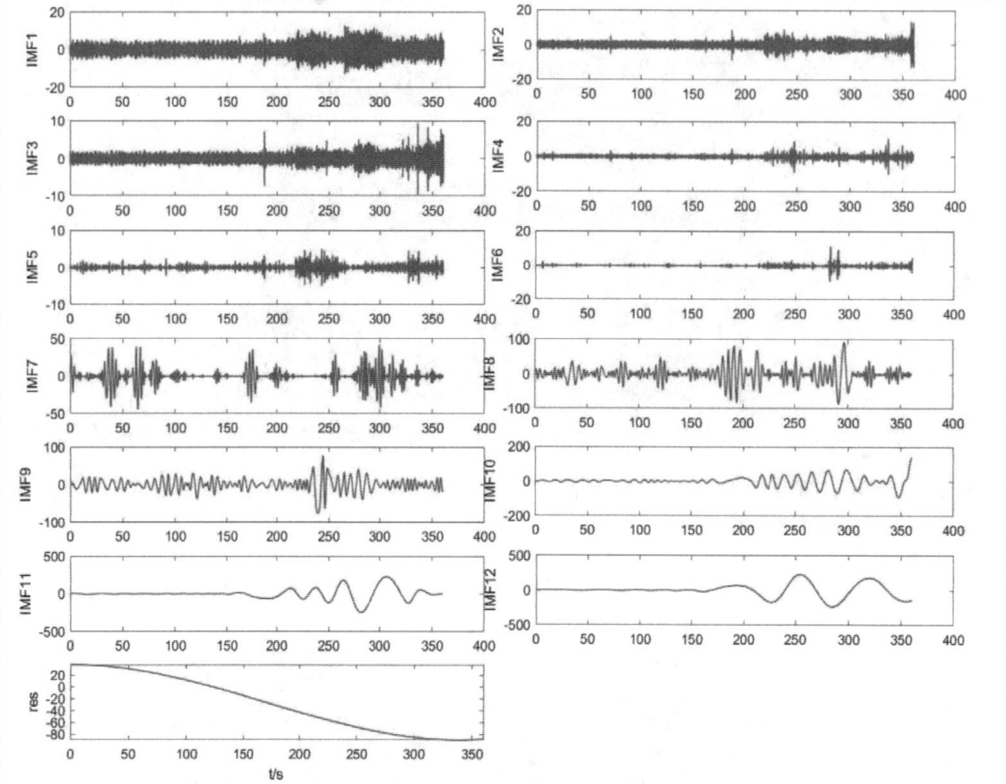


Figure 4. Statistical curve of IMF components after EMD decomposition.

For each IMF component, the correlation coefficient of each IMF is calculated according to formula (6). The result is shown in Figure 5:

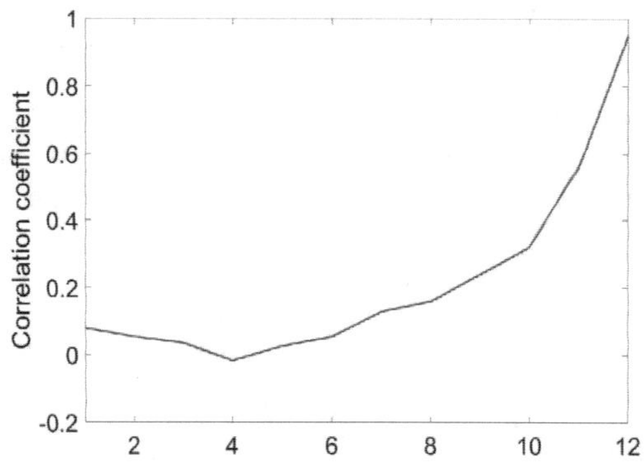


Figure 5. Correlation coefficient of each IMF component.

A minimum occurs at the 4th IMF. According to the method described in Section 1, IMF4 is defined as a transition IMF, IMF1, IMF2, and IMF3 are defined as noise IMFs, and the remaining IMFs are defined as useful IMFs. According to the EMD correlation coefficient reconstruction method, IMF1, IMF2, IMF3 are subjected to IIR low-pass filtering, and the transition IMFs, useful IMFs and the remainder are reconstructed, and the results are shown in Figure 6. The IIR low-pass filter uses the built-in toolbox FDATool toolbox of MATLAB software for filter design. The relevant parameters to be set are the pass-band cut-off frequency FP, the stop-band cut-off frequency FS, the maximum pass-band attenuation RP and the stop-band minimum attenuation RS.

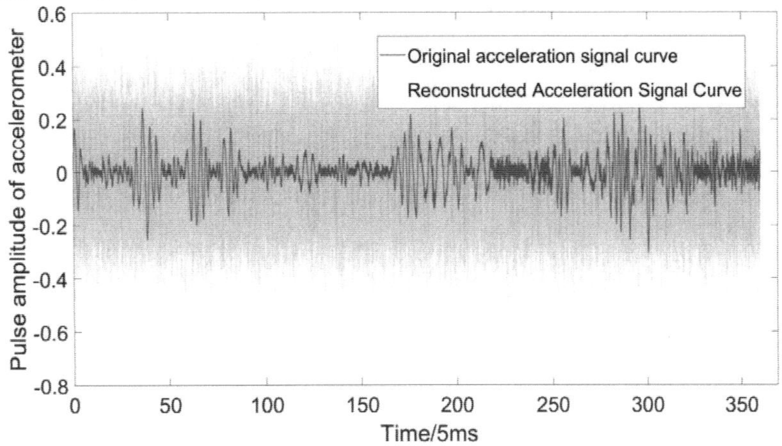


Figure 6. Y-axis accelerometer signal comparison before and after filtering.

It can be seen from Figure 6 that the noise reduction method has an obvious effect on the noise reduction of the acceleration signal. The high frequency part of the filtered accelerometer signal can be effectively filtered out, and the amplitude is also reduced. In the effect of the navigation experiment, the original signal and the reconstructed signal are imported respectively for pure inertial navigation calculation, and the horizontal positioning error of the navigation experiment is obtained as shown in Figure 7. The maximum horizontal positioning error of the navigation based on the original signal solution exceeds 0.42 nautical miles, while the maximum horizontal positioning error of the 8-hour navigation based on the reconstructed signal solution is better than 0.29 nautical miles, and the navigation accuracy is improved by 31%. It can be seen that the joint filtering based on EMD and IIR proposed in this paper can effectively improve the navigation accuracy of the SINS.

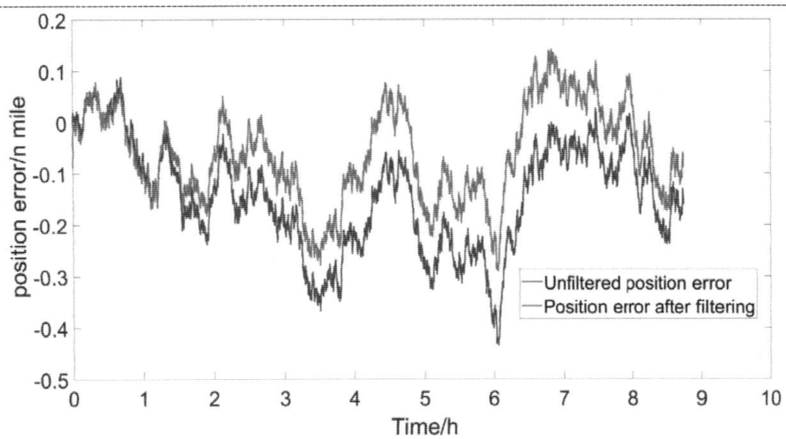


Figure 7. Comparison of SINS navigation accuracy before and after Y-axis accelerometer filtering.

## 6. Conclusions

Aiming at the problem that the noise of the accelerometer signal affects the navigation accuracy of the SINS in the vibration environment, a joint noise reduction method based on EMD decomposition and IIR low-pass digital filtering is proposed in this paper. On the basis of EMD decomposition, the IIR low-pass digital filter is used to filter the high-frequency IMF components, and the filtered signal and the low-frequency IMF components are reconstructed to obtain a noise reduction signal. The final results show that the joint noise reduction method based on EMD decomposition and IIR low-pass filtering proposed in this paper can effectively reduce the noise in the signal. After filtering, the horizontal positioning error of the inertial navigation system is better than 0.29 nautical miles, and the navigation accuracy is improved by 31%. It is proposed in this paper that the joint filtering based on EMD and IIR can effectively improve the navigation accuracy of SINS and provide a new idea for signal noise reduction of inertial devices.

**Acknowledgments**

This work was supported in part by the National Natural Science Foundation of Chinaunder Grant 61803015.

**References**

[1] David H Titterton, John L Weston. (2004) Strpdown Inertial Navigation Technology, Peter Peregrinus on Behalf of the Institute of Electrical Engineer, London.

[2] LIU Haijiang,LIU Shigao,LI Min. (2018) EMD and wavelet threshold noise reduction method for gear shift acceleration signals[J].Noise and Vibration Control, 38( 2): 198-203.

[3] WU Z, HUANG N E. (2004) A study of the characteristics of white noise using the empirical mode decomposition meth-od[J].Proceedings of the Royal Society A: Mathematical, Physical and Engineering Sciences, 460(2046):1597-1611.

[4] WU Z, HUANG N E. (2009) Ensemble empirical mode decompo-sition: anoise-assisted data analysis method [J]. Advances in Adaptive Data Analysis, 1(1):1-41.

[5] HUANG N E,SHEN Z,LONG S R,et al. (1998) The empirical mode decomposition and the hilbert spectrum for nonlinear and non-stationary time series analysis [J].Proceedings Mathematical Physical and Engineering Sciences, 454 ( 1971),903-995.

[6] WU Jun, ZHANG Yufeng. (2012) Comparison of filtering characteristics between empirical mode decomposition and wavelet decomposition[J]. Journal of Yunnan University (Natural Science Edition) ,2012,34( 3): 285 - 290.

[7] DONG Yawen,WANG Jianlin,WEI Qingxuan,et al. (2019) Accelerometer signal noise reduction based on sequence correlation and wavelet transform[J]. Journal of Transducer Technology, 32( 2): 225 - 230.

[8] QU Xing,TANG Ning,YAN Shu, et al. (2009) Design and Simulation of IIR Digital Filter Based on FPGA[J]. Computer Simulation, 26(8): 304 - 307.

# Ride comfort analysis of electric vehicle using hub motor as dynamic absorber

**Pingping Zhao[1], Zhimin Fan[2]**

[1] School of Mechanical and Electrical Engineering, Qingdao University of Science and Technology, Qingdao, Shandong, 266061, China

[2] Intelligent Manufacturing Institute, Weifang University of Science and Technology, Weifang, Shandong 262700, China

*Corresponding author's e-mail: zmfan@163.com

**Abstract.** To resolve the problems of adding unsprung mass and worsening ride comfort of hub motor drive electric vehicles, a new drive system with hub motor as dynamic absorber is studied and proposed. The mathematical models of the new hub motor drive system and the centralized motor drive system are established, and their ride comfort is compared and analyzed. The impact of dynamic absorber parameters on ride comfort of the new hub motor drive system is analyzed, and the parameter sensitivity is calculated. The results show that the body acceleration, suspension dynamic deflection, wheel dynamic load of the new hub motor drive system are better than those of the centralized motor drive system, and the dynamic absorber parameters have a great effect on the electric vehicle performance.

## 1. Introduction

The hub motor drive electric vehicle with the motor mounted directly in the wheel has many advantages: a high degree of integration of the in-wheel system, simplified chassis structure, increased interior space and drive flexibility. However, there are some problems with this type of drive, such as the increased unsprung mass of the vehicle caused by the hub motor, which affects the ride comfort.

To resolve the above problems, scholars at home and abroad have proposed some solution measures. In the literature [1,2], a dynamic absorber mechanism was designed to suppress vibrations and improve the ride comfort. In [3,4], an integrated structure was used to achieve the goal of decreasing the unsprung mass. The literature [5] proposes an electric wheel topology solution with a built-in suspension system. This solution improves the vehicle dynamics by separating the hub motor from the unsprung mass through the installation of rubber elements. The literature [6-10] improves the driving comfort of hub motor drive electric vehicles by designing control strategies and methods for active and semi-active suspensions. The literature [11-13] investigated control strategies to improve vehicle smoothness.

However, the above studies had some shortcomings in that they did not think about enhancing vehicle ride comfort and lessening motor vibration.

Therefore, this paper investigates and analyses a new drive system with a hub motor as a dynamic absorber. In the study of the ride comfort of this new system, four indicators of body acceleration, suspension dynamic deflection, dynamic wheel load and motor acceleration are analyzed, and not only the ride comfort but also the vibration of the motor is investigated.

## 2. Comparison of the ride comfort of the two models

### *2.1. Building the model*

In previous research work, the authors designed a new drive system with a hub motor as a dynamic absorber [14], a special structure that allows the hub motor to transform from an unsprung mass to a sprung mass. Conventional electric vehicles basically use a centralized motor drive. The two 1/4 vehicle vertical vibration models are shown in Figure 1.

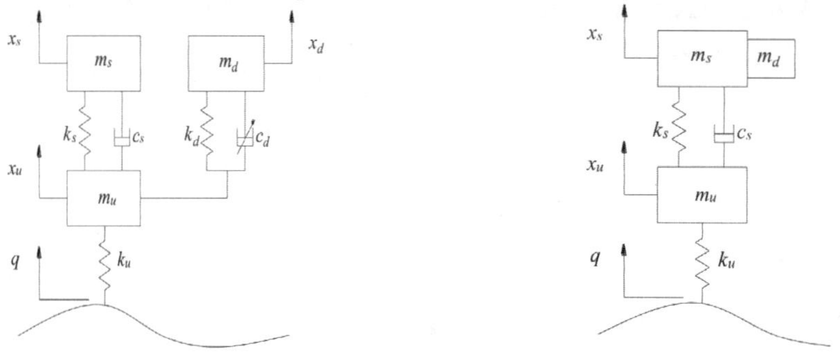


(a) Model of the new hub motor drive system (b) Model of a centralized motor drive system

Figure 1. 1/4 vehicle vertical vibration model

The vibration differential equations for the new hub motor drive system model and the centralized motor drive system model are

$$\begin{cases} m_s\ddot{x}_s + c_s(\dot{x}_s - \dot{x}_u) + k_s(x_s - x_u) = 0 \\ m_d\ddot{x}_d + c_d(\dot{x}_d - \dot{x}_u) + k_d(x_d - x_u) = 0 \\ m_u\ddot{x}_u - c_s(\dot{x}_s - \dot{x}_u) - k_s(x_s - x_u) - \\ c_d(\dot{x}_d - \dot{x}_u) - k_d(x_d - x_u) + k_u(x_u - q) = 0 \end{cases} \tag{1}$$

$$\begin{cases} (m_s + m_d)\ddot{x}_s + c_s(\dot{x}_s - \dot{x}_u) + k_s(x_s - x_u) = 0 \\ m_u\ddot{x}_u + k_u(x_u - q) - c_s(\dot{x}_s - \dot{x}_u) - \\ k_s(x_s - x_u) = 0 \end{cases} \tag{2}$$

where $m_s$ is the body mass, $m_d$ is the motor mass, $m_u$ is the wheel mass, $k_s$ is the suspension stiffness, $k_u$ is the tyre stiffness, $c_s$ is the suspension damping, $x_s$ is the body vertical vibration displacement, $x_u$ is the wheel vertical vibration displacement, $x_d$ is the motor vertical vibration displacement, $q$ is the road surface unevenness function, $k_d$ is the dynamic absorber stiffness, $c_d$ is the dynamic absorber damping. The vehicle parameter values are shown in Table 1.

Table 1. Vehicle parameters

| Parameters | Centralized | New |
|---|---|---|
| $m_s$ / kg | 350 | 350 |
| $m_d$ / kg | 35 | 35 |
| $m_u$ / kg | 40 | 40 |
| $k_s$ / (N·m$^{-1}$) | 15000 | 15000 |
| $k_u$ / (N·m$^{-1}$) | 200000 | 200000 |
| $c_s$ / (N·s·m$^{-1}$) | 1300 | 1300 |
| $c_d$ / (N·s·m$^{-1}$) | —— | 300 |
| $k_d$ / (N·m$^{-1}$) | —— | 15000 |

### *2.2. Comparative analysis of vehicle ride comfort*

The evaluation indicators for vehicle ride comfort include body acceleration, wheel dynamic load and suspension dynamic deflection [15]. As hub motors operate in a complex environment, motor vibration can directly affect the performance and lifetime of the motor, so it is necessary to evaluate motor acceleration when evaluating the ride comfort of hub motor drive electric vehicle.

In this section, a simulation program is designed in Matlab/Simulink to compare the ride comfort of the new hub motor drive system and the centralized motor drive system.

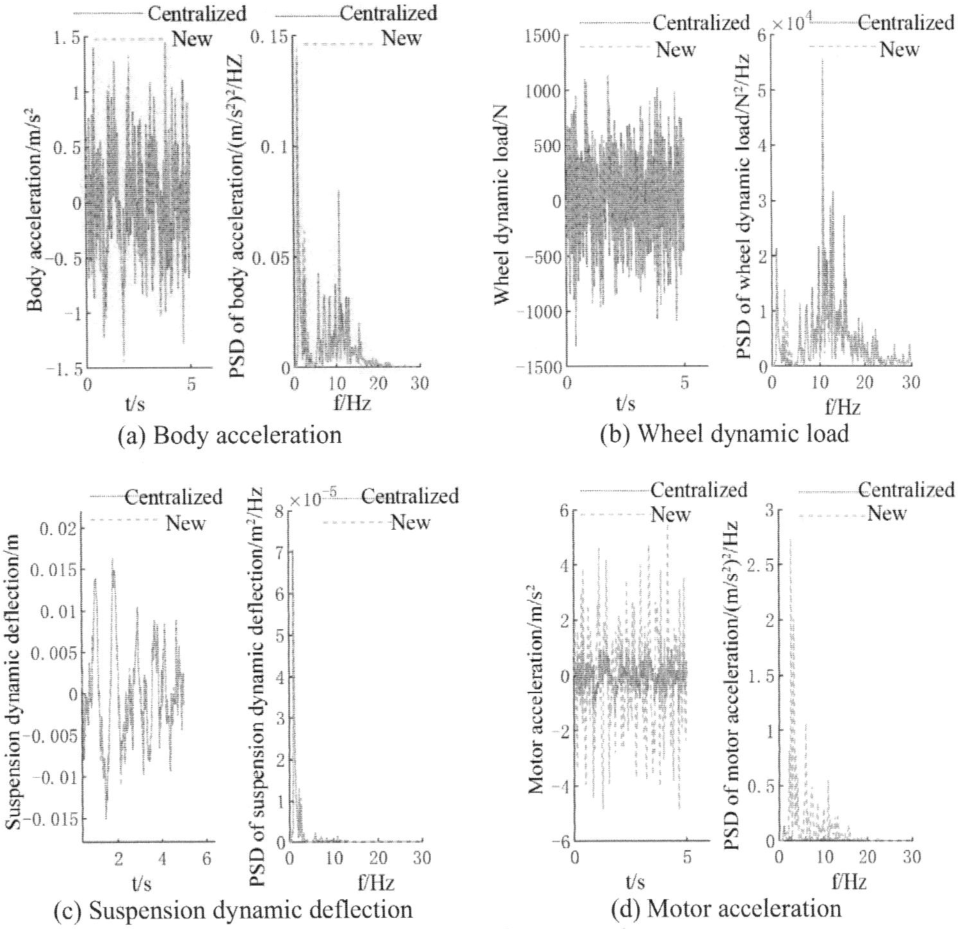


Figure 2. Ride comfort comparison

From Figure 2(a) to (c), it can be seen that the overall vibration amplitude of the new hub motor drive system model is less than that of the centralized motor drive system model in both the frequency and time domains, indicating that the ride comfort of the new hub motor drive system model is improved compared.

As seen in Figure 2(d), the fluctuations in the motor acceleration profile of the new hub motor drive system model are significantly greater than those of the centralized motor drive system model, both in the frequency and time domains. For comparison purposes, the maximum values and standard variances of motor acceleration for both models are listed in Table 2. As can be seen from Table 2, in the time domain, the maximum value of motor acceleration for the new hub motor drive system model is 3.21 times greater than that of the centralized motor drive system model, and the standard variance is 8.77 times higher. In the frequency domain, the maximum value of motor acceleration for the new hub motor drive system model is 24.96 times greater than that of the centralized motor drive system model, with a standard variance of 18.58 times. This indicates that the motor vibration acceleration of the new hub motor drive system model deteriorates, which directly affects the motor's performance.

Table 2. Motor acceleration

| | Time domain $/m/s^2$ | | Frequency domain $/(m/s^2)^2/Hz$ | |
|---|---|---|---|---|
| | Centralized | New | Centralized | New |
| Maximum peak | 1.719 | 5.52 | 0.17777 | 4.437 |
| Standard variance | 0.5175 | 1.95 | 0.01462 | 0.2717 |

## 3. Parameter sensitivity calculation

To analyze the influence of the dynamic absorber parameters, i.e. motor mass, dynamic absorber stiffness and dynamic absorber damping, on the ride comfort of the new hub motor drive system, the sensitivity values of body acceleration, suspension dynamic deflection, wheel dynamic load and motor acceleration to the dynamic absorber parameters are calculated in this section.

Sensitivity is defined as follows

$$S_{\sigma\_x_{ij}} = \lim_{\Delta x_{ij}\to 0} \frac{(\sigma(x_{ij}+\Delta x_{ij})-\sigma(x_{ij}))/\sigma(x_{ij})}{\Delta x_{ij}/x_{ij}}$$
$$S_{\sigma\_x_i} = \sqrt{\frac{1}{l}\sum_{j=1}^{l} S_{\sigma\_x_{ij}}^{\ 2}} \qquad (i=1,2,3,4) \tag{3}$$

where $\sigma$ a is the RMS value of the vibration response ( $\sigma_{\ddot{x}_c}$ 、 $\sigma_{F_d}$ 、 $\sigma_{f_d}$ 、 $\sigma_{\ddot{x}_d}$ ), $S_{\sigma\_x_{ij}}$ is the sensitivity of $\sigma$ to parameter $x_i$ at $x_{ij}$ , $S_{\sigma\_x_i}$ is the sensitivity of $\sigma$ to parameter $x_i$ over the entire design interval, $l$ is the total number of segments of parameter $x_i$ in the design interval.

The parameter sensitivity values of the new hub motor drive system at low frequency (0-4Hz), medium frequency (4-5Hz) and high frequency (10-20Hz) are calculated according to equation (3) and see Table 3 for the results.

Table 3. Sensitivity values

| Parameters | Body acceleration $S_{\sigma_{\ddot{x}_c}}$ | | | Wheel dynamic load $S_{\sigma_{F_d}}$ | | |
|---|---|---|---|---|---|---|
| | Low | Medium | High | Low | Medium | High |
| $m_d$ | 0.39 | 3.44 | 0.73 | 6.33 | 15.21 | 0.23 |
| $k_d$ | 0.94 | 4.70 | 0.65 | 4.63 | 17.89 | 1.57 |
| $c_d$ | 0.15 | 3.18 | 5.13 | 3.07 | 8.83 | 2.73 |
| Parameters | Suspension dynamic $S_{\sigma_{f_d}}$ | | | Motor acceleration $S_{\sigma_{\ddot{x}_d}}$ | | |
| | Low | Medium | High | Low | Medium | High |
| $m_d$ | 0.20 | 3.63 | 0.76 | 7.07 | 47.92 | 32.90 |
| $k_d$ | 0.33 | 4.60 | 1.11 | 22.46 | 45.79 | 11.03 |
| $c_d$ | 0.01 | 3.27 | 5.30 | 16.95 | 0.53 | 16.56 |

From the above table, it can be seen that the dynamic absorber parameters have different effects on the ride comfort performance in different frequency domains, which are summarised in Table 4 for ease of analysis.

Table 4. Effect of different frequency domains

| Indicators | Low frequency | Medium frequency | High frequency |
|---|---|---|---|
| Body acceleration | Little impact | Larger effect, dynamic absorber stiffness > motor mass > power absorber damping | High influence of dynamic absorber damping |
| Wheel dynamic load | Greater influence, motor mass > dynamic absorber stiffness > dynamic absorber damping | Significant effect, dynamic absorber stiffness > motor mass > dynamic absorber damping | No significant impact |
| Suspention dynamic deflection | Essentially no impact | Larger effect, dynamic absorber stiffness > motor mass > dynamic absorber damping | High influence of dynamic absorber damping |
| Motor acceleration | Significant effect, dynamic absorber stiffness > dynamic absorber damping > motor mass | Significant effect, motor mass > dynamic absorber stiffness | Significant effect, motor mass > dynamic absorber damping > dynamic absorber stiffness |

As can be seen from the table above, the parameters of the dynamic absorber, i.e. motor mass, dynamic absorber stiffness and damping, not only have a significant effect on the motor acceleration, but also on the ride comfort.

## 4. Conclusion

To address the problem of excessive unsprung mass and deteriorating ride comfort of hub motor drive electric vehicles, this paper investigates and analyzes a new drive system with a hub motor as a dynamic absorber. The results show that the body acceleration, suspension dynamic deflection and wheel dynamic load of the new hub motor drive system are better than those of the centralized motor drive system, but the motor acceleration is significantly higher. The effect of dynamic absorber parameters on vehicle ride comfort is investigated and the sensitivity is calculated. The simulation results show that the dynamic absorber parameters have different effects on vehicle ride comfort in different frequency domains, and the dynamic absorber parameters,i.e. motor mass, dynamic absorber stiffness and damping, make a great impact on the motor acceleration and vehicle ride comfort.

In summary, this study provides some basis and reference for enhancing the ride comfort of the hub motor drive electric vehicle.

**Acknowledgments**

This research is supported by Key Research and Development Program of Shandong Province of China (Grant No. 2019GGX104088).

## References

[1] Nagaya,G. (2003) Development of an in-wheel drive with advanced dynamic-damper mechanism. JSAE Review, 24(4): 477–481.

[2] Liu, M., Gu, F., Huang, J., et al. (2017) Integration Design and Optimization Control of a Dynamic Vibration Absorber for Electric Wheels with In-Wheel Motor. Energies, 10(12): 2069.

[3] Li, Y., Xue, N., Luo, F., et al. (2018) Design of a direct wheel drive system utilizing silent chain integrated with semi-trailing suspension. Transactions of Beijing Institute of Technology, 38(A01): 192–194.
[4] Yin, G., Yuan, Y. (2015) An integrated unequal length double wishbone suspension wheel drive system for reducing unsprung mass. China, CN2015102836671.
[5] Tan, D. (2013) Dynamics and structure optimization of the in-wheel motor system with rubber bushing. Guangzhou: South China University of Technology.
[6] Tan, D., Lu, C., Zhang, X. (2016) Dual-loop PID control with PSO algorithm for the active suspension of the electric vehicle driven by in-wheel motor. Journal of Vibroengineering, 18: 3915–3929.
[7] Katsuyama, E., Omae, A. (2016) Improvement of Ride Comfort by Unsprung Negative Skyhook Damper Control Using In-Wheel Motors. SAE International Journal of Alternative Powertrains, SAE International, 5(1): 214–221.
[8] Wang, R., Jing, H., Yan, F., et al. (2015) Optimization and finite-frequency H∞ control of active suspensions in in-wheel motor driven electric ground vehicles. Journal of the Franklin Institute, 352(2): 468–484.
[9] Shao, X., Naghdy, F., Du, H. (2017) Reliable fuzzy H∞ control for active suspension of in-wheel motor driven electric vehicles with dynamic damping. Mechanical Systems and Signal Processing, 87: 365–383.
[10] Hu, Y., Li, Z., Yang, C., et al. (2018) Analysis and optimization of suspension configuration of in-wheel motor electric vehicle. In: 2018SAECCE. Shanghai. pp. 952-956.
[11] Zhao, J., Wong, P. K., Xie, Z., et al. (2019) Design and Control of an Automotive Variable Hydraulic Damper Using Cuckoo Search Optimized Pid Method. International Journal of Automotive Technology, 20(1): 51–63.
[12] Jin, T., Liu, Z., Sun, S., et al. (2020) Theoretical and experimental investigation of a stiffness-controllable suspension for railway vehicles to avoid resonance. International Journal of Mechanical Sciences, 187: 105901.
[13] Lai, F., Huang, C. (2012) An investigation into vehicle active suspension with electromagnetic actuator. Automotive Engineering, 34(02): 170–174.
[14] Zhao, P., Fan Z. (2019) Electric wheel system with hub motor serving as mass of dynamic vibration absorber. China, 2019110230163.
[15] Yu, Z. (2009) Automobile theory. China Machine Press, Beijing.

Journal of Physics: Conference Series **2264** (2022) 012018 doi:10.1088/1742-6596/2264/1/012018

# Research on speed control of permanent magnet synchronous motor based on RBF neural network tuning PID

**Wanli WANG[1*], Han PANG[1], Xincheng LI[2], Yipeng WU[2], Xiufu SONG[2]**

[1]China University of Petroleum (East China)

[2]Yantai Jereh Petroleum Equipment & Technologies Co., Ltd.

[*]Corresponding author: wwli0919@163.com

Abstract: In order to solve the problem of insufficient speed regulation performance of conventional PID controller in multi gear switching of electric drive fracturing PMSM, RBF neural network parameter identification is used to set the PID parameters and improve the control effect of conventional PID controller. By analyzing the application characteristics of PMSM in electric drive fracturing truck, the RBF-PID controller and the speed control system model of PMSM are established based on s-function in MATLAB/SIMLINK, and the simulation research is carried out in combination with the application requirements. The results show that compared with the conventional PID controller, the RBF-PID controller has better real-time performance and robustness. The real-time response characteristics of the PMSM speed regulation control system are improved by 84% and the robustness is improved by 62%. The RBF-PID controller can effectively improve the working stability of PMSM.

## 1. Introduction

With the progress of science and technology, electric drive fracturing truck has become the main fracturing equipment for shale gas exploitation because of its characteristics of green environmental protection and good economic and social benefits. As the core equipment of the power system of electric drive fracturing truck, the stability and accuracy of PMSM directly affect the exploitation of shale gas. Due to the nonlinear and strong coupling characteristics of PMSM, the motor drive control system needs more advanced intelligent control algorithm [1].

At present, PID controller has been widely used in many fields of engineering control because of its simple structure and easy implementation. However, with the improvement of the complexity of the control object and the defect of its self-adaptive ability, PID controller can not meet the application scene of high-precision control. In order to improve the stability and robust characteristics of PMSM in complex environment, research teams at home and abroad have studied the composite control method of a variety of control strategies combined with PID controller. Wenbao Hu [2] proposed an improved PID control method to improve the dynamic response characteristics of the control system, but it has not been widely used because it relies too much on manual assistance and is easy to make mistakes. Chao Tian [3] and others proposed the method of using fuzzy control to adjust PI controller parameters online, which effectively reduces the overshoot and regulation time of the control system and improves the dynamic response characteristics of the control system. However, this method can not make the controlled system have strong speed tracking and robustness at the same time.

In this paper, Radial Basis Function (RBF) neural network is combined with PID control. By training neural network, PID parameters are adjusted according to the output of controller, which can effectively solve the influence of motor parameter changes and external disturbances.

## 2. Construction of mathematical model of PMSM

The stator side of PMSM is connected with three-phase symmetrical current. Because the three-phase stator has a difference of 120° in space, the three-phase stator current generates a rotating magnetic field in space, and the rotor moves under the action of electromagnetic force in the rotating magnetic field. Therefore, the vector control system must be transformed into a scalar system similar to DC motor by coordinate transformation method. For most PMSMs, there is no damping winding in the rotor, so the voltage, flux and torque equations of the coordinate transformed PMSM in the rotating d-q coordinate system are shown in equations (1) ~ (3) [4-5].

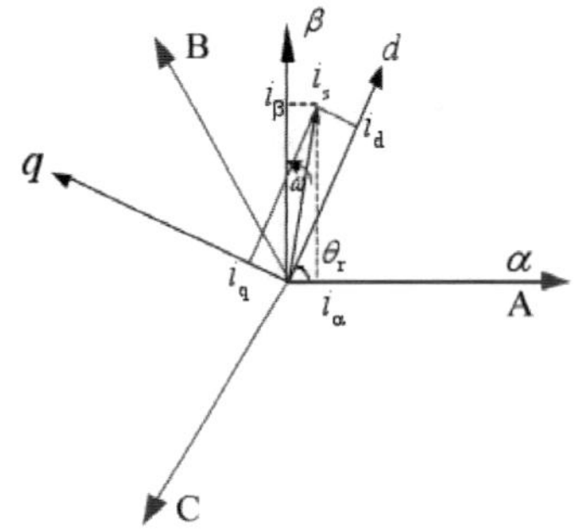


Fig. 1 Schematic diagram of coordinate transformation

Voltage equation:

$$\begin{cases} u_d = R_s i_d + \dfrac{d\psi_d}{dt} - \omega_r \psi_q \\ u_q = R_s i_q + \dfrac{d\psi_q}{dt} - \omega_r \psi_d \end{cases} \tag{1}$$

Flux linkage equation:

$$\begin{cases} \psi_d = L_d i_d + \psi_f \\ \psi_q = L_q i_q \end{cases} \tag{2}$$

Torque equation:

$$\begin{cases} \psi_d = L_d i_d + \psi_f \\ \psi_q = L_q i_q \end{cases} \tag{3}$$

By introducing the flux linkage equation into the torque equation:

$$T_e = p_n \left[ \psi_f i_q + (L_d - L_q) i_d i_q \right] \tag{4}$$

When the d-axis is 0, the following can be obtained:

$$T_e = p_n \psi_f i_q \tag{5}$$

The mechanical motion equation is:

$$T_e = T_L + B\omega + J\frac{d\omega}{dt} \tag{6}$$

In the above formula, $R_s$ is the stator resistance of the motor; $u_d$ 、 $u_q$ is the voltage of D and Q axes of stator winding; $i_d$ 、 $i_q$ is the current of D and Q axes of stator winding; $L_d$ 、 $L_q$ is the direct axis and quadrature axis synchronous inductance of stator winding; $\psi_d$、$\psi_q$ is the flux linkage of stator D and Q axes; $\psi_f$ is permanent magnetic flux; $p_n$ is the number of poles of the motor; $\omega$ is the electrical angular speed of motor rotor; $T_e$ is the electromagnetic torque of the motor; $T_L$ is the motor load torque; J is the moment of inertia of the rotor; B is the friction coefficient of the motor.

## 3. Design of PID controller based on RBF neural network

### 3.1. PID control algorithm

PID control usually adopts incremental PID algorithm, which uses the increment of control quantity Δu(k)=u(k)−u(k−1) to replace the original control quantity u(k), and the increment of the control quantity is only related to the latest sampling value. The formula of incremental PID algorithm is:

$$\Delta u(k) = K_P[e(k)-e(k-1)] + K_I e(k) + K_D[e(k)-2e(k-1)+e(k-2)] \tag{7}$$

### 3.2. RBF neural network algorithm

RBF neural network is a three-layer feedforward network with single hidden layer. It has fast learning speed and can effectively avoid the problem of falling into local minimum in the solution process. The general RBF neural network is composed of input layer, hidden layer and output layer as shown in Figure 2.

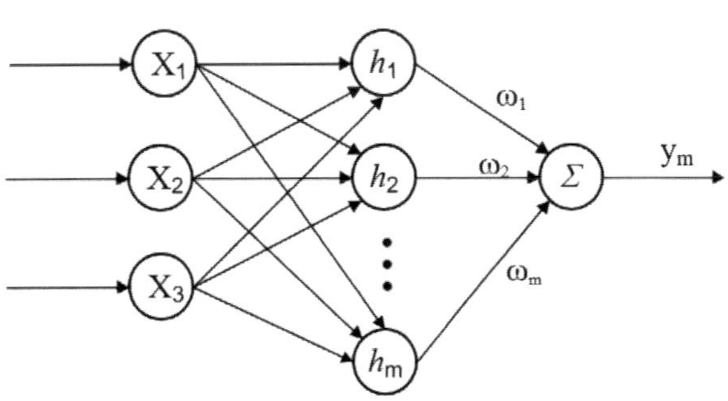


Fig 2 RBF neural network structure

The hidden layer of RBF neural network is composed of radial basis functions. The commonly used radial basis functions include thin plate spline function, inverse multiquadratic function, Gaussian function and so on. In this paper, Gaussian function is selected as the basis function, which has the characteristics of simple form, symmetrical structure and good smoothness [6-7], and its expression is:

$$h_j = \exp\left(-\frac{\|x-c_j\|^2}{2b_j^2}\right), \quad j=1,2,\cdots,m \tag{8}$$

Where: $\|x-c_j\|$ is Euclidean norm; $c_j$ is the center vector of the j-th hidden layer neuron node; $b_j$ is the width of the j-th hidden layer neuron node; $\omega_j$ is the weight vector of the output layer.

The gradient descent method is used as the learning algorithm of RBF neural network to correct the network center, width and weight of RBF. Because the conventional gradient descent method only depends on the error gradient direction of the current time and does not take into account the gradient of the previous historical time, it is easy to cause turbulence and local optimization problems in the training process. Therefore, this paper adds a momentum term to the adjustment formula of hidden layer output weight, center value and base width. The modified formula of RBF neural network parameters after adjustment is as follows:

Weight correction formula:

$$\begin{cases} \Delta\omega_j(k) = -\eta \dfrac{\partial J}{\partial \omega_j} = \eta[y(k) - y_m(k)]h_j \\ \omega_j(k) = \omega_j(k-1) + \Delta\omega_j(k) + \alpha[\omega_j(k-1) - \omega_j(k-2)] \end{cases} \tag{9}$$

Correction formula of hidden layer center:

$$\begin{cases} \Delta c_{ji}(k) = -\eta \dfrac{\partial J}{\partial c_{ji}} = \eta[y(k) - y_m(k)]\omega_j h_j \dfrac{x_j - c_{ji}}{b_j^2} \\ c_{ji}(k) = c_{ji}(k-1) + \Delta c_{ji}(k) + \alpha[c_{ji}(k-1) - c_{ji}(k-2)] \end{cases} \tag{10}$$

Base width correction formula:

$$\begin{cases} \Delta b_j(k) = -\eta \dfrac{\partial J}{\partial b_j} = \eta[y(k) - y_m(k)]\omega_j h_j \dfrac{\left\| x_j - c_{ji} \right\|^2}{b_j^3} \\ b_j(k) = b_j(k-1) + \Delta b_j(k) + \alpha[b_j(k-1) - b_j(k-2)] \end{cases} \tag{11}$$

Where: $\eta$ is the learning rate, which is generally selected between 0 and 1 to ensure the convergence of the iterative algorithm; $\alpha$ is the momentum coefficient, which is generally taken between 0 and 1.

### 3.3. PID parameter tuning based on RBF network

RBF neural network identifies the parameters according to the actual operation of PMSM, and obtains the Jacobian determinant information of the control input to output sensitivity of the controlled object, so as to adjust the parameters of conventional PID controller online. The error of the control system is:

$$e(k) = r - y(k) \tag{12}$$

Where: R is the given value, y(k) is the actual speed.

The error input of incremental PID is:

$$\begin{cases} x(1) = e(k) - e(k-1) \\ x(2) = e(k) \\ x(3) = e(k) - 2e(k-1) + e(k-2) \end{cases} \tag{13}$$

The performance index function of RBF neural network for PID controller tuning is:

$$E_k = \frac{1}{2} e^2(k) \tag{14}$$

The adjustment amount of PID parameters adjusted by gradient descent method are:

$$\begin{cases} \Delta K_P = -\eta \dfrac{\partial E_k}{\partial K_P} = -\eta \dfrac{\partial E_k}{\partial y(k)} \eta \dfrac{\partial y(k)}{\partial \Delta u} \eta \dfrac{\partial \Delta u}{\partial K_P} = \eta e(k) \dfrac{\partial y(k)}{\partial \Delta u} x(1) \\ \Delta K_I = -\eta \dfrac{\partial E_k}{\partial K_I} = -\eta \dfrac{\partial E_k}{\partial y(k)} \eta \dfrac{\partial y(k)}{\partial \Delta u} \eta \dfrac{\partial \Delta u}{\partial K_I} = \eta e(k) \dfrac{\partial y(k)}{\partial \Delta u} x(2) \\ \Delta K_D = -\eta \dfrac{\partial E_k}{\partial K_D} = -\eta \dfrac{\partial E_k}{\partial y(k)} \eta \dfrac{\partial y(k)}{\partial \Delta u} \eta \dfrac{\partial \Delta u}{\partial K_D} = \eta e(k) \dfrac{\partial y(k)}{\partial \Delta u} x(3) \end{cases} \tag{15}$$

Jacobian determinant information matrix algorithm is:

$$\frac{\partial y(k)}{\partial \Delta u(k)} \approx \frac{\partial y_m(k)}{\partial \Delta u(k)} = \sum_{j=1}^{m} \omega_j h_j \frac{c_{ji} - \Delta u(k)}{b_j^2} \tag{16}$$

### 3.4. Modeling and Simulation of RBF-PID control system

#### 3.4.1. Control system modeling

According to the structural block diagram of the speed control system, the model of PMSM is built in MATLAB/SIMLINK. Due to the complexity of RBF neural network tuning PID parameters, the application of s function in SIMLINK to write the control program can speed up the efficiency of building the system model. The simulation model of PMSM speed control system is shown in Figure 3. The RBF-PID controller constrains the speed loop and forms a closed-loop control with the output feedback of the control object. The SVPWM algorithm is introduced to obtain the voltage vector as the input of the PMSM.

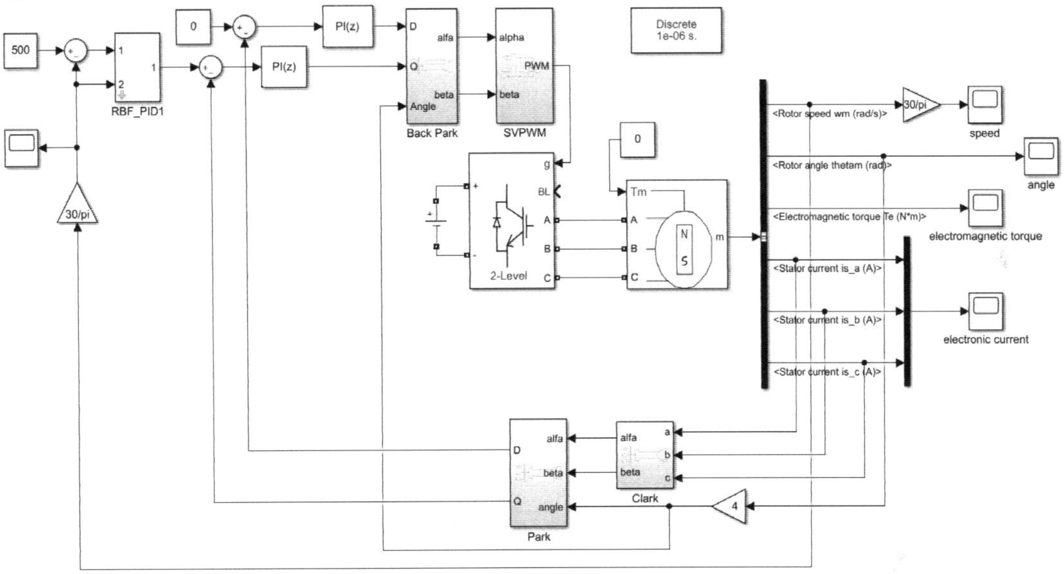


Fig. 3 simulation model of control system based on RBF-PID controller

The parameters of PMSM in the simulation model are shown in Table 1.

Table 1 Parameters of PMSM

| Name | Value | Unit |
|---|---|---|
| Rated voltage | 560 | V |
| Stator inductance | 5.26 | mH |
| Effective flux linkage of permanent magnet | 0.191 | Wb |
| Inertia coefficient | 0.01 | kg•m$^2$ |
| Polar logarithm | 4 | - |
| Static friction | 0 | N•m |

#### 3.4.2. Simulation analysis of control system

In order to verify the control effect of RBF-PID controller, the control effects of RBF-PID controller and PID controller are analyzed through system simulation. Set the PMSM of the simulation system as no-load start, and the given speed is 500r / min. The simulation results of speed tracking control are shown in Figure 4.

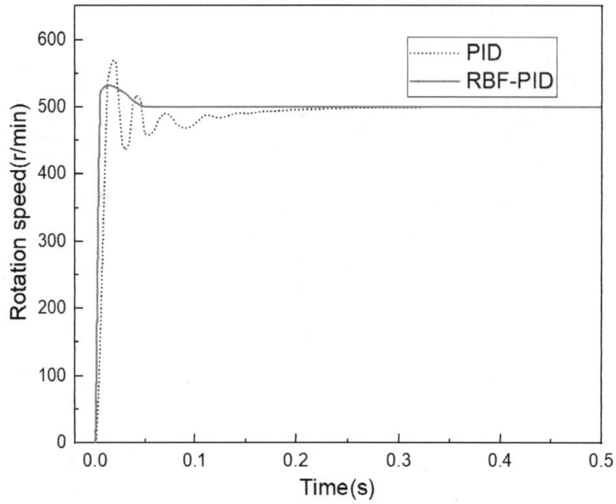


Fig. 4 speed response curves of two controllers

It can be seen from Figure 4 that during no-load start-up, the adjustment time of the speed response curve of PMSM based on traditional PID control is 0.265s, and the speed overshoot is 7.6%; The adjustment time of the speed response curve of PMSM based on RBF-PID controller is 0.042s, and the speed overshoot is 3.5%. Compared with the traditional PID controller, the adjustment time of RBF-PID controller is reduced by 84%. This shows that RBF tuning PID control is more efficient than conventional PID control.

Under the condition of no-load start, the load suddenly changes to 10N•m at 0.5s. The system speed response curve under the control of RBF-PID controller and PID controller after sudden load is shown in Figure 5.

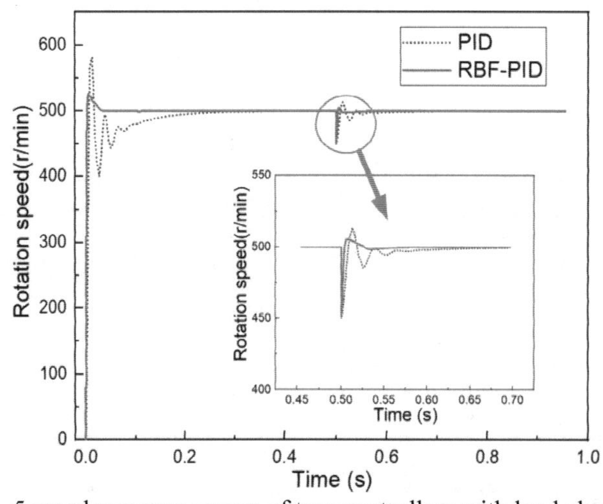


Fig. 5 speed response curves of two controllers with load changes

It can be seen from Fig. 5 that when the PMSM is suddenly loaded during operation, the adjustment time for the RBF-PID controller to adjust the motor speed to the target speed is 0.062s, the adjustment time of the traditional PID controller is 0.164s, and the adjustment time of the RBF-PID controller is 62% less than that of the traditional PID controller. This shows that RBF-PID controller has stronger anti-interference ability.

## 4. Conclusion

Aiming at the speed control system of PMSM of electric drive fracturing truck, this paper establishes the SIMLINK model of speed current closed-loop vector control, and compiles the program of RBF neural network tuning PID parameters on-line by using s-function. In this paper, the speed response time and anti-interference ability of the system model controlled by RBF-PID controller and traditional PID controller are compared and analyzed. The results show that RBF-PID controller has better control effect and can improve the real-time and anti-interference ability of PMSM speed control.

**Acknowledgments**
Fund project： Shandong Province key RESEARCH and development plan: key technology and industrialization of efficient all-electric shale gas stimulation equipment (2019TSLH0307); High technology Ship Research Project of Ministry of Industry and Information Technology: Research on safety assessment technology of underwater oil and gas production system (2018GXB01-02-002)

**Reference**

[1] Wenqiang Shao. PMSM control system based on RBF neural network[D]. Harbin University of Technology,2019.
[2] Wenbao Hu, Ruifeng Guo, Zhicheng Wang. Research on speed control of servo system based on Expert PID controller[J]. Computer engineering and design, 2013,34(07):2372-2376.
[3] Chao Tian, Yue Wei, Huanming Chen. Vector control of surface mounted PMSM based on Fuzzy PI[J]. Journal of Qingdao University, 2021,36(04):8-15.
[4] Qipeng Nie, Mingxin Tang. Application of RBF tuning PID in PMSM control[J]. Automation instrument, 2019,40 (01): 1-5.
[5] Yunbo He, Jiajun Chen. Research on compound control and parameter self-tuning of PMSM[J]. Electromechanical engineering, 2019,36 (09): 995-1000.
[6] Peilong Li, Yong Huang, Xianbo Sun, etc. Modeling and Simulation of rbf-pid control system for PMSM[J]. Journal of Hubei University for Nationalities, 2021,39(04):462-466.
[7] Qipeng Nie. Design and Simulation of electric vehicle drive control system based on RBF [D]. Dalian Jiaotong University, 2019.

# A method of estimation for the area and direction of pipe crack

**Haichao Wu[a], Yingchun Lv[b]* and Yan Chen[c]**

College of Electrical and Power Engineering, Taiyuan University of Technology, Taiyuan 030024, China

[a]email: tlwsuper@163.com· *Corresponding author: lvyingchun@tyut.edu.cn

**Abstract.** Crack direction will affect the ultrasonic wave reflected from it. Firstly, the correlation between crack direction and the reflected energy of ultrasonic is analyzed. Then, a mathematical model of crack direction and its reflected energy is developed due to the theory and methods of function approximation. Finally, aiming at the problem that the crack area must be known when using the model to estimate the crack direction, By observing the probability distribution of reflected wave energy of cracks with different areas and directions at different measurement distances, Using its law and mathematical model, an algorithm for estimating crack area and crack direction is designed. According to the related experiments, the study confirmed that both the estimation errors of the crack area and direction by the method are 5.25% and 8.21%, respectively.

## 1. Introduction

The ultrasonic flaw detection technique is a major type of pipe testing. Some studies [1-5] have shown that the technique has great potential for non-destructive testing applications. When ultrasound is traveling through the tested material, parts of the wave is reflected back by the flaw, and the waveform is changed. Some studies using the information of flaws presented by the wave variations are commonly used to discover, localize, and identify flaws [6-8]. Ongoing research continues to reveal how the information of flaws carried by the reflected wave extends the quantitative detection of flaws' dimensions and geometric shapes. Liang et al. [9] offered an estimation method for crack dimensions quantization with reflected signals from the front and rear borders of the crack, according to the interaction process between ultrasonic waves and pipe cracks. But the proposed method only works to quantitative assessment of axis dimension of pipelines cracks. Yang et al. [10] offered a probability quantization method of crack dimensions to correlate damage characteristics with crack dimensions due to a multiple regression model. But the method did not consider the influences of crack orientation and variation distance between measures. However, once the corresponding error was compensated by a new method, the effect was un-ideal as well. Zhong et al. [11] offered a method to approximately estimate the crack area by crack reflection coefficient, but there was no detailed algorithm presented. According to dozens of traction experiments of full-scale pipelines and actual results, Sha et al. [12] established a dimensions quantization model for welding cracks, but a poor estimation accuracy was found. Chao et al. [13] offered a diagnostic imaging method based on probability to indicate the orientation of the damaged edges clearly. However, the method needs a network combined with many sensors, including a suitable excitation frequency. Furthermore, there is only a single orientation of the damaged edge can be indicated.

The above research shows utilizing the information of flaws carried by the reflected wave to extend the quantitative detection of flaws' dimensions and geometric shapes is a feasible yet challenging solution. The dimensions and orientation of the cracks have been significant issues in pipelines flaw testing, particularly the hidden flaws, e.g., the cracks on or in the pipes. The dimensions and orientation of those cracks are treated as difficulties in non-destructive testing applications always. To solve the issues, this paper offers a method of estimation for the area and orientation of pipe cracks based on probability distribution and function approximation theory, according to correlation analysis of parameters. And the feasibility and availability of the proposed method are verified by experiments.

## 2. Treatment of Ultrasonic Echo

A flaw echo-signal treatment is proposed in this paper to test crack flaws. However, when cracks exist in the pipelines, not all of the ultrasonic echo signals are cracks reflected signals, which include the transmitted pulse and bottom echo inside. Fast Fourier Transform (FFT) lacks time and frequency positioning functions. It only shows the energy distribution of full-range frequency in the ultrasonic echo period, rather than just analyzing the crack signals in the echo. Wavelet transform is used for analyzing the crack signals in the echo through translation and scaling selectivities. Moreover, the Wavelet spectrum is a joint function of time and frequency to describe the energy distribution of crack reflected signals in different periods and frequencies and quantitatively analyze the cracks. In this paper, a Morlet wavelet with better features in part-time & frequency was provided as a continuous wavelet transform (CWT) to deal with crack echo.

$$W_{\psi}f(a,b) = < f, \psi_{a,b} > = |a|^{-1/2} \int_{-\infty}^{+\infty} f(t)\psi^*(\frac{t-b}{a})dt \qquad a \neq 0 \tag{1}$$

Where, $\psi(t)$ is the Morlet wavelet, "*" is the conjugate operation, $b$ is the time-shift factor which corresponds to time $t$, $a$ is the scale factor that corresponds to $f_a$ and the relation is:

$$f_a = \frac{f_c \times f_s}{a} \tag{2}$$

Where, $f_c$ is the center frequency of the Morlet wavelet. $f(t)$ is an ultrasonic echo signal in Eq.(1), that selectively analyzes the reflected signals of cracks in $f(t)$ due to scale factor and time-shift factor. According to the above analysis, the CWT energy time-frequency spectrum of crack reflected echo can be defined as:

$$S_{\psi}(t, f_a) = |W_{\psi}f(f_a, t)|^2 \tag{3}$$

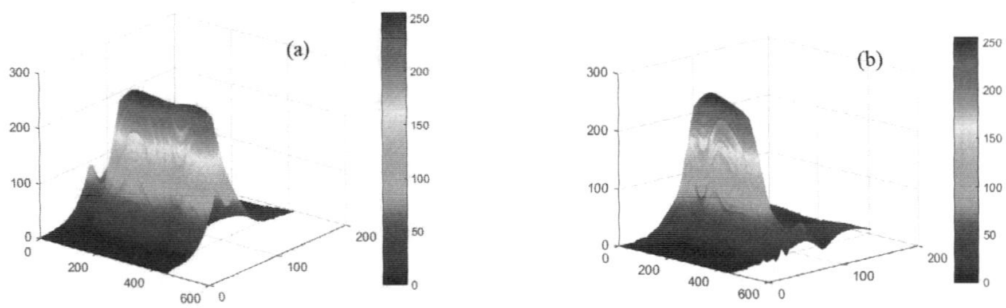


(a) crack orientation is equal to 0 degrees. (b) crack orientation is equal to 30 degrees.

Figure 1. Crack area = 8.2 πmm2, crack reflected wave time-frequency map.

3D surface graph conducted by CWT energy time-frequency spectrum can directly indicate the energy time-frequency distribution of crack echo. Fig.1 shows the time-frequency map conducted by CWT. It is easy to find the dominant of the crack reflected energy in the graph. When the angle of crack direct-ion is equal to 0 degrees, the crack area relates with the dominant volume of the crack reflected energy. When the crack direction changes, the effective area of the reflected ultrasonic wave decreases and the dominant volume of the reflected wave energy decreases, so the corresponding relationship between the dominant volume of the reflected wave energy and the crack direction can be analyzed. This paper extracts the dominant volume of the crack reflected energy by image processing technique, segments CWT time-frequency map by Otsu thresholding technique, and the obtained energy dominant geometry corresponds to the effective area of ultrasonic reflection from the crack. Therefore, the obtained energy dominant geometry can be used for quantitative analysis of cracks.

Assuming that the gray image function of the extracted energy dominant feature is f (x, y), the volume of the energy dominant feature is defined as:

$$V = \iint_{(x,y)\epsilon D} f(x,y)\, dxdy \tag{4}$$

Where, D=max $D = max\{D_i | f(x,y) \geq \xi\}$shows under threshold $\xi$.The region of crack reflected wave energy in the time-frequency map indicates its distribution in the map. $f$(x,y) shows the energy distribution of crack reflected wave in time $x$ and frequency y. Threshold $\xi$ meets the maximum variance between the energy dominant object and the background region, shows the best separation condition of both regions and explains the energy dominant object, which can indicate the energy of the crack reflected wave. In this paper, the correlation between crack orientation and the energy of the crack reflected wave would be analyzed due to the CWT time-frequency spectrum of crack.

## 3. Correlation analysis between crack orientation and crack reflected wave energy

### *3.1. Analytical method*

COMSOL program is used to analyze the correlation between crack orientation and its reflected wave energy. Fig.2. shows the pipe model designed from COMSOL.T24 steel is used in the pipe, and the dimensions are found in Table 1. The parameters of pipe material are default initialized. One end of the pipe is unrestrained, and the other is fully restrained. There are eight piezoelectric ceramic sensors deployed in the unrestrained end, where four sensors are transmitting and the rest receiving. The transmitting sensors send excitation ultrasonic signals; the excitation frequency is 140 kHz, as shown in Eq. (5). The receiving sensor receives the echo signal, and then uses Morlet wavelet to analyze the time-frequency of the crack signal in the echo,and studies the correlation between the energy dominant volume of the crack signal in the time-frequency diagram and the crack direction.

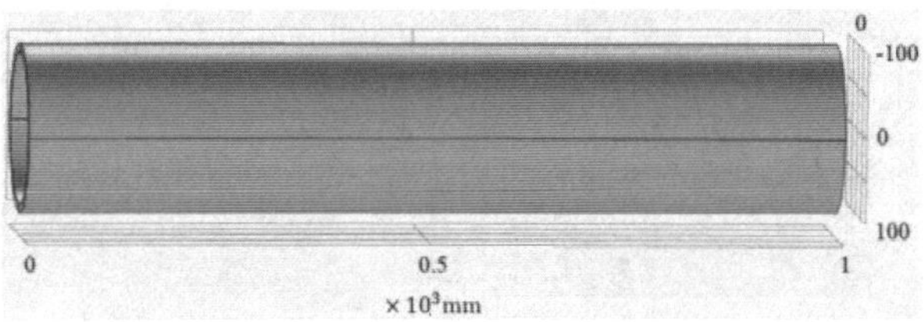


Figure 2. COMSOL pipe model.

Table 1. Model dimensions.

| Parameter | The parameter value(mm) |
|---|---|
| The length | 1000 |
| The outer diameter | 100 |
| The wall thickness | 10 |

$$y = \frac{1}{2}[1 - \cos\left(\frac{2\pi f t}{n}\right)] \cdot \sin(2\pi f t) \tag{5}$$

Where n is the number of cycles, $f$ is the center frequency of excitation signal, $t$ is time.There is a certain orientation when the crack in the pipe occurs. This paper defines crack orientation as the angle between the lengthwise and circumferential directions, as shown in Fig.3.

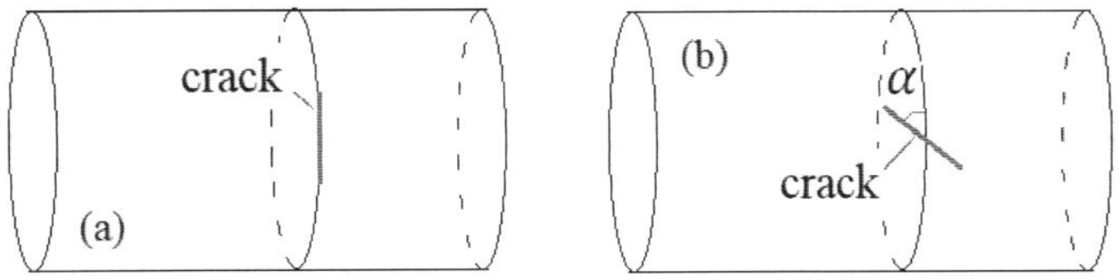


(a) Crack orientation at 0 degrees. (b) Crack orientation at $a$ degree.
Figure 3. Definition of crack orientation.

The reflected wave energy is related to the crack area. In the simulation, the calculation method of crack area is shown in Figure 4, that is, the large sector area where the crack is located minus the small sector area without crack.The reflected energy of crack is also related to its orientations. According to the rotating function of COMSOL, the study of various orientations with crack is available.

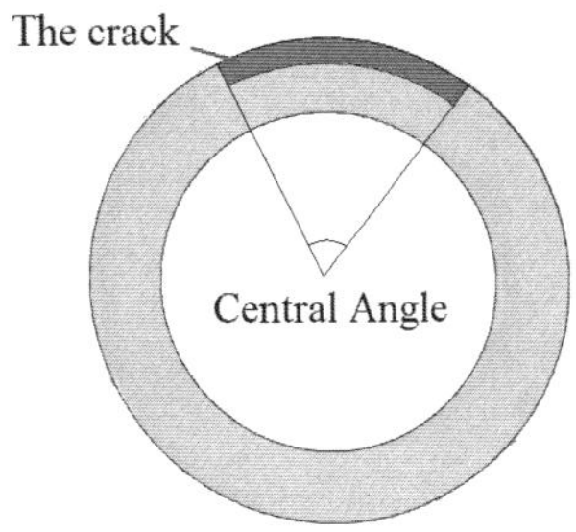


Figure 4. Cross-section of the pipe.

During the test, when the orientation is less than 45 °, there is an excellent regularity between the orientation and the reflected wave energy. Otherwise, an exception will occur. In this paper, an algorithm is developed to estimate the crack area and crack orientation based on samples of 0 ° ~ 45 ° crack orientation and its reflected wave energy.

In this paper, there are three cracks with various areas provided by the COMSOL pipe model. During the test, it is found the measured value of the minimum crack no longer varies with the crack orientation if the measurement distance is greater than 550 mm. When the distance between the crack and the transducer is less than 100 mm, the signals will be overlapped and confused. In this case, 100 mm is fixed as the nearest distance in crack detection.550mm as the maximum measurement distance.

*3.2. The corresponding relationship between crack orientation and reflected wave energy at various measure distances*

A circumferential crack with a width of 0.5 mm, a depth of 2 mm, and a central angle of 15 ° is made 100 mm away from the transducer. According to the method proposed, the crack area is 8.2 πmm2 after calculation. The measured distances were set as 100mm, 250mm, 400mm and 550mm, and then the reflected wave energies of the cracks in different directions at these distances were measured respectively. The reflected wave energy is indicated by the energy dominant volume of the time-frequency map. that is, CWT transformation is performed on the received crack reflected wave signal, and then the energy dominant volume is calculated according to the time-frequency diagram and Formula (4) of the reflected crack wave.Finally, the measurement data was obtained after the normalized process of calculated energy dominant volume through Eq. (6), as shown in Table 2.

$$V_i = \frac{v_i - \min}{\max - \min} \tag{6}$$

Where, $v_i$ is a measured value of energy dominant volume, min and max are the minimum and maximal measured value of energy dominant volume, $V_i$ is the value of energy dominant volume after the normalized process.

Table 2. The correlation between the crack orientation under various measure distances and energy dominant volume.

| Crack direction $\alpha/(°)$ | 0 | 15 | 30 | 45 | distance (mm) |
|---|---|---|---|---|---|
| Normalized energy dominant volume (V) | 1 | 0.939 | 0.866 | 0.781 | 100 |
| | 0.489 | 0.453 | 0.417 | 0.327 | 250 |
| | 0.161 | 0.137 | 0.120 | 0.069 | 400 |
| | 0.064 | 0.052 | 0.034 | 0.000 | 550 |

Table 2 shows the corresponding relationship between crack direction and energy dominant volume at different measured distances with crack area of 8.2πmm$^2$.

*3.3. The correlation between crack orientation and reflected wave energy at various measure distances*

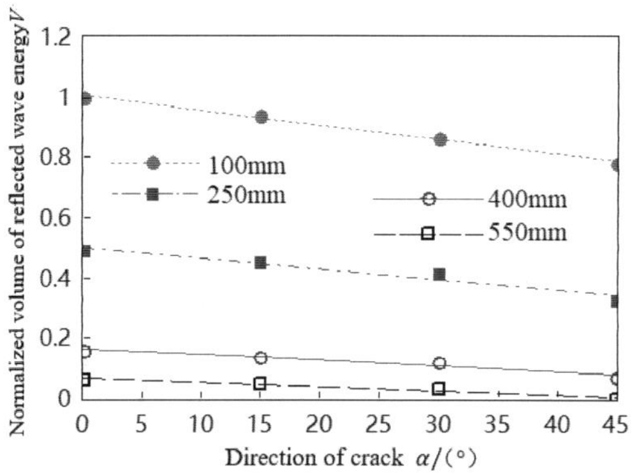


Figure 5. Scatter graph of measured values.

Fig 5 is a scatter diagram drawn from the data in Table 2.It can be seen from the figure that for the determined distance, the change of energy dominated volume with the crack trend is approximately linear.The slope of this linear relationship decreases with the increase of measurement distance.When the measurement distance is greater than 550mm, the energy dominated volume basically does not change with the change of crack trend.

In order to reveal the correlation between the crack direction and its reflected wave energy,According to the characteristics of scatter diagram in Figure 5.It is assumed that the sample data $(V_{i,}\alpha_i)_{S_j}$ observed at different distances $s_j$ in the figure meet the linear probability model of equation (7).

$$V_i = \beta_0 + \beta_1\alpha_i + \varepsilon_i \qquad i = 1,\ 2,\ \ldots,\ k \tag{7}$$

Where $\alpha$ is the crack orientation, $V$ is the energy dominant volume normalized, $\beta_0$ and $\beta_1$ are parameters to be estimated, $\varepsilon$ is the random error to meet the independent identically distributed hypothesis of Eq.8.

$$\begin{cases} \mathrm{E}(\varepsilon_i) = 0 \\ \mathrm{var}(\varepsilon_i) = \sigma^2 \end{cases} i = 1,2,\cdots,k \tag{8}$$

Calculating the conditional expectation from both ends of Eq.7, we have

$$E(V_i) = \beta_0 + \beta_1 \alpha_i \quad i = 1, 2, \ldots, k \tag{9}$$

Equation (9) shows the statistical law of energy dominant normalized volume V and crack direction α in an average sense.

Table 3. Normalization equation of various distances, area: 8.2πmm2.

| Regression Equation | Distance (mm) |
|---|---|
| $\hat{V} = 1.00600 - 0.004867\alpha$ | 100 |
| $\hat{V} = 0.49980 - 0.003480\alpha$ | 250 |
| $\hat{V} = 0.16560 - 0.001943\alpha$ | 400 |
| $\hat{V} = 0.06893 - 0.001401\alpha$ | 550 |

According to sample data$(V_i, \alpha_i)_{S_j}$ in FIG. 5,*Linear Fit.vi* provided by LabVIEW is used to estimate the parameters of Equation (9).Table 3 shows the calculation results of *Linear Fit.*vi.Table 4 is the evaluation index of the regression equation in Table 3.

Table 4. Evaluating goodness of fit of normalization equation.

| Distance(mm) | | 100 | 250 | 400 | 550 |
|---|---|---|---|---|---|
| Evaluating indicator | R2 | 0.9946 | 0.9400 | 0.9428 | 0.9481 |
| | SSE | $1.44 \times 10^{-4}$ | $8.74 \times 10^{-4}$ | $2.57 \times 10^{-4}$ | $1.21 \times 10^{-4}$ |

R^2 and SSE are evaluation indexes frequently used in current references to evaluate the goodness of fit of the normalization equation, as defined below.

$$SSE = \sum_{i=1}^{k} (V_i - \hat{V}_i)^2 \quad i = 1,2,\cdots,k \tag{10}$$

$$R^2 = 1 - \frac{SSE}{\sum_{i=1}^{k} (V_i - \overline{V})^2} \quad i = 1,2,\cdots,k \tag{11}$$

Where, $V_i$ is the measured value, $\hat{V}_i$ is the estimated value of the normalization equation, $\overline{V}$ is the mean of the measured value. Table 5 shows the four grades of the fitting results of the normalization equation given in reference [14]. According to the criteria in Table 5, the crack orientation and its energy dominant volume of echo follow the restriction correlation provided by the normalization equation in Table 3.

Table 5. Grades of the fitting result of the normalization equation.

| $R^2$ | 1 ~ 0.9 | 0.9 ~ 0.8 | 0.8 ~ 0.6 | < 0.6 |
|---|---|---|---|---|
| Fitting effect | excellent | good | Medium | Poor |

## 4. Modeling of the correlation between crack orientation and reflected wave energy

To realize the quantitative calculation between crack orientation and reflected wave energy, a mathematical model between both of them is established based on the above research.

### *4.1. Modeling*

As mentioned before, crack orientation with an area of 8.2 πmm2 and its reflected wave energy follow the restriction correlation provided by the normalization equation in Table 3. Therefore, this paper introduces the equation in Table 3 to establish the mathematical model of crack orientation and reflected wave energy. The equations of various distances in Table 3 are shown in a general formula.

$$\hat{V} = A - B\alpha \tag{12}$$

Eq.12 is the mathematical model of the correlation between crack orientation and reflected wave energy due to random measure distances when the crack area is 8.2 πmm2. Where, $\widehat{V}$ is the energy dominant volume of the time-frequency diagram of the reflected wave of the crack, $a$ is the crack direction, and A, B are the parameters to be estimated.

### *4.2. Model parameter estimation*

Table 3 shows that the values variation of A and B in Eq.12 correspond to the variation of measure distance s, so the formula can be built through A, B, and s. The measure distance s is a continuous variable. The parameters A and B always have values corresponding to them, according to each value on the measuring zone [100, 550] (unit mm) provided in this paper. Therefore, the parameters A and B and the measure distance s can be shown as the following functional relationship.

$$A = f(s) \quad B = \varphi(s) \tag{13}$$

Table 3 shows the value A = $f$ (s) with s values of 100 mm, 250 mm, 400 mm, and 550 mm are 1.006, 0.4998, 0.1656, and 0.06893, respectively, although the equation of it is unknown. Taking 100, 250, 400, and 550 as interpolation nodes, a cubic polynomial is established through the Lagrange interpolation algorithm.

$$\mathcal{A}(s) = 1.407 - 4.178 \times 10^{-3}s + 1.395 \times 10^{-6}s^2 + 3.236 \times 10^{-9}s^3 \tag{14}$$

Equation (14) conforms to $\mathcal{A}(100) = 1.006$, $\mathcal{A}(250) = 0.4998$, $\mathcal{A}(400) = 0.1656$, $\mathcal{A}(550) = 0.06893$.Therefore, Eq. (14) is an approximate expression of the function A=$f$(s).Equation (14) can be used to solve for the estimate $\mathcal{A}(s)$ of the function A=$f$(s).Similarly, the approximate expression of B=$\varphi$(s) function in Equation (13) can also be obtained by Lagrange interpolation algorithm.

$$\mathcal{B}(s) = -5.143 \times 10^{-3} - 1.25 \times 10^{-6}s + 4.574 \times 10^{-8}s^2 - 5.654 \times 10^{-11}s^3 \tag{15}$$

For the convenience of application, Eqs. (14), (15) are shown below.

$$\mathrm{A} = 1.407 - 4.178 \times 10^{-3}s + 1.395 \times 10^{-6}s^2 + 3.236 \times 10^{-9}s^3 \tag{16}$$

$$\mathrm{B} = 5.143 \times 10^{-3} + 1.25 \times 10^{-6}s - 4.574 \times 10^{-8}s^2 + 5.654 \times 10^{-11}s^3 \tag{17}$$

Where A in Eq. (16) and B in Eq. (17) are the parameters of the model in Eq. (12). The parameters A and B of any measured distance can be solved by the measured value of the distance.

### *4.3. Mathematical models of cracks with various areas*

Keep the depth and width of the crack unchanged, and expand the above cracks in sequence along the extension direction, so that the center angle of the expanded crack to the pipeline is 30 ° (area of 16.3 πmm2) and 45 ° (area of 24.5 πmm2).Then, according to the above method, the approximate expression of the mathematical model and parameters of the direction of the two area cracks and their reflected wave energy is as follows:

$$\hat{V} = C - D\alpha \tag{18}$$

$$\hat{V} = E - F\alpha \tag{19}$$

$$\mathrm{C} = 1.429 - 4.708 \times 10^{-3}s + 3.603 \times 10^{-6}s^2 + 8.168 \times 10^{-10}s^3 \tag{20}$$

$$\mathrm{D} = 7.956 \times 10^{-3} - 2.742 \times 10^{-5}s + 4.478 \times 10^{-8}s^2 - 2.948 \times 10^{-11}s^3 \tag{21}$$

$$\mathrm{E} = 1.35 - 3.563 \times 10^{-3}s + 8.985 \times 10^{-7}s^2 + 2.497 \times 10^{-9}s^3 \tag{22}$$

$$\mathrm{F} = 1.027 \times 10^{-2} - 4.579 \times 10^{-5}s + 1.152 \times 10^{-7}s^2 - 1.095 \times 10^{-10}s^3 \tag{23}$$

## 5. Estimation method of crack area and crack direction

Since the reflected wave energy is only related to the effective reflection area of the crack and has nothing to do with the crack shape , Eq. (12) is a mathematical model with the crack area is 8.2 πmm$^2$. It shows that if the crack area is 8.2 πmm$^2$, the energy dominant volume of crack measured under various orientations and distances follows the restriction correlation given in Eq. (12). In other words, the probability distribution of crack energy dominant volume follows the restraint given by Eq. (12). Similarly, if the crack area is fixed to 16.3 πmm$^2$ or 24.5 πmm$^2$, the probability distribution of its energy dominant volume must follow the restriction given by Eq. (18) or Eq. (19).

This paper treats the three measured energy dominant volumes of the cracks mentioned above by a unified normalization method, as shown in Table 6. Table 6 shows that the distribution of energy dominant volume of cracks with various areas is different in the numerical range due to the same measure distance. According to the difference, to determine the area of the detected crack based on the measured values, and then select the model through the crack area to solve the crack orientation. The algorithm is as follows.

①Firstly, the measured values are obtained. Two measured values can be obtained when detecting cracks, i.e., the transmission time t of the ultrasonic waves and the echo reflected by cracks. ② The transmission time t of the ultrasonic waves is converted into measure distance $S_m$ according to Eq.(24), and the echo reflected by the crack is converted into energy dominant volume $V_m$ due to the above method.

$$S_m = \frac{1}{2} tc \tag{24}$$

Where c is the speed of the ultrasonic waves in the material. ③According to Eq.(12), Eq.(18), and Eq.(19), the energy dominant volumes V0and V45 with the crack orientation of 0° and 45° are calculated at the measure distance *S*. Therefore, the distribution intervals of the energy-dominant volume of cracks with the area of 8.2πmm2, 16.3πmm2 and 24.5πmm2 are $[V0_{8.2}, V45_{8.2}]$, $[V0_{16.3}, V45_{16.3}]$ and $[V0_{24.5}, V45_{24.5}]$ at the measured distance of S, respectively.④ If the measured energy dominant volume $V_m$ of measure distance $S_m$ is in one of the regions, it means that the detected crack area is the region of the crack area correspondingly, which is shown in Table 6. ⑤ $S_m$ and $V_m$ are input into the mathematical model of the crack in this region, the orientation of the detected crack can be obtained. ⑥ If $V_m < V45_{8.2}$, it shows that the area of the detected crack is less than 8.2 πmm$^2$; If $V_m > V0_{24.5}$, it shows that the area is bigger than 24.5 πmm$^2$. ⑦ If $V_m$ is in the range of 24.5 πmm$^2$ of Table 6, but not in the range $[V0_{24.5}, V45_{24.5}]$ of $S_m$, it shows the crack area is 24.5 πmm$^2$, but the crack orientation is bigger than 45°. Similarly, the results of the crack area of 8.2 πmm2, 16.3 πmm$^2$, and the crack orientation bigger than 45°can be judged.

Table 6. Probability distribution of energy dominant volume.

| crack orientation (°) | 0 | 15 | 30 | 45 | Distance (mm) | Crack area (mm$^2$) |
|---|---|---|---|---|---|---|
| Normalized energy dominant volume (V) | 1 | 0.9226 | 0.8232 | 0.7542 | 100 | 24.5π |
| | 0.6021 | 0.5524 | 0.4894 | 0.4319 | 250 | |
| | 0.3258 | 0.2595 | 0.2197 | 0.1888 | 400 | |
| | 0.1822 | 0.1601 | 0.1357 | 0.1136 | 550 | |
| | 0.7348 | 0.6851 | 0.6336 | 0.5735 | 100 | 16.3π |
| | 0.4019 | 0.3503 | 0.3228 | 0.2901 | 250 | |
| | 0.1784 | 0.1440 | 0.1243 | 0.1054 | 400 | |
| | 0.1011 | 0.0753 | 0.0650 | 0.0512 | 550 | |
| | 0.4881 | 0.4585 | 0.4229 | 0.3813 | 100 | 8.2π |
| | 0.2389 | 0.2211 | 0.2033 | 0.1597 | 250 | |
| | 0.0787 | 0.0669 | 0.0586 | 0.0339 | 400 | |
| | 0.0313 | 0.0253 | 0.0165 | 0 | 550 | |

## 6. Validity analysis of estimation methods

### *6.1. Experimental equipment*

Fig.6 is the schematic diagram of the experimental system in this paper. The theory is introduced as follows. The ultrasonic transmitting transducer converts the electric pulse transmitted by the ultrasonic equipment into the ultrasonic pulse. The ultrasonic pulse transmits on the pipe wall and is reflected when it encounters a crack. Then it is caught by the ultrasonic receiving transducer on that occasion. The received signals are handled and analyzed by the LabVIEW program when the signals are converted into inputs through ultrasonic equipment. Finally, two measurement data are obtained: measurement distance $S_m$ and energy dominated volume $V_m$ of crack reflected wave.

Fig.7 shows the equipment and test piece of the experimental system. The left figure of Fig.7 is the ultrasonic equipment, which is ZXUT-PCI multi-channel ultrasonic testing system developed by WHZX Technology Co., Ltd. It is installed in a PCI slot on the computer. The device provides SDK dynamic link library (DLL), which can communicate with the LabVIEW program. The transducer is provided by Changzhou Ultrasonic Electronics Co., Ltd. The right figure of Fig.7 is the test piece.

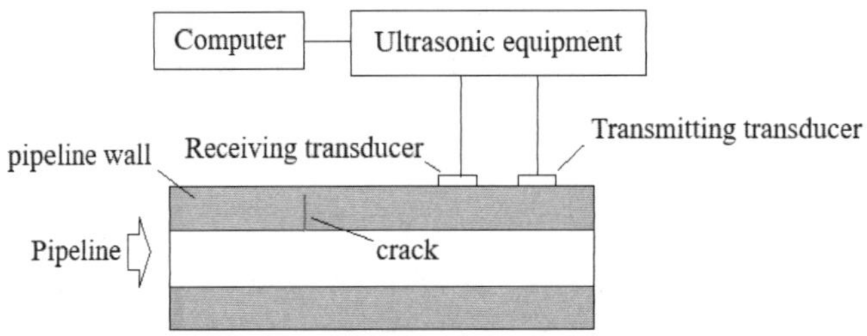


Figure 6. Schematic diagram of the experimental system.

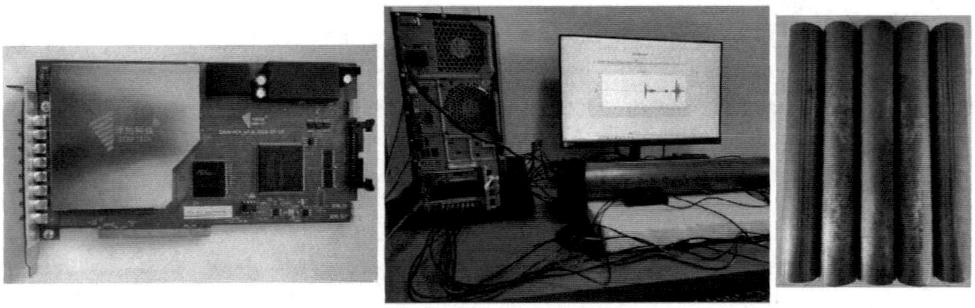

Figure 7. Device of experimental system and test-piece.

### *6.2. Tests methods*

There are four serial channels in ZXUT-PCI. During the tests, four transmitting transducers and four receiving transducers are installed alternately and evenly are placed at one end of the pipes, as shown in Fig. 7. Test-pieces are mainly used for the following analysis tasks. ① Error analysis of the model with distance variation. ② Error analysis of the model if the variation of crack orientation. ③ Error analysis of crack area estimation.

The specification of the test piece is as follows: the length is 650 mm, the outer diameter is 80 mm, and the wall thickness is 5 mm. Make the following cracks on the pipe: ① cracks with the same area and orientation, various measure distances; ② cracks with the same area and measure distance, and various orientations.

During the test, repeat the measurement for 15 times for each crack, then remove the maximum and minimum values, and take the average value of the remaining data as the measured value.

### 6.3. Test Results and Discussion

#### 6.3.1. Error analysis of model

In this paper, the estimation method shows the model is its core algorithm, which causes the estimation accuracy of the model to be the key to determining its accuracy. The crack direction can be estimated by substituting the measured energy dominated volume $V_m$ and the measured distance $S_m$ into the model.Then,the estimated value of the model approached in this paper compares with the measured value to evaluate the estimation accuracy of the model.

(1) The crack area(24.5πmm2) and orientation (30°) is kept unchanged, and the error is observed when comparing the estimated value of the model with the measured value at various measure distances. The results are shown in Figure 8.The error in the figure is the error after comparing the measured direction of the crack with the estimated direction of the model.There are three curves in the figure showing three cracks in different regions.Fig. 8 shows the increase of errors corresponding to the decrease of the area of the cracks.And with the increase of measurement distance, the error gap of cracks with different areas increases. Table 7 shows the maximum error, average error and relative error in Figure 8.

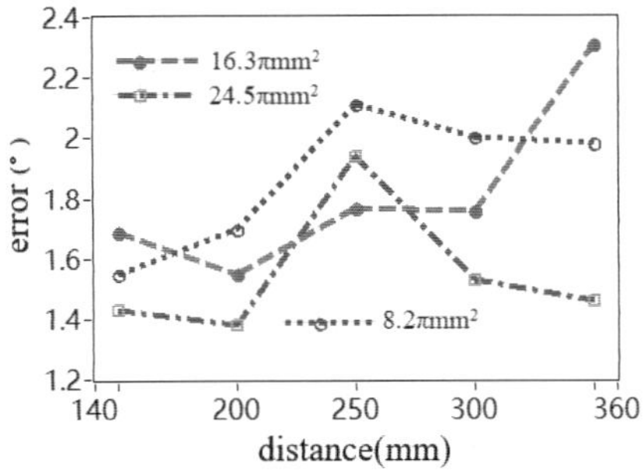


Figure 8. Model errors of various measure distances.

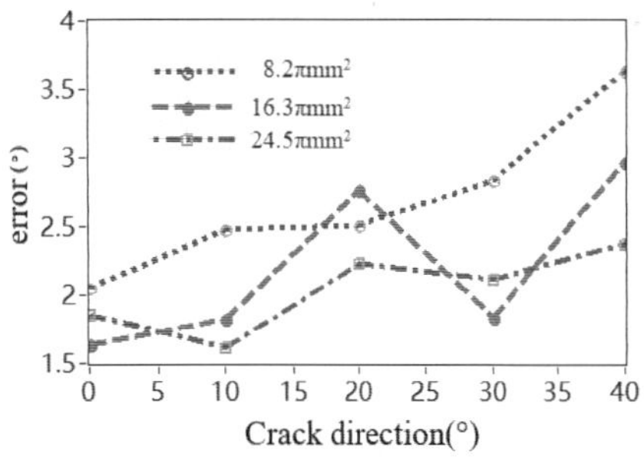


Figure 9. Model errors of various crack direction.

(2) The crack area(24.5πmm2) and measure distance(250mm) is kept unchanged, and the error is observed when comparing the estimated value of the model with the measured value at various crack direction. The results are shown in Fig. 9. As in Fig. 8, the error increases with the decrease of crack area;And with the increase of measurement distance, the error gap of cracks with different areas increases.Different from Fig. 8, as the degree of crack direction increases, the error of cracks in different areas shows an increasing trend. Table 8 shows the maximum error, average error and relative error in Figure 9.The relative error is the average.

Table 7. Error caused by distance change.

| Crack area ($mm^2$) | $8.2\pi$ | $16.3\pi$ | $24.5\pi$ |
|---|---|---|---|
| Maximum error(°) | 2.11 | 2.31 | 1.94 |
| Average error(°) | 1.87 | 1.82 | 1.55 |
| Relative error (%) | 6.23 | 6.07 | 5.17 |

Table 8. Error caused by crack direction change.

| Crack area ($mm^2$) | $8.2\pi$ | $16.3\pi$ | $24.5\pi$ |
|---|---|---|---|
| Maximum error(°) | 3.62 | 2.97 | 2.37 |
| Average error(°) | 2.69 | 2.21 | 2.03 |
| Relative error (%) | 8.21 | 7.13 | 5.74 |

Comparing Table 7 and Table 8, it has it has been found that the error caused by the crack direction variation is bigger than that caused by the measure distance variation.

*6.3.2 Error analysis of crack area estimation*

Based on the methods of references [10] and [11], the models are built according to the measured values of the pipe test pieces, as shown in Eqs.(25) and (26).

$$M = 62.46 + 28.79\theta - 64.57y + 55.08\theta y \tag{25}$$

$$M = (-0.4971 + 2.998x + 4.249x^2) \times M_0 \tag{26}$$

In Eq. (25), y is the normalized amplitude, $\theta$ is the phase angle. In Eq. (26), $x$ is the reflection coefficient of the crack, $M_0$ is the cross-sectional area of the pipeline.In Eq. (25), (26), M indicates the crack area. Obviously, Eqs. (25) and (26) are the correlation models between crack areas and damage parameters. Compare the methods in the references with those in this paper. ① the crack area(24.5πmm2) and crack direction(30°) are kept unchanged, and the errors in various distances are observed. The results are shown in Table 9. ② the crack area(24.5πmm2) and measure distance(250mm) are kept unchanged, and the errors in direction variation are observed. The results are shown in Table 10.

Table 9. Error analysis of various measure distances.

| Estimation method | This paper | Literature [10] | Literature [11] |
|---|---|---|---|
| Maximum error ($mm^2$) | 5.28 | 9.47 | 10.52 |
| Average error ($mm^2$) | 3.32 | 6.62 | 7.07 |
| Relative error (%) | 4.32 | 8.61 | 9.19 |

Table 10. Error analysis of crack direction change.

| Estimation method | This paper | Literature [10] | Literature [11] |
|---|---|---|---|
| Maximum error ($mm^2$) | 8.36 | 12.32 | 14.12 |
| Average error ($mm^2$) | 4.04 | 8.26 | 9.13 |
| Relative error (%) | 5.25 | 10.74 | 11.87 |

Table 9 and Table 10 show that all three errors are less than the methods of references. Because the model in this paper is built up due to the test data from various crack directions, the parameters of the model are estimated by the measured distances. The parameters of the model with various measure distances are different, which compensates for the error caused by the various crack directions and measure distances. However, the methods in the literature did not consider the influence of crack direction and measurement distance. Although Reference [10] offered a new method to compensate for the influence of crack direction, the effect is not obvious. Moreover, the parameters of the model are constant when the measurement distance changes, which can not compensate for the influence of the measurement distance change.

## 7. Conclusions

To solve the problems existing in the quantitative detection of pipe cracks, the analysis of simulation and experiments are conducted, and the conclusions follow.

(1) Quantitative detection of flaws has been affected by various orientations and measure distances. However, it has been ignored in current studies, which causes a large error in the quantitative results.

(2) The model of the relationship between crack direction and reflected wave energy constructed in this paper reveals the correlation between crack direction and reflected wave energy, and puts forward a new idea for the detection of crack direction.The estimation method of model parameters transforms the univariate linear model with crack direction as independent variable into a binary nonlinear model with crack direction and measurement distance as independent variables, so that the model can

effectively estimate the crack direction with different measurement distance,and improves the practicability of the model.

(3) By observing the probability distribution of reflected wave energy of cracks in different areas and directions at different measurement distances, an estimation method of crack area is proposed. As long as the number of data samples is increased, the accuracy of this method in estimating crack area can be greatly improved.

**Acknowledgments**
We are grateful to the financial support from the National Key R&d Program of China ( No.2018YFC1405902), as well as the critical and valuable comments from the editor and reviewers.

**References**

[1] Gao.F, Zhou.H, Huang.CH. (2020) Defect detection using the phased-array laser ultrasonic crack. Optics Communications,474:1-11.
[2] GAO.Y.D, GUAN.S.T, WANG.S.F,et al. (2020) Research on Focusing Energy-oriented Inspection for Non-through-thickness-Defects in Pipes. MACHINE TOOL &HYDRAULIC-S,48:175-180.
[3] DENG.X.Y, QIU.J.H, JI.H.l, et al. (2020) Delamination Damage Detection in Laminated Composite by LDR-Based Multi-frequency Method. Transactions of Nanjing University of Aeronautics and Astronautics,37:28-37.
[4] Andreades.CH, Fierro.G.P.M, Meo. Meo.M. (2022) A nonlinear ultrasonic modulation approach for the detection and localisation of contact defects. Mechanical Systems and Signal Processing,162:1-12.
[5] Vogelaar.B, Golombok.M. (2016) Quantification and localization of internal pipe damage.MechanicalSystemsandSignalProcessing,78:107-117.
[6] ZHOU.S.P, ZHANG.P.G, LV.W.C, et al. (2015) Detection of Cracks in Elbow Pipes Using Guided Waves. JOURNAL OF MECHANICAL ENGINEERING,51:58-65.
[7] Xu.Y, Luo.M.ZH, Liu.Q, et al. (2019) PZT transducer array enabled pipeline defect locating based on time-reversal method and matching pursuit de-noising. Smart Materials and Structures,28:1-16.
[8] HU.W.B, QIU.Sh, XU.X.Y, et al. (2021) Ultrasonic Detection and Classification for Internal Defect Of Rail Based on Deep Learning. JOURNAL OF THE CHINA RAILWAY SOCIETY,43:108-117.
[9] LIANG.B.A, ZHAO.K, WANG.X.J, et al. (2018) Simulation study of the characterization of pipe defect based on ultrasonic guided-waves. ]ournalof Xi'an University of Technology,34:489-495.
[10] Yang.J.S, He.J.J, Guan.X.F, et al. (2016) A probabilistic crack size quantification method using in-situ Lamb wave test and Bayesian updating. MechanicalSystemsandSignalProcessing,78: 118-133.
[11] Zhong.W.L,XU.J,XIE.T, et al. (2015) Defect identification of metal pipe based on guided wave dispersion characteristics.Journal of Central South University (Science and Technology), 46:3682-3688.
[12] SHA.S.Y, XIANG.X.Q, WU.X.Y, et al. (2018) Ultrasonic Ill signal identification technology for Girth weld defects of oil pipelines. Oil & Gas Storage and Transportation, 37: 757-762.
[13] Zhou.CH, Su.ZH.Q, Cheng.L. (2011) Quantitative evaluation of orientation-specific damage using elastic waves and probability-based diagnostic imaging. Mechanical Systems and Signal Processing,25:2135-2156.
[14] SUN.J,LIANG.J.Q,LIU K. (2013) A Combined Forecasting Model Based On Optimal Weighted Method And East-squares Method. BULLETIN OF SCIENCE AND TECHNOLOGY,29:10-12.

# Research and application of large size measurement calibration technology

**Chang'an Hu[1], Gang Li[1,*], Tao Jiang[2], Guosong He[3]**

[1]National Institute of Measurement and Testing Technology, Chengdu, Sichuan, 610021, China

[2]Erzhong (Deyang) Testing Technology Co. LTD., Deyang, Sichuan, 618099, China

[3]Sichuan Industrial Metrology and Testing Institute, Chengdu, Sichuan, 610199, China.

[*]Corresponding author's e-mail: jixie@nimtt.com

**Abstract:** Indoor large-length standard device is an integrated large-size measurement and calibration device, which plays an important role in large-size measurement and calibration technology. In this paper, an indoor standard device for large length based on linear guide is introduced. The uncertainty of the measurement result is 0.14 μm+$1\times10^{-7}$L. The device can be used to calibrate the laser interferometer and laser tracker located inside the triangle, the standard steel tape measure, special length scale and ground scanner located outside the triangle. Taking the calibration of ground scanner as an example, the application of large size measurement calibration technology is described in detail.

## 1. Introduction

Large size measurement is the measurement of the space coordinate (position), size, shape and motion track of objects within a range of several meters to several hundred meters. And large size measurement is a technology of large size measurement after quantification, can be said to be a special large size measurement, all measurement standards require traceability. When it comes to large size measurement, the indoor large size standard device has to be mentioned. The indoor large size standard device is also called the indoor large length standard device. Its main standard device is the optical path system composed of laser interferometer. The laser interferometer, the main standard of indoor large length standard device, the composition of indoor large length standard device and the application cases of large size measurement technology are introduced.

## 2. Laser interferometer

In 1960 Theodore Maiman made the world's first laser, the Ruby laser, which started the use of lasers. Up to now, there are many kinds of optical interferometers based on laser, including single-frequency interferometer, dual-frequency interferometer, Mach-Zehnder interferometer and grating interferometer. Dual-frequency interferometer has become the main research area because of its strong anti-interference ability, fast corresponding speed and high measurement accuracy.

Dual-frequency laser interferometer is improved on the basis of single-frequency laser interferometer. It introduces a certain frequency carrier into the signal of the interferometer, so as to realize AC amplification. The measured signal spreads through the carrier, which greatly reduces the trouble of signal being interfered. Dual-frequency laser interferometer is a heterodyne laser interferometer.

Dual-frequency laser interferometer system is mainly composed of laser head, computer, software measurement system, sensor and optical components, electrical box and tripod [1]. It has the measurement accuracy of micro and nano levels, so the measurement factors affecting the dual-frequency laser interferometer are also more, its measurement results are affected by the air environment, such as humidity, air pressure, temperature and so on, so many dual-frequency laser interferometers are equipped with temperature and air pressure sensors. Renissau's dual-frequency laser interferometer XL-80, for example, is equipped with a high-precision environmental sensor to compensate for changes in ambient temperature, which can be compensated at air pressure of 650-1150mpa, relative humidity of 0-95%RH, and air temperature of 0-40C. In the actual measurement process, the installation of dual-frequency laser interferometer is also affected by cosine error and Abbe error. Cosine error refers to the error caused by the Angle formed by the laser axis of the dual-frequency laser interferometer and the movement of the measured object. Cosine error is sometimes not easy to avoid and tends to increase with the increase of measurement distance. The mechanical design can be adjusted to make the movement direction of the measured object parallel to the laser direction as far as possible. Abbe error is the measurement error caused by the laser direction of the dual-frequency laser interferometer is not in the same line with the axis of the measured object. The Abbe error can be compensated by optical design principles or algorithms to minimize the error. In addition, the inherent error of dual-frequency laser interferometer itself is also one of the factors affecting the measurement accuracy, data drift in the measurement, polarization characteristics of laser light source, aging of optical and electrical devices, stability of laser frequency stabilizer and other aspects are some factors.

Currently, the most representative dual-frequency laser interferometers are mainly divided into two categories. One is the dual-frequency laser interferometer represented by the acoustoptic frequency shift method, which can modulated the laser interferometer of different frequency differences. The interferometer of Renissau Company in the UK and ZYGO Company in the United States. In 2017 Renishaw introduced an XM-60 multi-beam laser interferometer, which can directly measure six degrees of freedom (linearity, pitch, yaw, vertical and horizontal straightness, roll) on any form of linear axis with a single setting. The other type is the interferometer represented by American Tech. It is a dual-frequency laser formed under the common action of frequency traction effect, resonant cavity and Zeeman effect [2]. Dual-frequency laser interferometer is mainly used in high-precision measurement, and is often used in the debugging of machine tools, machining centers, coordinate measuring machines and precision guide rails. The field map of laser interferometer measurement is shown in Figure 1.

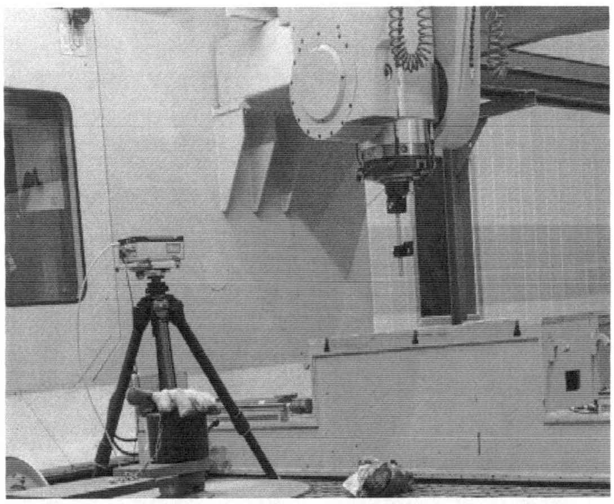

Figure 1. Dual-frequency laser interferometer.

## 3. Indoor large length standard device

This large length standard device is used for measuring the length from zero to 60 meters. Its resolution is 0.01μm. The indoor large length standard device is a necessary device for calibrating and verifying high precision and large range laser equipment and standard steel tape.

The large length standard device developed in this paper is composed of granite platform, metal linear guide rail, drive system, environmental detection system, granite platform adjustment device, standard set and data processing system [3]. Granite platform composed of multiple granite Mosaic, linear guide installed above the platform and arranged in a straight line, before and after the drive system of electric slide along the linear guide translation, environment detection system and standard for data processing system to provide the corresponding set of values and parameters, the algorithm conversion eventually get accurate displacement value of the corresponding position. In this device, three standard lasers are arranged in a triangular pattern. The abbe error is greatly reduced and cosine error is eliminated, and the uncertainty of measurement error is up to 0.14μm+1×10-7L. The measurement accuracy is much higher than that of the traditional measurement detection method, thus improving the accuracy of measurement and ensuring the reliability of the measurement value transfer.

The main standard device of indoor large length standard device is three-channel laser interferometer, three laser interferometers form an isosceles triangle. The displacement information of different spatial positions can be obtained by different specific algorithms. Therefore, this device can be used to verify the laser interferometer and laser tracker located inside the triangle, and the standard steel tape measure, special length ruler, ground scanner and other measuring equipment located outside the triangle.

## 4. Application of large size measurement technology

Indoor large length standard devices can calibrate floor scanners. Ground scanner is a spherical coordinate measurement system [4]. It forms the laser point cloud in space through the received laser beam, and then processes the laser point cloud through special software to form the three-dimensional morphology of the measured object [5][6]. Figure 2 shows a physical picture of the calibration of a ground scanner with a large indoor length standard device. The ground scanner measures the range by measuring the black and white target placed on the slide platform, and the measurement data of the three-way laser interferometer is the reference value. The measured data of the ground scanner are the measured values.

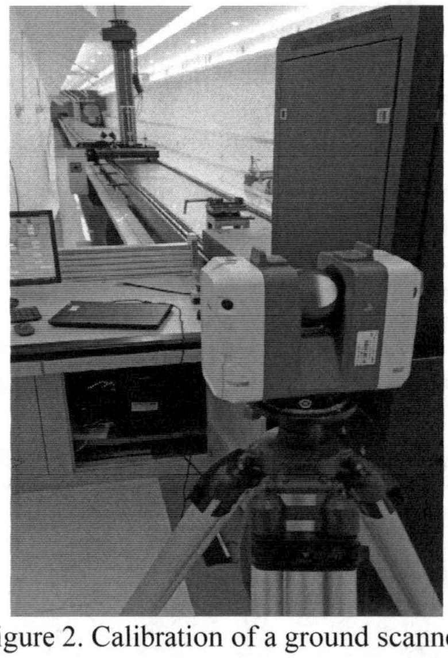

Figure 2. Calibration of a ground scanner.

Table 1 shows the radial distance error of the ground scanner. Radial measurement is also called co-measurement when the optical path of the ground scanner is parallel to that of the laser interferometer. It can be seen from the data in Table 1 that the maximum value error between the ground scanner and the large-length standard device is -0.818mm, and the reference value is 39999.1mm at this time.

Table 1. Radial distance error of ground scanner.

| Serial Number | Reference Value (mm) | The first time | | The second time | | The third time | |
|---|---|---|---|---|---|---|---|
| | | measured value (mm) | indication error (mm) | measured value (mm) | indication error (mm) | measured value (mm) | indication error (mm) |
| 1 | 4999.85 | 4999.991 | 0.141 | 5000.031 | 0.181 | 5000.074 | 0.224 |
| 2 | 9999.74 | 9999.739 | -0.009 | 9999.902 | 0.154 | 9999.918 | 0.170 |
| 3 | 14999.6 | 14999.462 | -0.168 | 14999.56 | -0.061 | 14999.61 | -0.018 |
| 4 | 19999.5 | 19999.249 | -0.290 | 19999.29 | -0.247 | 19999.28 | -0.251 |
| 5 | 24999.4 | 24998.961 | -0.490 | 24998.95 | -0.496 | 24999.01 | -0.437 |
| 6 | 29999.3 | 29998.632 | -0.739 | 29998.52 | -0.847 | 29998.56 | -0.809 |
| 7 | 34999.2 | 34998.615 | -0.674 | 34998.59 | -0.692 | 34998.66 | -0.620 |
| 8 | 39999.1 | 39998.377 | -0.818 | 39998.51 | -0.675 | 39998.46 | -0.718 |
| 9 | 44999.0 | 44998.293 | -0.793 | 44998.33 | -0.751 | 44998.36 | -0.718 |
| 10 | 49998.9 | 49998.261 | -0.653 | 49998.26 | -0.649 | 49998.32 | -0.591 |

Table 2 shows the measurement repeatability of radial distance of laser scanner. The data in the table can be obtained by 10 measurements at a measurement distance of 50 m. The point deviation is the measured value of the laser interferometer minus the measured value of the ground scanner. As can be seen from the table, the maximum value of point position deviation is 0.520 mm and the minimum value is 0.243 mm. The radial repeatability can be obtained from the formula of standard deviation.

Table 2. Repeatability of radial distance measurement.

| Serial number | Absolute Distance (mm) | Point Deviation (mm) |
|---|---|---|
| 1 | 49998.472 | 0.442 |
| 2 | 49998.415 | 0.499 |
| 3 | 49998.456 | 0.458 |
| 4 | 49998.394 | 0.520 |
| 5 | 49998.493 | 0.421 |
| 6 | 49998.671 | 0.243 |
| 7 | 49998.423 | 0.491 |
| 8 | 49998.472 | 0.442 |
| 9 | 49998.355 | 0.559 |
| 10 | 49998.464 | 0.450 |
| average | 49998.462 | 0.452 |
| Radial repeatability | 0.081 | |

## 5. Conclusions

With the development of manufacturing industry, the application of large size measurement is more and more extensive. As a typical representative of large-size measurement technology, indoor large-length standard device has been gradually applied in provincial metrology institutes. It is not only of high application value, but also of scientific research value.

## References

[1] Hu C.，Li W., Zhou Y. (2021) Application of the precision industrial measurement technology in geometric measurement. Journal of Physics: Conference Series,Vol.1885: 022021.
[2] Ding Z. (2021) Research on Environmental Error of Dual Frequency Laser Interferometer Measurement System. Guilin University of electronic technology.
[3] Hu C., Lv F., Luo S. (2021) Application of the Indoor Large-length Standard Device .Journal of Physics: Conference Series, Vol.1894:012017.
[4] Hu C., Kong L., Lv F. (2021) Application of 3D laser scanning technology in engineering field. E3S Web of Conferences, Vol.233:04014.
[5] Hu C., Peng D., Lv F. (2021) Application of Terrestrial Laser Scanner in engineering survey. IOP Conference Series: Earth and Environmental Science, Vol.783 (1):012084 (6pp).
[6] Hu C., Hu S., Kong, L. (2021) Application of ground laser scanner in surveying and mapping field. Journal of Physics: Conference Series,Vol.2029:012080.

# A Small Leak Detection and Localization Method for Oil Pipelines Based on Improved Robust Principle Component Analysis

**Haiou Shan, Yongqiang Zhu and Xianming Lang***

School of Information and Control Engineering, Liaoning Petrochemical University, Fushun, China

*E-mail: langxianming@mail.nwpu.edu.cn

**Abstract.** When there is a small leak in the oil pipeline, their susceptibility to the influence of external noise prevents their detection based on the robust principal component analysis (RPCA) method, which does not consider the matrix of dense noise. Thus, to solve this problem, leak detection and localization method based on an improved robust principal component analysis (IRPCA) for pipelines is proposed. By solving a convex optimization function, this method can remove the dense noise contained in the collected data and improve the efficiency of small leak detection. In addition, the collected leakage data is analyzed. Leak detection is monitored by the combined index $D_2$, which is a combined indicator that is composed of Hotelling's $T^2$ statistic and the SPE statistic. The experimental results show that the missing and false leak detection accuracies of the combined index $D_2$ are much higher than those of the Hotelling's $T^2$ statistic and the SPE statistic separately. Furthermore, it verifies that the small leak localization method proposed in this paper has a good effect.

## 1. Introduction

The pipeline is easily affected by corrosion, aging, etc., causing small leaks and slow leaks [1]. Thus, they have become an important hidden danger for the safe operation of pipelines. However, small leakages cause fluid parameters (e.g., flow, pressure, temperature, density and viscosity) to change at the ends of a pipeline, making it difficult to detect and locate the leakage position [2]. Determining an approach to detect and locate small leakages in long distance pipelines is one of the popular topics that concern many researchers.

Currently, the detection and localization of small leaks is mainly based on software-based methods [3]. Software-based methods mainly include pipeline model method, data-driven method and so on. Compared with the method using the mechanism, the data-driven method does not require the establishment of an accurate pipeline mechanism model and rich expert experience, and has been widely used in pipeline leak detection. Based on data collected during the pipeline's operation, it extracts the characteristics of the collected data and identifies the leakage condition of the pipeline [4]. At present, most of the data-driven methods use negative pressure waves or acoustic waves as leakage signals. Once the signal is collected, use support vector machines (SVM) [5], back propagation neural network (BPNN) [6] or convolution neural network (CNN) [7] to identify leaks. However, the above signal-based pipeline leakage detection methods only consider a single signal change and do not make full use of the data containing the pipeline operation status that results in the leakage.

Multivariate statistical process monitoring can extract real-time process state information on process data (e.g. principal component analysis (PCA) [8], independent principal component analysis (ICA) [9] and other methods that have been successfully applied to nonlinear and non-Gaussian industrial process fault diagnosis).

In addition, PCA has a defect of sensitivity to noisy observation data; however, robust principal component analysis (RPCA) is more robust to noisy observation data by introducing a sparse representation [10], where the training data is divided into two parts, namely a low rank matrix and a block sparse matrix by the original RPCA convex optimization functions. RPCA for fault diagnosis of a blast furnace process was separated from the row matrix and column matrix of the sparse matrix. In [11], a statistic suitable for process monitoring was designed for fault detection that is based on RPCA. In summary, for pipeline small leaks, an improved robust principal component analysis (IRPCA) method is proposed for leak detection and localization.

## 2. Robust principle component analysis

We choose a matrix $X \in R^{n\times m}$ in which each row is a sample. Besides, each sample contains $m$ variables. A low rank matrix $A$ and a sparse matrix $E$ of a robust matrix recovery problem with corrupted columns can be solved by the following model [12-14]:

$$\begin{aligned} &\min \ \|A\|_* + \lambda \|E\|_1 \\ &s.t. \ \ X = A + E \end{aligned} \tag{1}$$

where $\|A\|_*$ represents the nuclear norm. $\|E\|_1$ represents the $l_1$ norm. $\lambda$ is the parameter that balances these two terms with $\|A\|_* + \lambda \|E\|_1$. $\lambda$ can be found in [15], $\lambda = \max(n,m)^{-1/2}$.

In the RPCA algorithm, it assumes that $E$ consists of sparse errors. However, the dense noise always exists in measurement variables of a pipeline. Therefore, the dense noise will have a certain impact on RPCA-based methods.

Based on the above discussions, for oil pipelines, in order to achieve effective leak detection and localization based on RPCA, several issues need to be addressed.

## 3. Improved robust principle component analysis

It is natural that the performance of the RPCA algorithm would be degraded due to the inherent existence of external noise, which causes an inaccurate detection result of the oil pipeline leakage. To overcome this problem, an improved robust principal component analysis method is proposed.

### *3.1. Improved robust principle component analysis*

Here, the data matrix $X$ comprises of three parts:

$$\begin{aligned} &\min \ \|A\|_* + \lambda \|E\|_1 + \gamma \|N\|_F^2 \\ &s.t. \ \ X = A + E + N \end{aligned} \tag{2}$$

where $N$ is the matrix of dense noise, and $\gamma$ is a positive weighted parameter.

$$L(A,E,N,Y,\mu) = \|A\|_* + \lambda \|E\|_1 + \gamma \|N\|_F^2 + \langle Y, X - A - E - N \rangle + \frac{\mu}{2}\|X - A - E - N\|_F^2 \tag{3}$$

The augmented Lagrange function has more than one penalty term with a constraint on the Lagrange function. We can use an alternate updating method to minimize the augmented Lagrange function. First, we find an $A$ which minimizes the objective function $L$ by fixing $E$ and $Y$, and then fixing $A$ and $Y$ to obtain the $E$ which minimizes $L$. Thus, we can achieve the optimal solution to the problem by iteration.

Updating $A$

$$\arg\min \|A\|_* + \frac{\mu}{2}\|X - A - E - Y + \mu^{-1}Y\|_F^2 = D_{\mu^{-1}}(A - E - N + \mu^{-1}Y) \tag{4}$$

Updating $E$

$$\arg\min \lambda\|E\|_1 + \frac{\mu}{2}\|X - A - E - N + \mu^{-1}Y\|_F^2 = S_{\mu^{-1}\lambda}(X - A - E - N + \mu^{-1}Y) \tag{5}$$

Updating N

$$\arg\min \gamma\|N\|_F^2 + \frac{\mu}{2}\|X - A - E + \mu^{-1}Y\|_F^2 = \frac{\mu}{\mu + 2\gamma}(X - A - E + \mu^{-1}Y) \tag{6}$$

Alternate updates until convergence.

*3.2. Convergence*

The following formula can be solved by the augmented Lagrange multiplier (ALM) method.

$$\begin{aligned} &\min f(x) \\ &s.t. \quad h(x) = 0 \end{aligned} \tag{7}$$

where $f: R^n \to R, h: R^n \to R^m$.

The augmented Lagrange function of this optimization model can be expressed as follows.

$$L(x, Y, \mu) = f(x) + \langle Y, h(x) \rangle + \frac{\mu}{2}\|h(x)\|_F^2 \tag{8}$$

where $\mu$ is a positive scalar.

If the parameters increase gradually during the iteration and the sum of the functions is a continuous differential function, the optimal solution for the objective function can be obtained by ALM under certain conditions.

However, the objective function of the equation is not a continuous differential equation; therefore, the ALM method cannot be used directly. The convergence of convex optimization functions for the formula must be proved.

Matrix E and matrix N are sparse matrices and noise matrices, respectively. Hence, the training matrix X will be divided into the following three parts: the low rank matrix A, the noise matrix N and the sparse matrix E.

***Theorem 1**:* In the IRPCA algorithm, if the parameters increase gradually in the iterative process, then the convex optimization function formula in IRPCA can be solved by the algorithm, and an optimal solution can be obtained.

***Proof:*** $(A^*, E^*, N^*)$ is defined as a saddle point of the IRPCA algorithm, $X = A^* + E^* + N^*$.

$$L(A, E, N, Y) = \|A\|_* + \lambda\|E\|_1 + \gamma\|N\|_F^2 + \langle Y, X - A - E - N \rangle \tag{9}$$

Thus,

$$\begin{aligned} &\partial_A(A, E, N, Y) \in 0 \\ &\partial_E(A, E, N, Y) \in 0 \\ &\partial_N(A, E, N, Y) \in 0 \end{aligned} \tag{10}$$

$$\begin{aligned} &\partial\|A\|_* - Y^* \in 0 \\ &\partial\|\lambda E\|_1 - Y^* \in 0 \\ &\partial\|\gamma N\|_F^2 - Y^* \in 0 \end{aligned} \tag{11}$$

$$\begin{aligned} &Y^* \in \partial \|A\|_* \\ &Y^* \in \partial \|\lambda E\|_1 \\ &Y^* \in \partial \|\gamma N\|_F^2 \end{aligned} \tag{12}$$

A new sequence is defined from the algorithm.

$$\hat{Y}_k = Y_{k-1} + \mu_{k-1}(X - A_k - E_{k-1} - N_{k-1}) \tag{13}$$

$$X - A_k - E_{k-1} - N_{k-1} = \mu_k^{-1}(Y_{k+1} - Y_k) \tag{14}$$

Then, we deduce

$$\begin{aligned} &\|E_{k+1} - E^* + N_{k+1} - N^*\|_F^2 + \mu_k^{-2}\|Y_{k+1} - Y^*\|_F^2 = \|E_{k+1} - E^* + N_k - N^*\|_F^2 \\ &+ \mu_k^{-2}\|Y_{k+1} - Y^*\|_F^2 - \|E_{k+1} - E_k + N_{k+1} - N_k\|_F^2 - \mu_k^{-2}\|Y_{k+1} - Y_k\|_F^2 \\ &- 2\mu_k^{-1}(\langle Y_{k+1} - Y_k, E_{k+1} - E_k\rangle + \langle Y_{k+1} - Y_k, N_{k+1} - N_k\rangle\langle A_{k+1} - A^*, Y_{k+1} - Y^*\rangle \\ &+ \langle E_{k+1} - E^*, Y_{k+1} - Y^*\rangle + \langle N_{k+1} - N^*, Y_{k+1} - Y^*\rangle) \end{aligned} \tag{15}$$

$$\begin{aligned} &\mu_k^{-1}\langle Y_{k+1} - Y_k, E_{k+1} - E_k\rangle \\ &= -\langle A_{k+1} - A^*, Y_{k+1} - Y^*\rangle - \langle E_{k+1} - E^*, Y_{k+1} - Y^*\rangle - \langle N_{k+1} - N^*, Y_{k+1} - Y^*\rangle \\ &= \langle A_{k+1} - A^*, \hat{Y}_{k+1} - Y_{k+1}\rangle - \langle A_{k+1} - A^*, \hat{Y}_{k+1} - Y^*\rangle \\ &\quad - \langle E_{k+1} - E^*, Y_{k+1} - Y^*\rangle - \langle N_{k+1} - N^*, Y_{k+1} - Y^*\rangle \\ &= \mu_k\langle A_{k+1} - A^*, E_{k+1} - E_k\rangle - \mu_k\langle A_{k+1} - A^*, N_{k+1} - N_k\rangle \\ &\quad - \langle A_{k+1} - A^*, \hat{Y}_{k+1} - Y^*\rangle - \langle E_{k+1} - E^*, Y_{k+1} - Y^*\rangle \end{aligned} \tag{16}$$

$$\begin{aligned} &\|E_{k+1} - E^*\|_F^2 + \mu_k^{-2}\|Y_{k+1} - Y^*\|_F^2 = \|E_k - E^*\|_F^2 - \|E_{k+1} - E_k\|_F^2 + 2\langle E_{k+1} - E^*, E_{k+1} - E_K\rangle \\ &+ \mu_k^{-2}((\|Y_{k+1} - Y^*\|_F^2 - \|Y_{k+1} - Y_k\|_F^2) + 2\langle Y_{k+1} - Y^*, Y_{k+1} - Y_k\rangle) \\ &= \|E_k - E^*\|_F^2 + \mu_k^{-2}\|Y_k - Y^*\|_F^2 - \|E_{k+1} - E_k\|_F^2 - \mu_k^{-2}\|Y_{k+1} - Y_k\|_F^2 + 2\langle A_{k+1} - A^*, E_{k+1} - E_k\rangle \\ &- 2\mu_k^{-1}\left(\langle A_{k+1} - A^*, \hat{Y}_{k+1} - Y_{k+1}\rangle + \langle E_{k+1} - E^*, Y_{k+1} - Y^*\rangle\right) \\ &= \|E_k - E^*\|_F^2 + \mu_k^{-2}\|Y_k - Y^*\|_F^2 - \|E_{k+1} - E_k\|_F^2 - \mu_k^{-2}\|Y_{k+1} - Y_k\|_F^2 - 2\mu_k^{-1}(\langle A_{k+1} - A^*, \hat{Y}_{k+1} - Y^*\rangle \\ &+ \langle E_{k+1} - E_k^*, Y_{k+1} - Y^*\rangle + \langle N_{k+1} - N^*, Y_{k+1} - Y^*\rangle) \\ &= \|E_k - E^*\|_F^2 + \mu_k^{-2}\|Y_k - Y^*\|_F^2 - \|E_{k+1} - E_k\|_F^2 - \mu_k^{-2}\|Y_{k+1} - Y_k\|_F^2 - 2\mu_k^{-1}\langle Y_{k+1} - Y_k, E_{k+1} - E_k\rangle \\ &- 2\mu_k^{-1}(\langle A_{k+1} - A^*, \hat{Y}_{k+1} - Y^*\rangle + \langle E_{k+1} - E_k^*, Y_{k+1} - Y^*\rangle + \langle N_{k+1} - N^*, Y_{k+1} - Y^*\rangle) \end{aligned} \tag{17}$$

This completes the proof.

***Lemma 1:*** If there is no reduction in each iteration, the following sequence is nonnegative and its sum is bounded.

$$\begin{aligned} &\sum_{k=1}^{+\infty} u_k^{-1}(\langle Y_{k+1} - Y_k, E_{k+1} - E_k\rangle + \langle A_{k+1} - A^*, \hat{Y}_{k+1} - Y^*\rangle \\ &+ \langle E_{k+1} - E^*, Y_{k+1} - Y^*\rangle + \langle N_{k+1} - N^*, Y_{k+1} - Y^*\rangle) < +\infty \end{aligned} \tag{18}$$

***Proof:***
Therefore,

$$\begin{aligned} &\hat{Y}_{k+1} \in \partial \left\| A_{k+1} \right\|_* \\ &Y_{k+1} \in \partial \left\| \lambda E_{k+1} \right\|_1 \\ &Y_{k+1} \in \partial \left\| \gamma N_{k+1} \right\|_F^2 \end{aligned} \quad (19)$$

Then, we obtain

$$\begin{aligned} &\left\langle A_{k+1} - A^*, \hat{Y}_{k+1} - Y^* \right\rangle \geq 0 \\ &\left\langle E_{k+1} - E^*, Y_{k+1} - Y^* \right\rangle \geq 0 \\ &\left\langle E_{k+1} - E^*, Y_{k+1} - Y_k \right\rangle \geq 0 \\ &\left\langle N_{k+1} - N^*, Y_{k+1} - Y^* \right\rangle \geq 0 \end{aligned} \quad (20)$$

Therefore, $\left\| E_{k+1} - E^* \right\|_F^2 + \mu_k^{-2} \left\| Y_{k+1} - Y^* \right\|_F^2 \geq 0$ is obvious. As a result,

$$\begin{aligned} &2u_k^{-1}(\langle Y_{k+1} - Y_k, E_{k+1} - E_k \rangle + \left\langle A_{k+1} - A^*, \hat{Y}_{k+1} - Y^* \right\rangle \\ &+ \left\langle E_{k+1} - E^*, Y_{k+1} - Y^* \right\rangle + \left\langle N_{k+1} - N^*, Y_{k+1} - Y^* \right\rangle) \\ &\leq \left\| E_k - E^* \right\|_F^2 + \left\| E_{k+1} - E_k \right\|_F^2 + \mu_k^{-2} \left\| Y_k - Y^* \right\|_F^2 - u_{k+1}^{-2} \left\| Y_{k+1} - Y^* \right\|_F^2 \end{aligned} \quad (21)$$

Thus,

$$\begin{aligned} &\left\| E_{k+1} - E^* + N_{k+1} - N^* \right\|_F^2 + \mu_k^{-2} \left\| Y_{k+1} - Y^* \right\|_F^2 = \left\| E_k - E^* + N_k - N^* \right\|_F^2 + \mu_k^{-2} \left\| Y_{k+1} - Y^* \right\|_F^2 \\ &- \left\| E_{k+1} - E_k + N_{k+1} - N_k \right\|_F^2 - \mu_k^{-2} \left\| Y_{k+1} - Y_k \right\|_F^2 - 2\mu_k^{-1}(\langle Y_{k+1} - Y_k, E_{k+1} - E_k \rangle \\ &+ \langle Y_{k+1} - Y_k, N_{k+1} - N_k \rangle + \left\langle A_{k+1} - A^*, \hat{Y}_{k+1} - Y^* \right\rangle + \left\langle E_{k+1} - E^*, Y_{k+1} - Y^* \right\rangle + \left\langle N_{k+1} - N^*, Y_{k+1} - Y^* \right\rangle) \\ &\leq \left\| E_k - E^* + N_k - N^* \right\|_F^2 + \mu_k^{-2} \left\| Y_k - Y^* \right\|_F^2 \end{aligned} \quad (22)$$

$$\mu_k^{-2} \left\| Y_{k+1} - Y^* \right\|_F^2 \leq \left\| E_k - E^* + N_k - N^* \right\|_F^2 + \mu_k^{-2} \left\| Y_k - Y^* \right\|_F^2 - \mu_k^{-2} \left\| Y_{k+1} - Y^* \right\|_F^2 \quad (23)$$

$$\sum_{k=1}^{+\infty} \mu_k^{-2} \left\| Y_{k+1} - Y^* \right\|_F^2 < +\infty \quad (24)$$

Thus, $\left\| X - A_k - E_k - N_k \right\|_F = \mu_{k-1}^{-1} \left\| Y_{k+1} - Y^* \right\|_F \to 0$ exists. Therefore, $(A_k, E_k, N_k)$ can be used to resolve IRPCA issues. Suppose $(A^*, E^*, N^*)$ and the optimal solution are basically equal; therefore, $f^* = \left\| A^* \right\|_* + \lambda \left\| E^* \right\|_1 + \gamma \left\| N^* \right\|_F^2$ is the minimum value.

$$\begin{aligned} &\| A_k \|_* + \lambda \| E_k \|_1 + \gamma \| N_k \|_F^2 \\ &\leq \left\| A^* \right\|_* + \lambda \left\| E^* \right\|_1 + \gamma \left\| N^* \right\|_F^2 - \left\langle \hat{Y}_k, A^* - A_k \right\rangle - \left\langle Y_k, E^* - E_k \right\rangle - \left\langle Y_k, N^* - N_k \right\rangle \\ &= f^* + \left\langle Y^* - \hat{Y}_k, A^* - A_k \right\rangle - \left\langle Y^* - Y_k, E^* - E_k \right\rangle - \left\langle Y^* - Y_k, N^* - N_k \right\rangle \\ &\quad - \left\langle Y^*{}_k, A^* - A_k + E^* - E_k + N^* - N_k \right\rangle \\ &= f^* + \left\langle Y^* - \hat{Y}_k, A^* - A_k \right\rangle - \left\langle Y^* - Y_k, E^* - E_k \right\rangle - \left\langle Y^* - Y_k, N^* - N_k \right\rangle - \left\langle Y^*{}_k, X - A_k - E_k - N_k \right\rangle \end{aligned} \quad (25)$$

$$\begin{aligned} &\sum_{k=1}^{+\infty} u_k^{-1}(\langle Y_{k+1} - Y_k, E_{k+1} - E_k \rangle + \left\langle A_{k+1} - A^*, \hat{Y}_{k+1} - Y^* \right\rangle) \\ &+ \left\langle E_{k+1} - E^*, Y_{k+1} - Y^* \right\rangle + \left\langle N_{k+1} - N^*, Y_{k+1} - Y^* \right\rangle) < +\infty \end{aligned} \quad (26)$$

As $\sum_{k=1}^{+\infty} u_k^{-1} = +\infty$ , a subsequence $(A_{k_j}, E_{k_j}, N_{k_j})$ that satisfies $\left\langle A_{k_j} - A^*, \hat{Y}_{k_j} - Y^* \right\rangle + \left\langle E_{k_j} - E^*, \hat{Y}_{k_j} - Y^* \right\rangle + \left\langle N_{k_j} - N^*, \hat{Y}_{k_j} - Y^* \right\rangle \to 0$ exists.

Letting $k \to +\infty$ , we have that

$$\lim_{k \to +\infty} \left\| A_{k_j} \right\|_* + \lambda \left\| E_{k_j} \right\|_1 + \gamma \left\| N_{k_j} \right\|_F^2 \le f^* \tag{27}$$

Hence, $(A_{k_j}, E_{k_j}, N_{k_j})$ is an optimal solution to the IRPCA problem.

Because $u_k \to +\infty$ and $Y_k$ are bounded, it can be calculated

$$\left\| E_{k_j} - E^* + N_{k_j} - N^* \right\|_F^2 + \mu_{k_j}^{-2} \left\| Y_{k_j} - Y^* \right\|_F^2 \to 0 \tag{28}$$

In addition,

$$\left\| E_k - E^* + N_k - N^* \right\|_F^2 + \mu_k^{-2} \left\| Y_k - Y^* \right\|_F^2 \to 0 \tag{29}$$

Therefore, it can be concluded that

$$\lim_{k \to \infty} E_k + N_k = E^* + N^* \tag{30}$$

In addition,

$$\lim_{k \to \infty} X - A_k - E_k - N_k = 0 \tag{31}$$

Hence, $X = A_k + E_k + N_k$ .

Therefore, it can be seen that $\lim_{k \to \infty} A_k = A^*$ .

Thus, this completes the proof.

### *3.3. A small leak detection and localization based on IRPCA*

The data matrix of the normal condition $X \in R^{n \times m}$ is obtained, and the data matrix of the small leakage $Z \in R^{p \times m}$ is obtained. There is a sample in the data matrix of the normal condition $n$ , a sample $p$ in the data matrix of the small leakage condition and the number of variables is $m$ . The method of small leak detection and localization based on the proposed IRPCA is as follows:

1. Matrix decomposition

The training matrix $X$ is divided into three parts with the proposed IRPCA method: a low rank matrix $A$ , a sparse matrix $E$ with useful pipeline leak information and a matrix of dense noise $N$ .

$$\begin{aligned} &\min \left\| A \right\|_* + \lambda \left\| E \right\|_1 + \gamma \left\| N \right\|_F^2 \\ &\text{s.t.} \;\; X = A + E + N \end{aligned} \tag{32}$$

The mean values and the standard deviations of the variables in matrix E are calculated, and the singular value decomposition of the low rank matrix E is calculated.

2. Hotelling's $T^2$ and SPE statistics calculation

The control limit of the Hotelling's $T^2$ and SPE statistics that are achieved with the training matrix are calculated as follows:

$$T_\sigma^2 = \frac{G(n^2 - 1)}{n(n - G)} F_{G,n-G;\sigma} \tag{33}$$

where $\sigma$ is the significant level, and $G$ is the largest $G$ singular values of $E$. $n$ is the sample size.

$$SPE_\sigma = \theta_1 \left( \frac{c_\sigma \sqrt{2\theta_2 h_0^2}}{\theta_1} + \frac{\theta_2 h_0 (h_0 - 1)}{\theta_1^2} \right)^{1/h_0} \tag{34}$$

where $\theta_i = \sum_{j=1}^{m-G} \lambda_j^i$ , $(i=1,2,3)$ , $h_0 = 1-2\theta_1\theta_3 / 3\theta_2^2$ , $\lambda_j$ is the $j$-th largest eigenvalue of the covariance matrix of $E$, and $c_\sigma$ is the critical value of the normal distribution under significance level of $\sigma$ .

The SPE monitoring statistics and Hotelling's $T^2$ can be calculated as:

$$T^2 = A^T P \Lambda^{-1} P^T A \tag{35}$$

where $P$ is composed of the $G$-$th$ eigenvectors, $\Lambda = diag\{\lambda_1, \lambda_2, \cdots, \lambda_G\}$ .

$$SPE = e^T e \tag{36}$$

where $e$ represents the residuals of the monitoring model.

3. Combined index $D_2$

In statistical process monitoring, Hotelling's $T^2$ statistics and the SPE statistics are two commonly applied for statistical indicators. Hotelling's $T^2$ statistics are applied to describe normal data change measures, while SPE is used to describe abnormal data change measures. The combined index $D_2$ is composed of the Hotelling's $T^2$ and SPE statistical indicators with weighted terms, as shown in Eq. (37).

$$D_2 = \alpha \cdot T_\sigma^2 / T^2 + \beta \cdot SPE / SPE_\sigma^2, 0 < \alpha < 1, 0 < \beta < 1 \tag{37}$$

4. Online leak detection and localization

If the $D_2$ statistic of the test matrix calculated in step 3 is greater than the control limit calculated by the combined index $D_2$ under normal conditions, the leakage occurs.

$L_x$ represents the distance between the leak point and the head end of the pipeline, which can be calculated by the following formula:

$$L_x = \frac{|L - a_w \cdot t_D|}{2} \tag{38}$$

where $|\cdot|$ is absolute value, $L$ represents the total length of the oil pipeline, $a_w$ is the velocity of the negative pressure wave (NPW), (m/s), $t_D$ represents the estimated delay time.

The algorithm of the leak detection step based on the IRPCA method is shown in Table 1.

Table 1. Leak detection and localization based on IRPCA

| **Algorithm**: Leak detection and localization based on IRPCA |
| --- |
| 1. Data collection: collect measurement data from the pipeline operational process. |
| 2. Normalization: normalize the measurement data set $X$; the normalized data divides matrix $X$ into three parts: $A$, $E$, and $N$. |
| 3. Compute $D_2$: conduct an SVD on $A$; the monitoring statistics $D_2$ and the corresponding control limits can be calculated. |
| 4. Leak detection: If a new observation is obtained, it is normalized; then, the normalized new observation X is projected to the IRPCA. Afterwards, if the $D_2$ statistics exceeds the corresponding limit, a small leak is detected. |
| 5. Localization: Compute the leak position with Eq. (38) |

Calculate the fault alarm rate (FAR) and the fault detection rate (FDR) by Eq. (39) and Eq. (40).

$$\text{FAR}=\frac{\text{No. of normal samples identified as fault}}{\text{total No. of normal samples}}\times 100\% \tag{39}$$

$$\text{FDR}=\frac{\text{No. of fault samples detected}}{\text{total No. of fault samples}}\times 100\% \tag{40}$$

## 4. Simulation

The schematic diagram of the different leakage apertures in the pipeline leak simulator is shown in Figure 1 Water was used to simulate pipeline leaks in the experiment. The ultrasonic signal, the flow signal, the density signal, the viscosity signal, the pressure signal and the temperature signal at both ends of the pipeline are chosen for the test.

A pipeline with a length of 2 km and an inner diameter of 50 mm was used in the experiment. The velocity of water flow in pipeline is 12 m3/h, different small valve apertures were installed on pipeline to emulate leaks. The lift of upstream stations pump was 120 m. The density and the viscosity of water are 1000 kg/m3 and 0.8949×10-3 Pa·s, respectively. NPW has a wave speed of 1,000 m/s. Ultrasonic wave speed is 1520 m/s, $K = 2.13\times10^{-6}$. The ambient temperature is 25℃.

We obtained a database by National Instruments DAQ-9184. The computer and software used were: computer with Intel Core processor (2.8 GHz) and 8 GB RAM, MATLAB R2014a.

Figure 1. Different leakages

Flow ball valves are installed to emulate small leaks at 198.5 m upstream. The length of acquisition signal is 70 s with a sampling rate of 20 Hz. The measured flow rate, pressure, temperature density, viscosity, and the sound velocity of the ultrasonic signal are set as the data matrix $X$ .

There are three types working conditions in the pipeline: (1) 3 mm leakage aperture, (2) 4 mm leakage aperture and (3) the normal condition.

Therefore, the data acquisition equipment collects the signal when there without leakage and when the leak aperture is 3 mm, 4 mm. Four types of working condition samples are created, and 200 normal condition samples and 100 leakage aperture samples of 3 mm, 4 mm are chosen for training the IRPCA, and the RPCA. Thus, the normal condition samples are divided into three parts. The training matrix contains 100 normal condition samples. Each testing matrix contains 100 normal condition samples and 100 leakage samples of 3 mm, and 4 mm.

In the IRPCA algorithm, the parameter $\lambda = 1/\sqrt{100}$ is 0.1, and $\gamma$ is 0.1. The $\alpha$ and $\beta$ of the combined index $D_2$ are set to 0.8 and 0.2, respectively.

When the pipeline leakage is 2.5 L/min, the 4 mm leakage aperture of the measured Hotelling's $T^2$ statistic and SPE statistic and the 4 mm leakage aperture of the $D_2$ statistic are as shown in Figures 2-3. Meanwhile, the 3 mm leakage aperture of the measured Hotelling's $T^2$ statistic and SPE statistic and the 3 mm leakage aperture of the $D_2$ statistic are as shown in Figures 4-5.

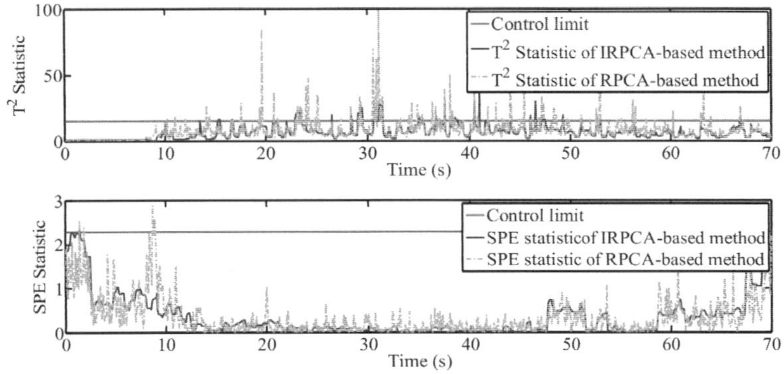


Figure 2. Hotelling's $T^2$ statistic and SPE statistic of IRPCA-based and RPCA-based methods for leak detection of 4 mm leakage aperture

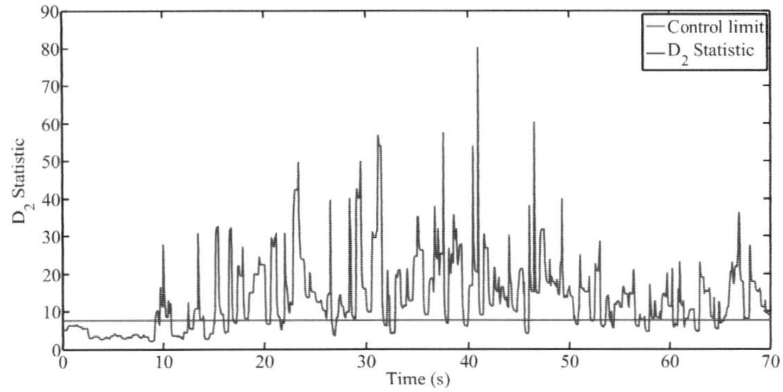


Figure 3. $D_2$ statistic of IRPCA-based method for leak detection of 4 mm leakage aperture

In Figure 2, the Hotelling's $T^2$ statistic result of the IRPCA-based method is much better than that of the RPCA-based method for the 4 mm leakage aperture. However, the SPE statistic results of the IRPCA-based method and the RPCA-based method are below the control limit; therefore, they are not accurate to detect leakage.

Figure 3 shows that the combined index $D_2$ has an obvious sharp transition upstream and downstream where the 4 mm leak aperture occurs.

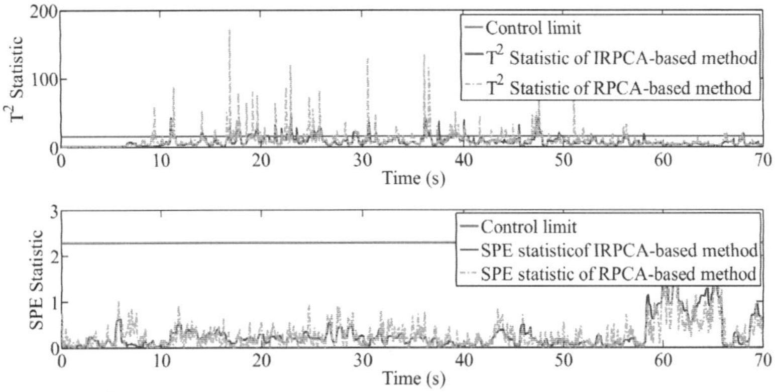


Figure 4. Hotelling's $T^2$ statistic and SPE statistic of IRPCA-based and RPCA-based methods for leak detection of 3 mm leakage aperture

 2264 (2022) 012021 doi:10.1088/1742-6596/2264/1/012021 

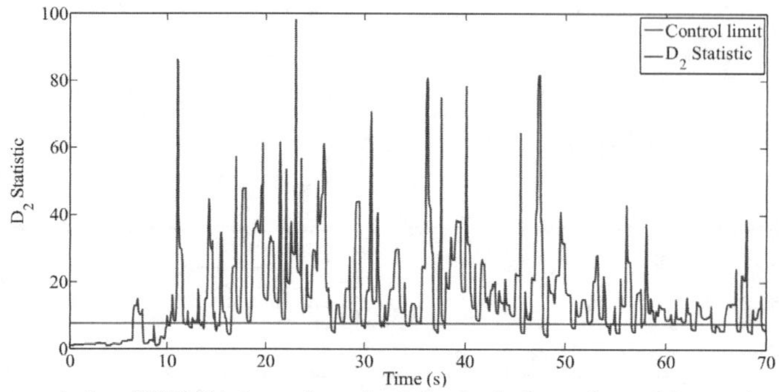

Figure 5. $D_2$ statistic of IRPCA-based method for leak detection of 3 mm leakage aperture

In Figure 4, the Hotelling's $T^2$ statistic result of the IRPCA-based method is much better than that of the RPCA-based method in the 3 mm leakage aperture. However, the SPE statistic results of the IRPCA-based method and the RPCA-based method are below the control limit; therefore, they are not accurate to detect leakage.

Figure 5 shows that the combined index $D_2$ has an obvious sharp transition upstream and downstream where the 3 mm leak aperture occurs.

In conclusion, Figure 5 shows that $D_2$ statistics at the ends of pipeline have obvious sharp transitions at both ends of the pipeline in the 4 mm leakage. However, as can be seen from Figure 5, the measured $D_2$ statistic also changes significantly, even when the leakage aperture is small.

The leak detection results by the IRPCA-based method and the RPCA-based method with the FDR and the FAR calculated by Eq. (39) and Eq. (40) are listed in Table 2.

Table 2. The far and FDR of IRPCA-based and PCA-based leak detection methods

| Leakage aperture | FDR-IRPCA | FAR-IRPCA | FDR-RPCA | FAR-RPCA |
|---|---|---|---|---|
| 4 mm | 100 | 0 | 92 | 1 |
| 3mm | 96 | 3 | 90 | 5 |

As seen from Figures 2-5 and Table 2, the leak detection result by the IRPCA-based method is better than that of the RPCA-based method. By solving a convex optimization function and separating the sparse matrix and noise matrix from the data matrix, the proposed method can effectively eliminate noise in a data matrix. This shows that IRPCA-based method is suitable to detect pipeline leakage.

At 198.5m of the pipeline, the method in this paper and the conventional method are used to locate the leakage and compare, and the comparison results are shown in Table 3: (1) the conventional cross-correlation of the pressure signals at both ends of the pipeline, (2) the conventional cross-correlation of the information fusion signals at both ends of the pipeline in [2] and (3) the proposed method. The calculated localization of these methods for the 20 groups is the highest possible error.

Table 3. Experimental results of different leakage holes

| Localization method | Leakage aperture | Leak position (m) | Calculated leak localization (m) | The highest possible error (m) |
|---|---|---|---|---|
| 1 | 4 mm | 198.5 | 178 | 20.5 |
| | 3 mm | 198.5 | 160 | 38.5 |
| 2 | 4 mm | 198.5 | 180 | 18.5 |
| | 3 mm | 198.5 | 218 | 19.5 |
| 3 | 4 mm | 198.5 | 212 | 13.5 |
| | 3 mm | 198.5 | 181 | 17.5 |

As can be seen form Table 3, compared with methods based on fusion signals and methods using pressure waves, the IRPCA-based method is more accurate in leak localization. When the leak aperture is 4mm, the method based on the pressure wave locates the leak position at 178m, and the positioning error is 20.5m. When using the fusion signal-based method for localization, the leak position at 180m, and the maximum possible localization error within the same leak aperture is 18.5 m. When using the method proposed in this paper to locate the leak, the leak location is 212 m, and the maximum possible positioning error is 13.5 m.

The inflection points of the flow, pressure, temperature, density and viscosity signal become indistinguishable with the decrease of the leakage aperture, which affects the leak localization.

## 5. Conclusion

Aiming at the problem that the leak signal contains noise, an improved robust principal component analysis method is proposed, which can be applied to pipeline leakage detection and localization. This method can recover low-rank matrices with important pipeline leak information in the original data matrix, and it is convergent. By solving a convex optimization function and separating the noise data from the training matrix, this method is proven to be more effective than the other RPCA approaches in leakage detection. Additionally, the combined index $D_2$ has obvious inflection points upstream and downstream, and the precision of the leak localization is more accurate than that of the other methods.

## Acknowledgments

This work was supported by the National Natural Science Foundation of China under Grant 61673199.

## References

[1] Datta, S., Sarkar, S. (2016) A review on different pipeline fault detection methods. J. Loss Prevent. Proc., 41: 91-106.
[2] Lang, X., Li, P., Cao, J. T., Li, Y. and Ren, H. (2018) A small leak localization method for oil pipelines based on information fusion. IEEE Sens. J., 18: 6115-6122.
[3] Lu, W., Liang, W., Zhang, L., Liu, W. (2016) A novel noise reduction method applied in negative pressure wave for pipeline leakage localization. Process Saf. Environ., 104: 142-149.
[4] Liang, W., Zhang, L. (2012) A wave change analysis (WCA) method for pipeline leak detection using Gaussian mixture model. J. Loss Prevent. Proc., 25: 60-69.
[5] Qu, Z., Feng, H., Zeng, Z., Zhuge, Jin, J. S. (2010) A SVM-based pipeline leakage detection and pre-warning system. Measurement, 43: 513-519.
[6] Zhang, H.G., Liu, J.H., Ma, D.Z., Wang, Z.H. (2011) Data-core-based fuzzy min–max neural network for pattern classification. IEEE T. Neural Network., 22: 2339-2352.
[7] Feng, J., Li, F., Lu, S., Liu, J., Ma, D. (2017) Injurious or noninjurious defect identification from MFL images in pipeline inspection using convolutional neural network. IEEE T. Instrum. Meas., 66: 1883-1892.
[8] Deng, P. C., Gui, W. H., Xie, Y. F. (2010) Latent space transformation based on principal component analysis for adaptive fault detection. IET Control Theory A., 4: 2527-2538.
[9] Deng, X., Zhong, N., Wang, L. (2016) Research on mixed signal separation method for pipeline leakage based on Robust ICA. In: International Conference on Robotics and Automation Engineering (ICRAE). Jeju. pp. 70-74.
[10] Han, N.N., Song, Y.M., Song, Z. J., (2017) Bayesian robust principal component analysis with structured sparse component. Computational Statistics and Data Analysis, 109: 144-158.
[11] Pan, Y., Yang, C., An, R., Sun, Y. (2018) Robust principal component pursuit for fault detection in a blast furnace process. Ind. Eng. Chem. Res., 57: 283-291.
[12] Bouwmans, T., Zahzah, E.H., (2014) Robust PCA via Principal Component Pursuit: A review for a comparative evaluation in video surveillance. Compu. Vis. Image Und., 122: 22-34.

[13] Lim, Y., Park, Y., Oh, H. (2014) Robust principal component analysis via ES-algorithm. J. Korean Stat. Soc., 43: 149-159.
[14] Pan, Y., Yang, C., An, R., Sun, Y. (2009) Robust PCA for skewed data and its outlier map. Comput. Stat. Data An., 53: 2264-2274.
[15] Vidal, R., Ma, Y., Sastry, S. S. (2016) Robust principal component analysis. Springer Publishing, New York.

# Research on Magnetron Fault Reasoning Based on Fault Tree Reasoning Model

**Weiyi WU [1], Lianjun XIE [2*], Bingpeng GAO [3]**

[1]College of Electrical Engineering, Xinjiang University, Urumqi, Xinjiang, China

[2]Xinjiang Urumqi Civil Aviation Aviation Administration Bureau, Urumqi, Xinjiang, China

[3]College of Electrical Engineering, Xinjiang University, Urumqi, Xinjiang, China

*Lianjun XIE:e-mail:383111731@qq.com

**Abstract:** Aiming at the characteristics of magnetron of airport surveillance radar, such as short operation time, complexity and nonlinear vibration signal, a fault reasoning method based on T-S fuzzy fault tree model was proposed to realize fault reasoning analysis of magnetron of radar receiving and receiving system. A T-S fuzzy fault tree fault inference model is constructed by using the field expert's opinion expressed by fuzzy number language set. The experimental results show that compared with the traditional manual fault detection method, this method can make more comprehensive and efficient inference analysis for the magnetron fault events. This method provides a new idea for the application of magnetron fault reasoning analysis in engineering.

## 1. Introduction

Airport scene surveillance radar is a large mechanical and electronic equipment, which plays a very important role in the operation of the airport and plays a very important role in the daily operation of the airport. The magnetron is one of the most important components in the transceiver system, which is one of the two important components of the head of airport scene surveillance radar. If it fails, the receiving and sending signal process can not be carried out normally, and the air traffic at the airport will be seriously affected. Here we take the magnetron as the research object of this paper. There are many reasons for the failure of the magnetron, which may be caused by the internal problems of the magnetron, or it may be caused by the external factors suddenly encountered in the process of operation. Therefore, the research on fault diagnosis and reasoning of magnetron is of great significance in engineering application.

Le Chen et al [1], using T-S fuzzy fault tree model, carried out corresponding research on the fault analysis of important parts of CNC lathe. Wu Chen et al [2], analyzed and evaluated the probability of tunnel collapse by using T-S fuzzy fault tree. Na Deng et al [3], in the aspect of biased tunnel, used fault tree to carry out reasoning analysis on its collapse risk rate. Feng Li et al [4], the T-S fuzzy fault tree model is used to analyze the reliability of truck crane. Haiming Zhao et al [5], analyzed the failure probability of hydraulic cylinder by using fuzzy set language and fault tree analysis method. Yuhong Nan et al [6], made a risk assessment on tunnel collapse risk by using accident tree analysis method. Lu Sun [7], explained the application of scene surveillance radar in airport. Gang He [8], elaborated on the daily troubleshooting process of important components of scene surveillance radar in Urumqi.

To sum up, in this paper, we propose a fault diagnosis reasoning analysis of the magnetron based on the established T-S fuzzy door fault tree structure model.

## 2. Theoretical analysis of magnetron fault reasoning

### *2.1. Fuzzy set theory concept*

Scene in the actual engineering application, airport surveillance radar transceiver magnetron components in the system failure occurs, the state of the parts we use fuzzy language to say here, because there is no precise numerical description, so we in order to achieve the analysis of fault tree is more convenient, here we will be the first to put forward the concept of gradient function, as shown in figure 1.

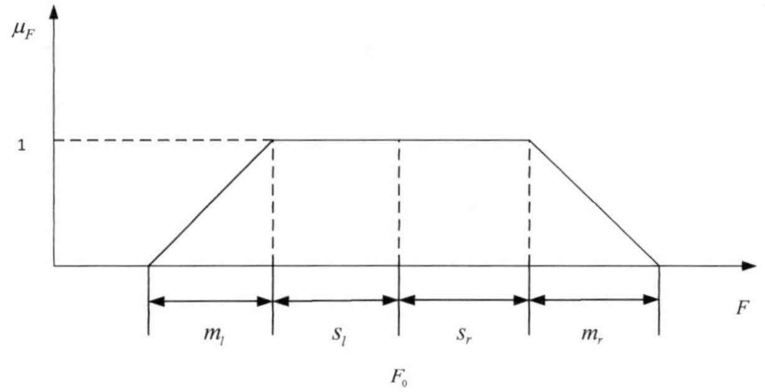


Figure 1. Fuzzy set trapezoidal membership function diagram.

In the concept, $F_0$ is the center that the fuzzy data set supports, $m_l$ as the left fuzzy region, $m_r$ as the right fuzzy region, and $s_l$ as the left supporting radius, $s_r$ as the right supporting radius. As can be seen from Figure 1, the specific expression of membership function $\mu_F$ is:

$$\mu_F=\begin{cases} 0 & 0 \le F \le F_0 - s_l - m_l \\ \dfrac{F-(F_0 - s_l - m_l)}{m_l} & F_0 - s_l - m_l \le F \le F_0 - s_l \\ 1 & F_0 - s_l \le F \le F_0 + s_r \\ \dfrac{F_0 + s_r + m_r - F}{m_r} & F_0 + s_r \le F \le F_0 + s_r + m_r \\ 0 & F_0 + s_r + m_r < F \le 0 \end{cases} \tag{1}$$

When $s_l=s_r=0$, in the trapezoidal membership function, the original membership function will become triangular membership function, and when the left and right supporting radius becomes 0 and the value of the left and right fuzzy area also becomes 0, the fuzzy number will correspondingly change to a certain value. In normal practical applications, the degree of failure of the important components diagnosed is represented by fuzzy values of [0,1] this interval. Then, in the process of detection and diagnosis of our engineering, we usually divide the fault degree of components into three grades, that is, normal (no fault), general fault and major fault. Next, the three fault degrees are represented by 0, 0.5 and 1 fuzzy values respectively. Where, the parameter $s_l=s_r$, $m_l=m_r$.

### *2.2. Language fuzzy processing*

Here, we use Wicken's 7-level judgment theory to express the fuzzy language. We divide the probability of an event into 7 levels, and then combine it with the trapezoidal membership function to express the fuzzy number of the event. Therefore, the fuzzy number corresponding to the fuzzy language is shown in Table 1.

Table 1. Fuzzy numbers corresponding to event fuzzy language

| Probability level | Fuzzy language | Fuzzy number | Probability level | Fuzzy language | Fuzzy number |
|---|---|---|---|---|---|
| VL | Very low | 0,0.1 | FH | Higher | 0.7 |
| L | Low | 0.2 | H | High | 0.8 |
| FL | The lower | 0.3 | VH | Very high | 0.9,1 |
| M | Medium | 0.4, 0.5,0.6 | | | |

## 3. Fault tree analysis method

In the field of Fault diagnosis, Fault Tree Analysis (FTA) is an important research method of reliable Fault diagnosis. It has the characteristics of clear logical reasoning, so it is often widely used at Fault reasoning of important electronic and mechanical equipment system. The biggest advantage of fault tree analysis method is that it can carry out qualitative analysis and quantitative calculation for the faulty equipment components. Because it has the characteristics of clear logical reasoning, it can use reasoning to get the minimum cut set that causes the failure of equipment components, and it can also get the probability of the event of the minimum cut set through mathematical calculation. After the analysis stage of fault tree analysis method, we can provide more rigorous and reliable fault reasoning theory for the subsequent fault reasoning stage, so as to shorten the time of fault reasoning with engineering, so as to achieve the goal of improving the efficiency of fault diagnosis and identification.

### *3.1. The basic concept of fault tree and the representation method of fault tree*

Fault tree is a structure diagram that can represent the causal relationship contained in an event. Fault tree analysis method is an analysis method that uses the model of fault tree and then uses logic to represent the causal relationship in the event in the form of tree structure diagram. The relevant expressions are shown in Table 2.

Table 2. Corresponding symbol representation of fault tree

| Professional term | Symbol | Instructions |
|---|---|---|
| The top event |  | Fault events that need to be diagnosed when a system fault occurs |
| Intermediate events |  | An event that lies between a low-level event and a top-level event |
| The underlying event |  | The basic events that occur in the system |
| With the door |  | All of the underlying basic events happen |
| Or gate |  | There are 1 or more underlying minimum events occurring |

The structural relation diagram of the fault tree is shown in Figure 2. In Figure 2, $X_1$, $X_2$, and $X_3$ low-level events, A is intermediate event, and T represents top-level event. According to the concepts corresponding to gates and or in Table 2, when $X_2$ and $X_3$ in the diagram of fault tree structure occurs at the same time, the intermediate event A will occur. Similarly, when any of the sum events $X_1$ and A occurs, the top event T occurs. Therefore, the faults that need to be treated by fault reasoning are used as the top events of the fault tree. The occurrence of the top events is associated with many intermediate events, and the generation of intermediate events is determined by a large number of low-level events. Therefore, layer after layer of logical reasoning can be completed from top to bottom, and then the necessary and sufficient conditions leading to the occurrence of top events can be finally obtained, so as to form a complete tree structure diagram of logical reasoning relations.

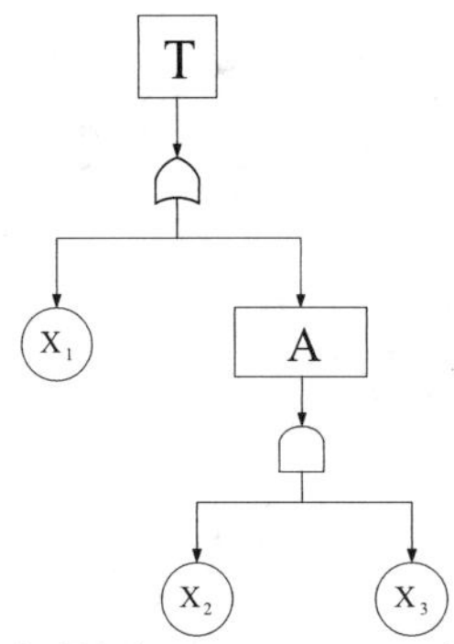


Figure 2. Fault tree structure diagram.

### *3.2. The detailed description of the structural relationship of fault tree*

In the process of equipment fault reasoning, we assume that there are only two operating states in the airport scene monitoring , that is, normal state and fault state. Then there are only two cases of occurrence and non-occurrence of top-level event T and low-level events $X_i$, which are defined as follows:。

$$T=\begin{cases} 1 & \text{Top-level event occurred (failure state)} \\ 0 & \text{Top-level event did not occur (normal state)} \end{cases} \tag{2}$$

$$X_i=\begin{cases} 1 & \text{An underlying event occurs (fault state)} \\ 0 & \text{The underlying event does not occur (Normal state)} \end{cases} \tag{3}$$

In this way, the top-level event T will serve as a function of $X_i$, and the relationship between the top-level event T and the low-level event $X_i$ can be expressed as:

$$\begin{cases} T = T(X) \\ X = X_i \quad i = 1,2,...,n \end{cases} \tag{4}$$

With the gate and or gate are the two logical structures used most often in the fault tree model. The meaning represented by and gate logical structure is that only when all the bottom events happen, the top event will happen; The meaning represented by the or gate logical structure is that the top-level event will occur if one or more low-level events occur among all low-level events. In the fault tree model, the logical relationship between the logical structure of the gate and or the logical structure of the gate can be expressed as:

$$\begin{cases} T(X)=\prod_{i=1}^{n} X_i & (\text{With the door}) \\ T(X)=\sum_{i=1}^{n} X_i & (\text{Or gate}) \end{cases} \tag{5}$$

## 4. T-S model is combined with fuzzy theory

The traditional fault tree analysis method commonly used by us has some defects in itself, which may not to get the desired results when we use it. For example, in the process of analysis and diagnosis, it often fails to accurately describe the fault state and fault degree, the causal relationship between faults cannot be accurately determined, and the probability of occurrence of underlying events is inaccurate, etc. Because of these defects, the traditional fault tree model focuses on qualitative analysis, but it is difficult to achieve ideal results in quantitative calculation. In view of this, here, we combine T-S model and fuzzy theory to model, and then integrate the new model into the fault tree analysis method, optimize the traditional fault tree analysis method, so as to achieve the goal of using fuzzy number to represent the fault degree of components. T-S fuzzy gates are used to replace and gates and or gates mentioned above to represent the relationship between each level of faults, so as to complete the qualitative analysis and quantitative calculation of fault reasoning by fault tree analysis method.

### *4.1. Establish T-S fuzzy door fault tree model*

Given the degree of fault condition and fault accurately describe, failure and failure of cause and effect relationship between each other could not be determined accurately, the underlying event probability problems such as inaccurate, and above, using fuzzy possibility, we can be a description of component failure probability with fuzzy possibility to say, at the same time, Fuzzy number is used to represent the fault degree of components, and then fuzzy gates are used to replace the traditional fuzzy gates mentioned above to represent the relationship between each level of events, so as to establish a new fault tree analysis method structure -- T-S fault tree structure. In this way, we can achieve the goal of complete reasoning from the occurrence of low-level events to the occurrence of top-level events, as shown in FIG. 4.3, $G_1$ and $G_2$ are T-S fuzzy gates, $X_1$, $X_2$ and $X_3$ represent low-level events.

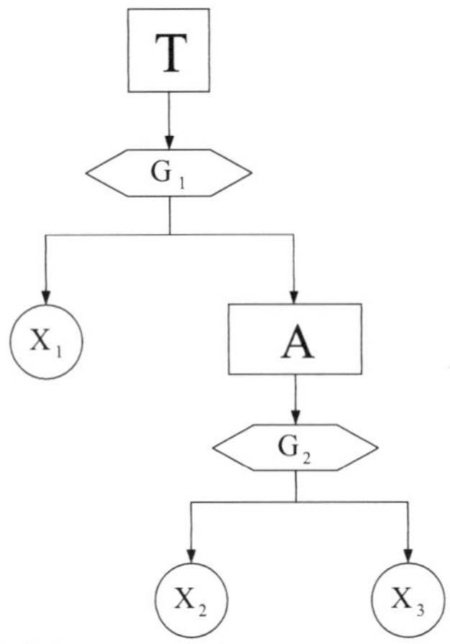


Figure 3. T-S fuzzy gate fault tree structure diagram.

T-S model can be used to represent the relationship between events, which is similar to a kind of approximator, including a series of fuzzy language rules, which are used to describe the relationship between events. T-S fuzzy gate is described as follows:

Rule $l$=($l$=1,2,…,$m$), if $x=\{x_1,x_2,\ldots,x_n\}$ is the pre-event variable, y as the post-event variable, $A_{lj}$ ($j$=1,2,…,$n$)is the fuzzy set, if $x_1$ is $A_{l1}$, $x_2$ is $A_{l2}$, $x_n$ is $A_{ln}$, then y is $y^l$. If the membership function of $A_{lj}$ is $\mu_{A_{lj}}\left(x_j\right)$, then the output of T-S model is:

$$y=\prod_{j=1}^{n}\mu_{A_{lj}}\left(x_{lj}\right)/\sum_{j=1}^{n}\prod_{j=1}^{n}\mu_{A_{lj}}\left(x_j\right)y^{(l)} \tag{6}$$

### *4.2. Possibility analysis of T-S fuzzy door fault tree*

Set $x=\{x_1,x_2,\ldots,x_n\}$ as the basic event, $y$ as the superior event, and then apply the fuzzy set $\left\{x_1^{(1)},x_1^{(2)},\ldots,x_1^{(k_2)}\right\},\ldots,\left\{x_1^{(1)},x_1^{(2)},\ldots,x_1^{(k_n)}\right\}$ and $\left\{x_1^{(1)},x_1^{(2)},\ldots,x_1^{(k_y)}\right\}$ to describe the fault degree of $x$ and $y$, among which:

$$\begin{cases} 0\le x_1^{(1)}<x_1^{(2)}<\ldots<x_1^{(k_1)}\le 1 \\ 0\le x_2^{(1)}<x_2^{(2)}<\ldots<x_2^{(k_2)}\le 1 \\ \qquad\ldots \\ 0\le x_n^{(1)}<x_n^{(2)}<\ldots<x_n^{(k_n)}\le 1 \\ 0\le y^{(1)}<y^{(2)}<\ldots<y^{(k_n)}\le 1 \end{cases} \tag{7}$$

The relevant rules of fuzzy gates are described as follows:

Rules $l$=($l$=1,2,…,$m$), if $x_1$ is $x_1^{(i_1)}$, $x_2$ is $x_2^{(i_2)}$ ,..., $x_n$ is $x_n^{(i_n)}$, then the possibility of $y$ is $y^{(2)}$ is $p^{(l)}\left(y^{(2)}\right)$,..., the possibility of y is $y^{(k_y)}$ is $p^l\left(y^{(k_y)}\right)$, where $i_1$=1,2,…,k$_1$, $i_2$=1,2,…,k$_2$; ... ; $i_n$=1,2,…,k$_n$.

The total number of rules is:

$$m=\prod_{i=1}^{n} k_i \tag{8}$$

If the probability of everything happens in terms of $p\left(x_1^{(i_1)}\right),\dots,p\left(x_n^{(i_n)}\right)$ then the possibility of the rule $l$ is:

$$p_0^l=p\left(x_1^{(i_1)}\right)p\left(x_2^{(i_2)}\right)\dots p\left(x_n^{(i_n)}\right) \tag{9}$$

So, the $y$ 's fuzzy possibility is:

$$\begin{cases} p\left(y^{(1)}\right)=\sum_{l=1}^{m} p_0^{(l)} p^{(l)}\left(y^{(1)}\right) \\ p\left(y^{(2)}\right)=\sum_{l=1}^{m} p_0^{(l)} p^{(l)}\left(y^{(2)}\right) \\ \dots \\ p\left(y^{(k_y)}\right)=\sum_{l=1}^{m} p_0^{(l)} p^{(l)}\left(y^{(k_y)}\right) \end{cases} \tag{10}$$

If the state of the antecedent $x=\{x_1,x_2,\dots,x_n\}$ is known as $x'=\{x_1',x_2',\dots,x_n'\}$, then the $y$ 's fuzzy possibility we can obtain according to the T-S fuzzy structure model is:

$$\begin{cases} p\left(y^{(1)}\right)=\sum_{l=1}^{m} \beta_l^*\left(x'\right) p_0^{(l)} p^{(l)}\left(y^{(1)}\right) \\ p\left(y^{(2)}\right)=\sum_{l=1}^{m} \beta_l^*\left(x'\right) p_0^{(l)} p^{(l)}\left(y^{(2)}\right) \\ \dots \\ p\left(y^{(k_y)}\right)=\sum_{l=1}^{m} \beta_l^*\left(x'\right) p_0^{(l)} p^{(l)}\left(y^{(k_y)}\right) \end{cases} \tag{11}$$

$$\beta_l^*\left(x'\right)=\prod_{j=1}^{n}\mu_{x_j}\left(i_j\right)\left(x_j'\right)/\sum_{i=1}^{r}\prod_{j=1}^{n}\mu_{x_j}\left(i_j\right)\left(x_j'\right) \tag{12}$$

In the formula, $\mu_{x_j}\left(i_j\right)\left(x_j'\right)$ represents the fuzzy set membership degree corresponding to the fault degree of the fi $j$ component in the $l$ rule.

If we know the basic events of all kinds of fault degree, and the corresponding fuzzy possibility, then we can of the t-s fuzzy fault tree and the corresponding T-S fuzzy rules to door by formula (10) and formula (11) is calculated to get the correlation among the various events of all kinds of fuzzy probability, then, in the same way, We can then combine the respective fuzzy possibilities corresponding to various fault degrees of intermediate events with T-S fuzzy gate rules, and then use Formulas (10) and (11) to calculate the respective fuzzy possibilities corresponding to various fault degrees of top events.

## 5. Reasoning analysis of magnetron fault diagnosis for scene surveillance radar

*5.1. General flow of magnetron fault reasoning*

FIG. 4 shows the general flow chart of reasoning analysis for magnetron fault diagnosis of scene surveillance radar, and the specific steps are as follows:

(1) Firstly, the fault tree of magnetron fault inference event is established;

(2) The magnetron fault event is established as the top event;

(3) T-S fuzzy fault tree structure is constructed according to each event code;

(4) Input the expert opinions obtained in the project site and convert them into fuzzy number language set;

(5) Establish fuzzy membership function;

(6) Through T-S fuzzy fault tree, the fuzzy possibility is calculated from bottom to top by using Formula (9) and Formula (10).

(7) Verify the calculated results manually;

(8) If there is any discrepancy with the actual fault result, remove it manually; Return to the fuzzy possibility calculation process again and repeat (6)-(8).

Next, we will introduce the specific process of constructing fault tree of magnetron fault inference event.

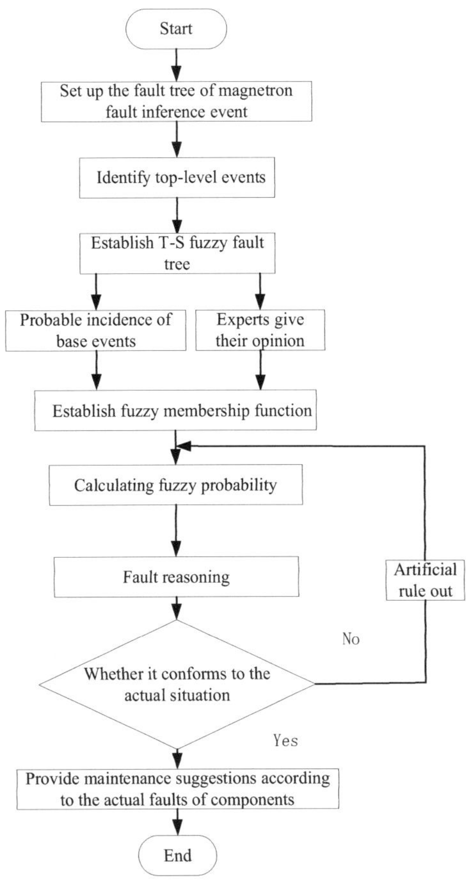


Figure 4. The fault reasoning flow chart of the airport surface surveillance radar magnetron based on T-S fuzzy fault tree.

### *5.2. Establish T-S fuzzy fault tree structure reasoning model*

#### *5.2.1. Formulate relevant rules for T-S fuzzy doors*

Here, we assume that the basic events only contain two failure states, fault and normal, and that each thing exists independently from each other and is not interfered by various external factors. We take it as a precondition, and T-S fuzzy door fault tree model is based on this precondition to determine the corresponding top event, intermediate event and basic event successively. The codes of each event are shown in Table 3.

Table 3. Magnetron fault tree event codes of airport surface surveillance radar

| Code | Events | Code | Events |
|---|---|---|---|
| T | The magnetron device is faulty | $X_1$ | The modulator is faulty |
| $A_1$ | Poor anti-high pressure impact device | $X_2$ | No high pressure impact protection device |
| $A_2$ | Signal failure | $X_3$ | Quality defects |
| $A_3$ | Short circuit leakage | $X_4$ | Error |
| $A_4$ | Waveguide damage | $X_5$ | Unqualified inspection |
| $A_5$ | Filament interrupt | $X_6$ | Poor contact |
| $A_6$ | Failure of high voltage impact protection device | $X_7$ | Failure of high pressure impact protection device |

Next, set the fuzzy number set of common faults of $A_1$~$A_6$ and $X_1$~$X_7$ as (0,0.5,1), membership function $s_l$=$s_r$=0.1, $m_l$=$m_r$=0.3， and according to the rules of T-S fuzzy gate, each line in table 4 and table 5 respectively represents a fuzzy rule, as the rule represented by line 6 in table 4 is: When $X_3$=0, $X_4$=0, the probability of the intermediate event $A_3$=0 is 0, the probability of $A_3$=0.5 is 0, and the probability of $A_3$=1 is 1. The rule represented in line 5 of Table 5 is: when the base event $X_7$=0.5, $X_2$=0.5, then probability $A_4$=0 is 0.3, probability $A_4$=0.5 is 0.6, and probability $A_4$=1 is 0.1. In this way, the corresponding rules of other T-S fuzzy gates can be set.

Table 4. T-S $G_3$

| Rules | $X_3$ | $X_4$ | $A_3$ | | |
|---|---|---|---|---|---|
| | | | 0 | 0.5 | 1 |
| 1 | 0 | 0 | 1 | 0 | 0 |
| 2 | 0 | 0.5 | 0.2 | 0.6 | 0.2 |
| 3 | 0 | 1 | 0 | 0 | 1 |
| 4 | 0.5 | 0 | 0.5 | 0.3 | 0.2 |
| 5 | 0.5 | 0.5 | 0.1 | 0.5 | 0.4 |
| 6 | 0.5 | 1 | 0 | 0 | 1 |
| 7 | 1 | 0 | 0 | 0 | 1 |
| 8 | 1 | 0.5 | 0 | 0 | 1 |
| 9 | 1 | 1 | 0 | 0 | 1 |

Table 5. T-S $G_4$

| Rules | $X_7$ | $X_2$ | $A_4$ | | |
|---|---|---|---|---|---|
| | | | 0 | 0.5 | 1 |
| 1 | 0 | 0 | 1 | 0 | 0 |
| 2 | 0 | 0.5 | 0.2 | 0.6 | 0.2 |
| 3 | 0 | 1 | 0 | 0 | 1 |
| 4 | 0.5 | 0 | 0.4 | 0.4 | 0.2 |
| 5 | 0.5 | 0.5 | 0.3 | 0.6 | 0.1 |
| 6 | 0.5 | 1 | 0 | 0 | 1 |
| 7 | 1 | 0 | 0 | 0 | 1 |
| 8 | 1 | 0.5 | 0 | 0 | 1 |
| 9 | 1 | 1 | 0 | 0 | 1 |

*5.2.2. Establish T-S fault tree structure diagram*

Build the structure diagram we need from the fault tree event code, as shown in Figure 5.

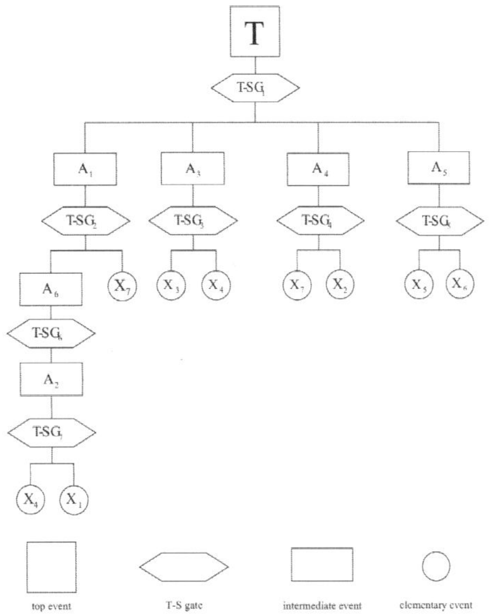


Figure 5. T-S fuzzy fault tree structure diagram.

*5.3. Analyze the fuzzy possibility*

The radar magnetron of an airport in Urumqi, Xinjiang is taken as the research object. Assume that all basic events of $X_1 \sim X_7$ the fault degree and fuzzy number is 0.5 for all the fuzzy number 1, according to a xinjiang urumqi airport scene surveillance radar transceiver system instructions as well as the related data of the magnetron, coupled with the scene of the two experienced experts on-site operation and maintenance for all of our basic events happened to check the possibility of scoring, Finally, the possible incidence rate of basic events with failure degree 1 is shown in Column 2 of Table 4.6. In order to improve the accuracy of the data we obtained, fuzzy numbers are used here to represent the failure rate of all events, as shown in column 3 of Table 6.

Table 6. Failure rates for basic events

| Basic event | Failure rate | Fuzzy language membership |
|---|---|---|
| $X_1$ | 0.01462 | (0.00147,0.00136,0.00164,0.00153) |
| $X_2$ | 0.00357 | (0.00171,0.00182,0.00176,0.00211) |
| $X_3$ | 0.00832 | (0.00154,0.00165,0.00177,0.00186) |
| $X_4$ | 0.02790 | (0.00401,0.00470,0.00530,0.00609) |
| $X_5$ | 0.00421 | (0.00702,0.00810,0.00970,0.02001) |
| $X_6$ | 0.00090 | (0.00209,0.00610,0.00122,0.00123) |
| $X_7$ | 0.03168 | (0.00483,0.00360,0.00640,0.00716) |

Then, the fault incidence rate of basic events obtained in Table 6 was substituted for the T-S fault tree model established by us. Then, the fuzzy possibility of the event $A_1 \sim A_6$ was calculated according to formula (9) and Formula (10) introduced in the paper, and the fuzzy possibility was used to represent the fault occurrence probability of the event. Here, We take the specific calculation process of fuzzy possibility corresponding to each fault degree of intermediate event $A_3$ as a specific example, as shown in Formula (13) :

$$\left.\begin{array}{l}P\{A_3=0\}=\sum_{l=1}^{9}P_0^lP^l(A_3=0)=\\P_0^1P^1(A_3=0)+P_0^2P^2(A_3=0)\\+P_0^4P^4(A_3=0)+P_0^5P^5(A_3=0)=\\(0.81042,0.67385,0.70286,0.76345)\\P\{A_3=0.5\}=\sum_{l=1}^{9}P_0^lP^l(A_3=0.5)=\\P_0^2P^2(A_3=0.5)+P_0^4P^4(A_3=0.5)\\+P_0^5P^5(A_3=0.5)=\\(0.08103,0.02161,0.12331,0.16120)\\P\{A_3=1\}=\sum_{l=1}^{9}P_0^lP^l(A_3=1)=\\P_0^2P^2(A_3=1)+P_0^3P^3(A_3=1)\\+P_0^4P^4(A_3=1)+P_0^5P^5(A_3=1)\\+P_0^6P^6(A_3=1)+P_0^7P^7(A_3=1)\\+P_0^8P^8(A_3=1)+P_0^9P^9(A_3=1)=\\(0.11486,0.21563,0.14355,0.16658)\end{array}\right\}\tag{13}$$

According to the above fuzzy possibility results, the fuzzy possibility corresponding to various faults of intermediate events $A_3$ is a trapezoidal fuzzy value, which is very small in the case of general faults or major faults, and very high in the case of normal state. In comparison, the fuzzy possibility of major faults is higher than that of general faults. Therefore, the conclusion obtained by T-S fuzzy fault tree inference is the maximum possibility that this intermediate event $A_3$ does not occur. Similarly, fuzzy possibilities corresponding to various fault states of other intermediate events can be obtained.

Then we can deduce and calculate from bottom to top by using this method, and the fuzzy possibilities corresponding to various fault states of magnetron fault events as top events can be obtained as follows:

$P(T=0)=(0.81742,0.68270,0.617828,0.58251)$

$P(T=0.5)=(0.01694,0.01671,0.016338,0.01413)$

$P(T=1)=(0.16351,0.20146,0.24701,0.28403)$.

It can be seen from the above fuzzy possibility results that the failure of the magnetron during operation is a small probability event, but if the failure occurs, the probability of major failure is higher than the probability of general failure. Based on the inference results we can obtain the real-time fault degree of the magnetron and the intermediate and basic events which are most likely to cause the fault. Finally, it is found that the result of inference calculation meets the requirements through manual investigation, which saves for half an hour than the usual sequential investigation process, and verifies the effectiveness of the method.

## 6. Conclusion

After analyzing the characteristics of radar component magnetron with short operation time, this paper proposes a magnetron fault reasoning model based on T-S fuzzy fault tree. The test verifies that this method can reduce the time cost of traditional fault elimination methods of engineering. The main conclusions are as follows:

(1). Fuzzy language is used to replace the fault data onto traditional fault reasoning, which reduces the need for a large number of historical fault data and makes the fault reasoning process of components more convenient and efficient;

(2). The T-S fuzzy fault tree reasoning model can be used to analyze the fault events of the magnetron more comprehensively and effectively. Not only can the maximum probability cause of the fault be obtained, but also the fault degree of the maximum probability of the magnetron can be obtained;

(3).The proposed model is applied to the magnetron of a scene surveillance radar in Urumqi, Xinjiang, for fault reasoning. The reasoning results prove the effectiveness, comprehensiveness and high efficiency of the proposed method.

**References**

[1] Chen Le, Wang Xian-lin, Li Wei-fei, WANG Kai, Liu Kai-le. Tool holder system based on t-s fuzzy fault tree reliability analysis [J]. Journal of combination machine tools and automatic processing technology, 2019 (02) : 143-146. The DOI: 10.13462 / j.carol carroll nki mmtamt. 2019.02.038.

[2] Wu Chen, Guohua Zhang, Hao Wang, Guoqiang Zhong, Chengtang Wang. Based on t-s fuzzy fault tree of drilling and blasting method construction of tunnel collapse probability evaluation [J]. Rock and soil mechanics, 2019, 40 (S1) : 319-328. The DOI: 10.16285 / sm j.r. 2018.1712.

[3] Deng Na, Wen Hai. Tree Analysis of Landslide risk in Topographic Biased Tunnel Construction [C]// Proceedings of the 2015 National Engineering Geology Academic Conference.,2015:351-358.

[4] Li Feng, YUAN Zhikai, He Zhenxin, Wang Hao. Reliability Analysis of Truck Crane Outrigger Hydraulic Circuit Based on T-S Fuzzy Fault Tree Model [J]. Machine tool & hydraulics,2018,46(07):160-163.

[5] Zhao Haiming, Xiong Zhihong, Zeng Lei, Zhang Huan. Research on fault tree analysis method of hydraulic cylinder based on fuzzy set theory [J]. Journal of hefei university of technology (natural science),2016,39(02):150-155.

[6] Nan Yuhong, Zhao Song, Yang Zhengmao. Risk assessment of tunnel collapse based on accident tree [J]. Road machinery & construction mechanization,2017,34(07):106-110.

[7] Sun Lu. Application of field surveillance radar in airport [J]. Electronic technology and software engineering, 2019 (18): 93-94

[8] He Gang. Case study of Terma surface surveillance radar maintenance in Urumqi [j]. Scientific and technological innovation and application, 2019 (26): 58-60

[9] Li Xinxi. Overview of the civil aviation airport surface monitoring system [j]. China new communication, 2020,22 (09): 67

[10] Yang Baojun, Hong Rongjing, Pan Yubin. Fault diagnosis of wind turbine Gearbox based on CEEMdan-EFica denoised [J]. Modular Machine Tool and Automatic Processing Technology , 2020 (02) :115-118+122.

# Research and Implementation of Containerization for Virtual Instrumentation Measurement

**Wang Guo**[1, a]

[1]School of Physics and Electronics, Central South University, Changsha, Hunan, China

[a]email: vanguo@csu.edu.cn

**Abstract.** The instrumentation measurement system has fruitful achievements in various fields with cloud computing technology. However, the instrumentation cloud needs more PaaS (Platform-as-a-Service) capabilities to improve development and IT operations efficiency by expanding distributed measurement systems. In recent years, container and cloud-native technology growth have brought developing chances for instrumentation cloud. We propose a novel container and instrumentation communication architecture to apply container technology to virtual instrumentation. Firstly, we analyze the critical technologies of Container-Instrumentation communication architecture and give the solution of its implementation. Secondly, we develop and deploy the web backend and frontend to visualize measurement data with Kubernetes. Finally, completing the IV sweep measurement on the web demonstrates the feasibility of building the measurement system in a cloud-native way. So, the container-instrumentation architecture provides practical significance for constructing measurement systems and improving the instrumentation cloud capabilities with cloud-native architecture in the future.

## 1. Introduction

With the continuous development of cloud computing technology, networked virtual instrumentation measurement systems and remote labs have various applications in several fields[1-3]. However, with the progressive expansion of the virtual instrumentation measurement system scale, the cloud platform of the system will face more challenges, which need to have wider compatibility and more scalability with fast-responding and occupy fewer computing resources at the same time. He Hengjing proposed system architecture for an instrumentation virtualization technology based on the IaaS (Infrastructure-as-a-Service) model, which can not meet the requirements of low resource consumption and high response time with virtual machine (VM)[4]. Because of the lightweight and finer-grained resource isolation, container technology enables a PaaS (Platform-as-a-Service) cloud platform for efficient resource scaling and system maintenance with low resource consumption, which are the instrumentation cloud features delicate to optimize[5-7].

To enhance the PaaS capabilities such as resource scheduling, resource monitoring, and container orchestration, we committed to giving a solution to apply the container technology to virtual instrumentation. Here we analyze the communication methods between measurement service (MS) and instrumentation in networked instrumentation virtualization and propose the Container-Instrumentation communication architecture. Besides, we use Golang and React as the web backend and frontend to build the online measurement web for verifying the architecture and attempts to implement the measurement system in a cloud-native way. After the declarative API of Kubernetes is used to deploy the system components, users can make measurements work on the web, and developers manage the system on the Kubesphere console. Based on the work of this paper, we hope future work can make a better user experience for users and more efficient in developing and managing the measurement system for developers with a cloud-native PaaS platform.

## 2. The Proposal of Container-Instrumentation communication architecture

### *2.1 Instrumentation Virtualization*

The emergence of Virtual Instrumentation (VI) technology is a milestone in the progress of instrument virtualization from traditional to intelligent[8]. With the advancement of computer network technology, instrumentation can be further abstracted as "instrumentation on the Internet." based on the remote procedure call (RPC) protocol. As is shown in figure 1-a, and the physical image is shown in figure 1-b, the client sends remote connection requests for instrumentation through the VISA library, which reaches the instrumentation server according to the IP address. In short, the Internet decoupled the MS and instrumentation, and the networked instrumentation is more flexible and efficient than traditional measurement.

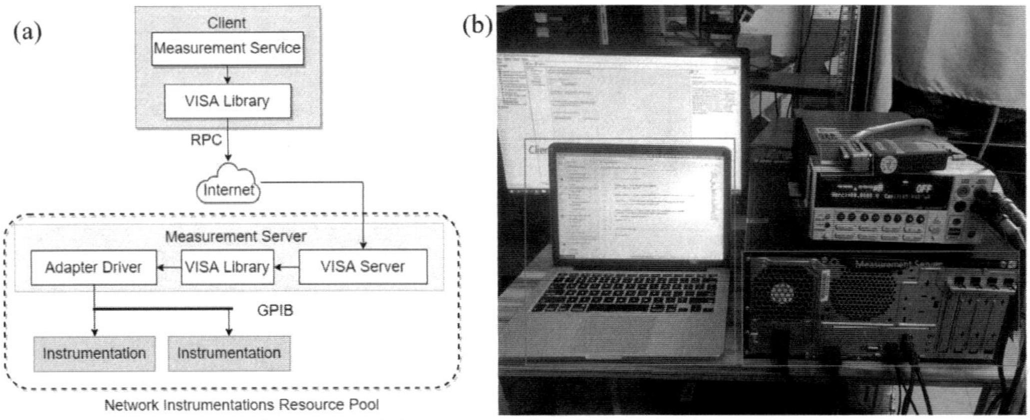


Figure 1 RPC instrument virtualization architecture

### *2.2. Analysis of Container-Instrumentation communication architecture*

Based on the RPC instrument virtualization architecture, a client connecting to the instrument is abstracted as the MS connecting to the VISA library. Therefore, the critical technology of communication between MS container and instrumentation is that the container accesses the VISA library and the VISA driver in the host[9]. So, there are two conditions for MS to communicate with instruments, *condition 1*: The MS in the container calls the VISA library in the host OS, *condition 2*: The container accesses the kernel module through the I/O driver interface in the VISA library to send the connection request. The following subsection will provide MS container and instrumentation communication principles and critical technologies, including container data volumes and container kernel permissions.

#### *2.2.1. container data volumes.*

The Docker data volume is designed to mount files or folders from the host computer to the container independently of the container union file system. To meet *condition 1*, the MS container can call the VISA library of the host with mounting the directory */usr/lib/x86_64-linux-gnu* of the VISA library and */etc* of the VISA passport configuration in the host to the same path of the container.

### *2.2.2. container kernel permissions*
Because of the Linux capabilities mechanism, the container kernel permissions have more fine-grained division. Therefore, the mechanism can significantly reduce the system kernel's security provisioning surface. To satisfy *condition 2*, the MS container needs to add the capacity of load/unload kernel modules.

### *2.3. Container-Instrumentation communication architecture*
Summary, the container-instrument communication architecture on the Kubernetes is shown in figure 2-a and its physical image is shown in figure 2-b. The pod is the most basic control object of the Kubernetes and is a group of containers with a specific IP address. From this perspective, a pod is considered a "logical host" in the Kubernetes, and the MS containers are processes running in the "logical host". The following paper adopts a "bottom-up" approach to implement the container-instrument communication architecture.

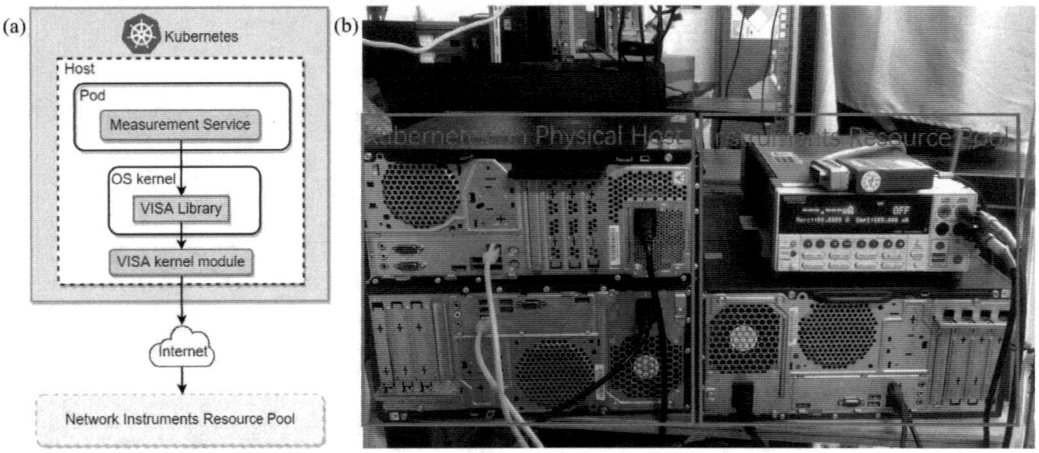

Figure 2 Container-Instrumentation communication architecture

### *2.4 Building web for testing the architecture*
To visualize the measurement data from measurement service, gRPC and WebSockets protocol is for bidirectional data transmission in real-time. Considering building the system more cloud-native way, Golang is used to create the web backend and React for the web frontend.

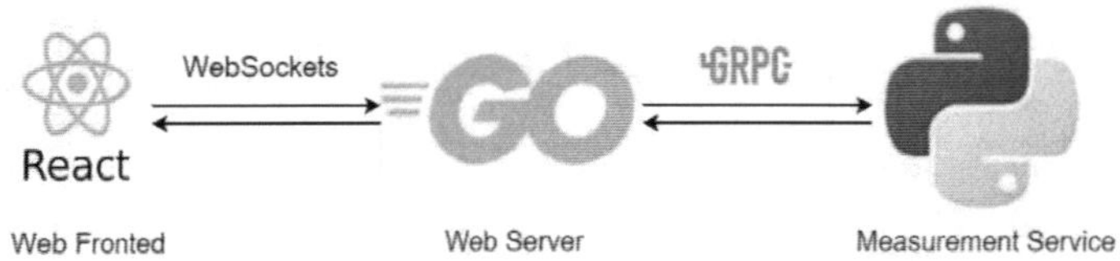

Figure 3 The solution for testing the architecture

## 3. The Implementation of Container-Instrumentation communication architecture

### *3.1 Construction of infrastructure*
The infrastructure includes hardware and software. The hardware is Keithley 2400 source meter and 10U and 20G computer for the cloud platform. The software is CentOS-7.9.2009 and the 3.10.0 kernel version for the cloud platform operating system, where all install the 2022 Q1 version of the NI-VISA kernel module. The instrument server installs the NI-VISA suite and starts VISA Server as the instrumentation server. As is shown in figure 4, the instrumentation container cloud platform is built on

Journal of Physics: Conference Series **2264** (2022) 012023 doi:10.1088/1742-6596/2264/1/012023

Kubesphere, a distributed operating system for cloud-native application management based on Kubernetes.

Figure 4 The Kubesphere console of instrumentation cloud

*3.2 Construction of MS container*

The MS is built on Python 3.6.8, PyVISA 1.11.3, and Docker 20.10.12. the Dockerfile is shown in Figure 5. Docker builds the MS as a container image in the development environment and pushes it to the image registry. After completing the first preparation stage, the next part uses Kubernetes to achieve the deployment and the electronic device measurement.

```
FROM centos:centos7.9.2009
WORKDIR /app
COPY . .
RUN yum install -y python3 python3-devel && \
   pip3 install --upgrade setuptools && \
   pip3 install --upgrade pip && \
   pip3 install -r requirements.txt
CMD [ "python3", "server.py"]
```

Figure 5 The MS Dockerfile

*3.3 Deploy the MS using Declarative API*

Figure 6-a is the specific content of the MS Kubernetes Deployment API. The key fields that determine the MS operation will be emphasized. *spec.container.security.Context.capabilities.add* field is the list of added container capabilities. Adding *SYS_MODULE* allows the container to load the host kernel module. *spec.container.volumeMounts* field is the storage volume mounted by the container, where *visa-lib* and *visa-config* represent the VISA library and VISA passport configuration, respectively. *spec.container.volumes* field defines the kind of data volumes mounted by the container. The data volume type is *hostPath* because the driver files of the host are to be shared. Golang web server Deployment is shown in figure 6-b, almost the same as the React frontend Deployment.

After all pods are in Running status and successfully assigned cluster IP addresses, the container data volume of *hostPath* type is successfully created and mounted in the /usr/lib/x86_64-linux-gnu and /etc directories. The container can access the VISA library in the host.

a)

```yaml
apiVersion: apps/v1
kind: Deployment
metadata:
  ...
spec:
  replicas: 2
  ...
  template:
    ...
    spec:
      containers:
        - name: container-t7sxes
          imagePullPolicy: IfNotPresent
          image: registry.gitlab.com/vanguo996/mes-backend-pyvisa
          volumeMounts:
            - mountPath: /usr/lib/x86_64-linux-gnu
              name: visa-lib
              readOnly: true
            - mountPath: /etc
              name: visa-config
              readOnly: true
          securityContext:
            capabilities:
              add: ["SYS_MODULE"]
      imagePullSecrets:
        - name: gitrepo
      volumes:
        - hostPath:
            path: /usr/lib/x86_64-linux-gnu
            type: Directory
          name: visa-lib
        - hostPath:
            path: /etc
            type: Directory
          name: visa-config
```

b)

```yaml
apiVersion: apps/v1
kind: Deployment
metadata:
  ...
spec:
  replicas: 1
  ...
  template:
    ...
    spec:
      containers:
        - name: container-t7sxes
          imagePullPolicy: IfNotPresent
          image: registry.gitlab.com/vanguo996/mes-backend-go
      imagePullSecrets:
        - name: gitrepo
```

Figure 6 The MS and web server Deployment

*3.4 The result of IV sweep measurement*

The Device Under Test (DUT) is a 10K resistor with IV sweep mode, which has a starting voltage of -2 V, an abort voltage of 2 V, a step of 0.5 V, current compliance of 0.1 A, and measurement speed of 5. As shown in figure 7, the User inputs the parameters for measurement and hits the start button, and then Keithley 2400 starts the IV Sweep Mode. Simultaneously, the web camera shows the real-time status of instrumentation.

*3.5 Discussion*

From the above, we can conclude that developers need to pay attention to the OS and kernel version selection of the container cloud platform host due to the NI Linux driver support limitation in version 2022Q1[10]. The container will be more lightweight because of mounting */etc/ni-visa* instead of the entire */etc* directory from the host. However, it causes the measurement system to throw the *VI_WARN_CONFIG_NLOADED* warning, which cannot find the VSIA Passport module. The IV sweep mode on Kubernetes demonstrates that we can build a containerized instrumentation measurement system based on Container-Instrumentation architecture and use measurement service on the container PaaS cloud platform.

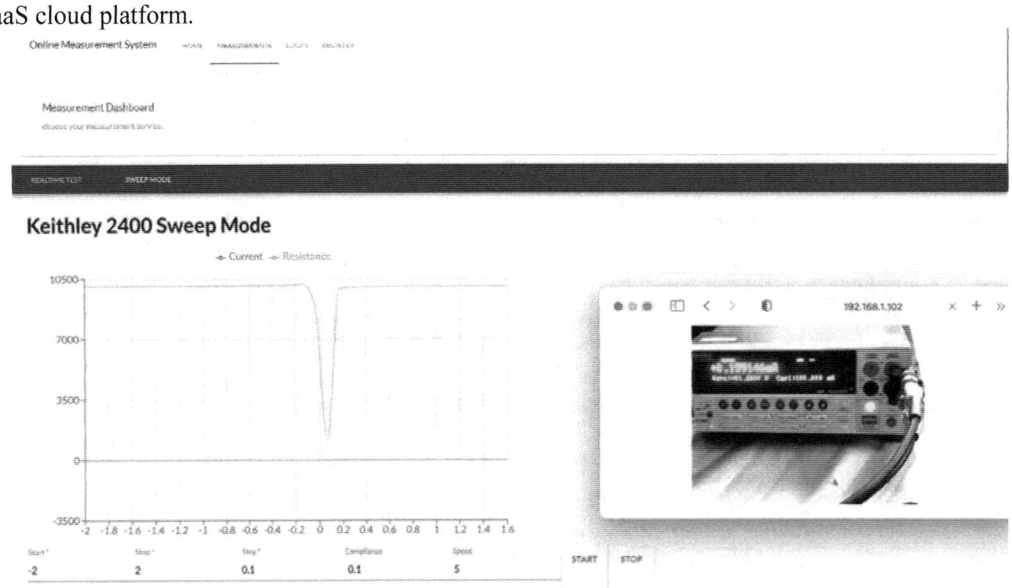


Figure 7 The IV Sweep measurement of 10K resistance on React web

## 4. Conclusion

We propose and implement a technical architecture for communication between MS container and instrumentation. The result shows that users can perform the measurement work online with time and location independent on the Kubernetes cloud platform. This work explains that researchers can build a cloud-native measurement system based on the novel container-instrumentation architecture and manage distributed measurement systems more efficiently using Kubesphere. In the future, developers can take full advantage of the capabilities of Kubesphere to make the instrumentation PaaS cloud more efficient. Such as DevOps[11] to accelerate the software delivery and sidecar model to realize noninvasive microservice governance[12].

## References

[1] Wu, S.C., Liu, T., Xiao, Y., Zhao, W., Zhang, Z.H., Huang, S.L. (2017) A remote on-line detection system for electronic Watt-hour meter based on instrumentation and measurement cloud. Electrical Measurement & Instrumentation, 54: 80-85.

[2] Sierra-Fernández, J.M., Florencias-Oliveros, O., Espinosa-Gavira, M.J., Palomares-Salas, J.C., Agüera-Pérez, A., González-de-la-Rosa, J.J. (2020) Reconfigurable web-interface remote lab for instrumentation and electronic learning. In: 2020 IEEE Global Engineering Education Conference. Porto. pp. 713-717.

[3] Khanafer, M., Shirmohammadi, S., (2020) Applied AI in instrumentation and measurement: The deep learning revolution. IEEE Instrumentation & Measurement Magazine, 23: 10-17.

[4] He, H.J., Zhao, W., Huang, S.L. (2014) Research on the Instrumentation Virtualization in Cloud Computing Environment. Electrical Measurement & Instrumentation, 51: 11-16.

[5] Zhejiang University Software Engineering Lab. (2016) Docker and Kubernetes under the Hood. Post & Telecom Press, Beijing.

[6] Yang, Q.B., Chen, Z.Y., Liu, D., Xia, W.Y., Huang, Y.H., Li, D.P., Di, F.C., Gao, X.Y. (2020) Design and Implementation of Dispatching and Control Cloud PaaS Platform Based on Container. Power System Technology, 44: 2030-2037.

[7] Zhao, Y.F., Lei, S., Zhang, G.G., Geng Y.S. (2021) Design and Development of Container-based Cloud Platform for Power Equipment Simulation. Computer Engineering, 47: 171-177.

[8] Grimaldi, D., Marinov, M. (2001) Distributed measurement systems. Measurement, 30: 279-287.

[9] Wei, H.C., Yang, H.Q., Li, Y.M. (2015) Principles Study on Virtual Instrument Software Architecture (VISA) Resource Management. Measurement & Control Technology, 34: 96-99.

[10] NI (2022) NI Linux Device Drivers Readme. https://www.ni.com/pdf/manuals/378353f.html.

[11] Ebert, C., Gallardo, G., Hernantes, J., Serrano, N. (2016) DevOps. Ieee Software, 33: 94-100.

[12] Song, M., Liu, Q., Haihong, E. (2019) A Mirco-Service Tracing System Based on Istio and Kubernetes. In: 2019 IEEE 10th International Conference on Software Engineering and Service Science. Beijing. pp. 613-616

# Analysis of surveillance system under epidemic conditions

**Quanying Guo[1,a], Mingjia Yang[2,b], Qiang Peng[1*,c] and Zhuoer Wang[4]**

[1] School of Marxism, Wuhan Technology of University, Wuhan, Hubei, 40070, China

[2] School of Marxism, Wuhan Technology of University, Wuhan, Hubei, 40070, China

[1*] Modern Educational Technology and Service Center, Wuhan Technology of University, Wuhan, Hubei, 40070, China

[4] School of Information Engineering, Wuhan Technology of University, Wuhan, Hubei, 40070, China

[a] email: Guo.414@whut.edu.cn

[*] Corresponding author's e-mail: peng@whut.edu.cn

**Abstract.** Under the situation of normalized epidemic prevention and control, how to establish a perfect and intelligent epidemic monitoring system has become an important prerequisite for epidemic prevention and control. The epidemic detection system reminds people to wear masks correctly through target detection technology to reduce the possibility of cross infection in public places; Through the use of original temperature measurement technology, the body temperature of personnel can be detected more quickly and conveniently to isolate patients from the source of epidemic infection; Through developed communication technology and communication means, the information identified by the system is retained and retrieved, and the identification accuracy and detection effect are continuously enhanced. This study systematically analyzes and summarizes the existing epidemic detection systems. According to the different key technologies and the emphasis of design schemes, this study summarizes the design types of "target detection", "original temperature measurement technology", "IOT communication technology", and is committed to exploring an optimization scheme with parallel economic benefits and technical efficiency.

## 1. Introduction

Since the outbreak of the epidemic in 2019, 31 provinces, autonomous regions and municipalities directly under the central government have launched first-class response to major public health emergencies. The spread of New Coronavirus's epidemic is diverse, long incubation period and strong camouflage, which continue to endanger people's lives and health. Aggregated public places and dense crowds are likely to be the channels for the rapid spread of the virus. Therefore, the supervision of the density of clustered people and the wearing of masks is an important way to control the spread of the virus [1]. With the external assistance of the epidemic monitoring system, people began to consciously wear masks, consciously measure body temperature and take the initiative to isolate themselves. With the increasingly severe epidemic situation, the traditional manual detection system has many shortcomings, such as high infection risk, high labor cost and low supervision efficiency. A more intelligent and automatic epidemic monitoring system has important theoretical significance and practical value for the normalized epidemic prevention and control management. People began to consciously look forward to the new answer of epidemic prevention and control. Many scholars at home and abroad have devoted themselves to the research in this field, and the research results are relatively remarkable.

At present, China's epidemic prevention and control war has made significant strategic achievements, and the epidemic prevention and control work has entered the state of normalized prevention and control from the state of emergency. However, China's import pressure of prevention and control abroad continues to increase, and the risk of rebound of domestic epidemic always exists. Therefore, we must constantly improve the normalized prevention and control mechanism. First, normalize the prevention and control work of professional institutions and improve the nucleic acid detection ability of medical and health institutions, disease prevention and control institutions and other professional institutions. Second, ensure the normalization of prevention and control work in key places and key groups [2]. On the basis of ensuring the treatment of patients and the supply of medical resources, the key to the work at this stage is to strictly prevent "human to human" and strictly control large-scale gathering people. How to effectively detect novel coronavirus pneumonia and patients, how to isolate, prevent and control such highly contagious people in advance, how to use technology instead of manpower is the main objective and important content of the epidemic surveillance system. The existing epidemic surveillance system has experienced many tests, and constantly improved its technology and upgrading in the challenges. It is generally equipped with key technologies such as "target detection", "original temperature measurement technology" and "IOT communication technology", and continues to develop in the direction of strong real-time and high accuracy.

## 2. Action mechanism and application effect of epidemic surveillance system

### *2.1. Action mechanism*

*2.1.1. Target detection and data acquisition.*The vision system of the epidemic monitoring system is composed of various sensors, face recognition and sensitive object recognition modules. Image acquisition equipment can carry out face recognition and information acquisition on the target, and obtain video images through cameras or other image acquisition equipment. With the development of intelligence, the vision system can also adjust the distance and angle according to the specific place and monitoring needs to obtain as much monitoring place information as possible [3].

*2.1.2. Information pre-treatment and transmission.* After the base obtains the image information, it often uses the cross platform computer vision library Opencv (Open source computer vision library) for image pre-treatment, and inputs the image as the target detection module. Then, the information is transmitted through IOT communication technology, and the collected data is transmitted to the processor by taking wide area communication in wireless communication as the main way [4].

*2.1.3. Model reasoning.* The core of epidemic surveillance system is the establishment of model algorithm. Take the information collected after image pre-treatment as the input to identify the "person" and "object" in the information. The novel coronavirus pneumonia patients were identified by whether they had a staff aggregation, whether the target personnel wore masks, and whether the temperature of the target personnel was abnormal. Compared with the data of the new crown pneumonia virus carrier in the database, the screening work was conducted to identify the risk persons who had contact with the designated personnel, so as to prevent the spread and spread of the epidemic situation of the high-risk population.

*2.1.4. Result analysis and response.* The output result obtained by substituting the picture information into the model. Often accompanied by data comparison and analysis, the analysis results are transmitted to the base display. When pathogen carriers or risk personnel are found, cooperate to send alarm information and carry out epidemic warning and dangerous personnel screening.

### *2.2. Application effect*
The emergence of epidemic monitoring system is conducive to the development of face recognition system, infrared imaging temperature measurement technology and can bus communication technology. It has derived a series of application products such as forehead temperature gun, face recognition thermometer, AI body temperature early warning system and so on. The epidemic situation monitoring system is conducive to the prevention and control of the emergence and spread of the epidemic situation, timely avoid sudden risks, intelligently monitor the risk of collective infection, improve the efficiency of manual monitoring in public places, and protect the life, health and safety of the people.

*2.2.1. Epidemic prevention and control.* The system monitors the identity of personnel through the face recognition system, compares the database for analysis, and screens and controls the access of dangerous personnel to public places.

*2.2.2. Risk assessment and early warning.* Various temperature measurement techniques judge the temperature data of the target personnel, and record the access time and place. The system will respond quickly after detecting abnormal body temperature data, quickly lock the flow track of "epidemic related" personnel, intercept abnormal personnel and report warnings.

*2.2.3. Information integration.* IOT communication technology can integrate the cloud platform, create an information visualization platform, predict and trace the development trend of the epidemic, locate the space-time collision point according to the diagnosis sequence of patients and close contacts, and then calculate the disease transmission path, so as to provide a theoretical basis for the traceability analysis of infectious diseases.

## 3. Design scheme of existing epidemic surveillance system

### *3.1. Mask wearing detection algorithm based on improved YOLOv3 in complex scenes.*
In terms of mask wearing detection, Wang Yihao and others pay attention to complex factors such as occlusion, dense crowd and small-scale targets in natural scenes. They combined the cross phase local network to improve the DarkNet53 backbone network to reduce the computing consumption and improve the training speed. The improved spatial pyramid pool structure is introduced into the YOLOv3 algorithm, and the multi-scale prediction network is optimized through the top-down and bottom-up feature fusion strategy, so as to realize feature enhancement. CloU is selected as the loss function, and the central point distance, overlap rate and aspect ratio information between the target and the detection frame are considered [5].

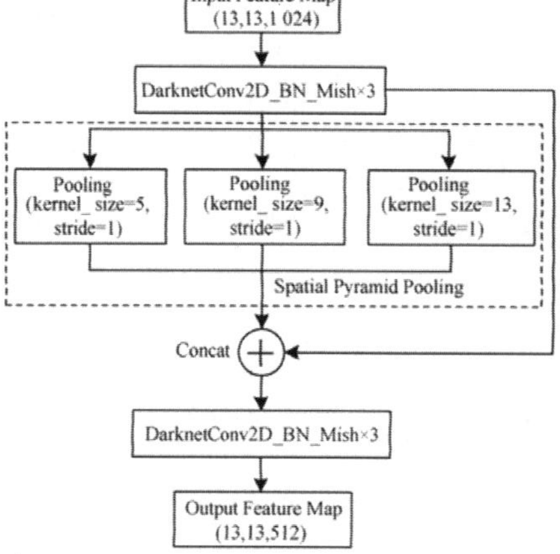


Figure.1 Improved network structure of spatial pyramid pooling

### *3.2. Mask wearing detection method based on YOLO-Mask Algorithm.*

Cao Chengshuo and Yuan Jie also pay attention to the difficulty of mask wearing detection. They draw attention to the mechanism, try to improve on the basis of YOLOv3 algorithm, and propose a detection algorithm for mask wearing detection task. They introduced senet into YOLOv3 basic network and proposed a network based on SEDarknet-53 to improve the channel weight of useful features and suppress the interference of useless features to the model. They take FPN with path aggregation strategy as the YOLO-Mask Algorithm of feature fusion network, strengthen the use of low-level features, and improve the detection effect of the network on small targets and occluded targets. The last test of the article establishes the data set of mask wearing in natural scenes, and carries out algorithm verification and comparative experiments on the data set to verify the feasibility of the detection method [6].

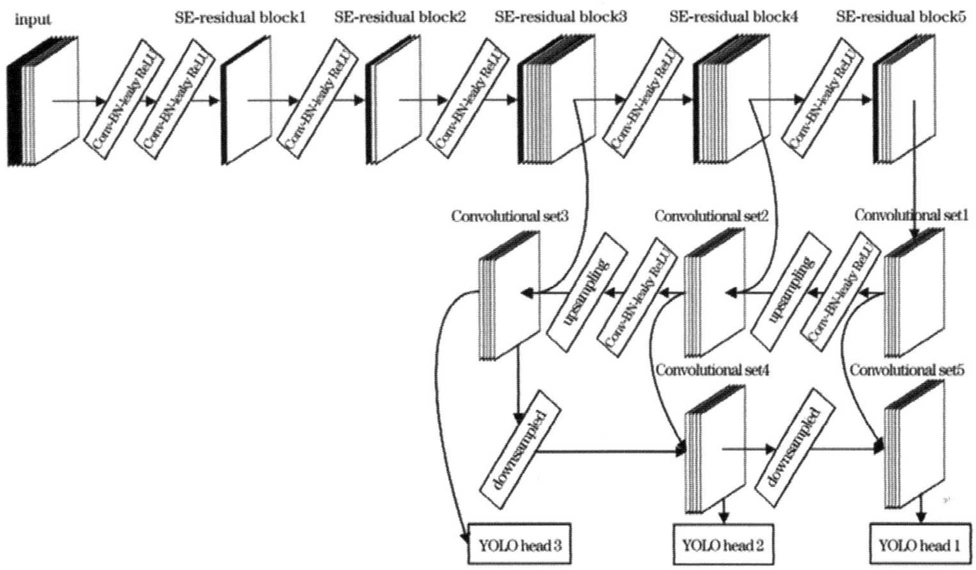


Figure.2 YOLO-Mask algorithm network structure

### *3.3. Improved RetinaFace natural scene mask wearing detection algorithm.*

Niu Zuodong and others believe that the deep learning model for face detection can be applied to the detection task of mask wearing. They improved the RetinaFace algorithm, added the face mask wearing detection task, and optimized the loss function. An improved self-attention mechanism is introduced into the feature pyramid network to enable the feature extraction network to extract effective target information and suppress useless information, so as to enhance the expression ability and context description ability of feature graph. In addition, they made a new data set and made a lot of manual annotation on the data for the training and testing of the model network [7].

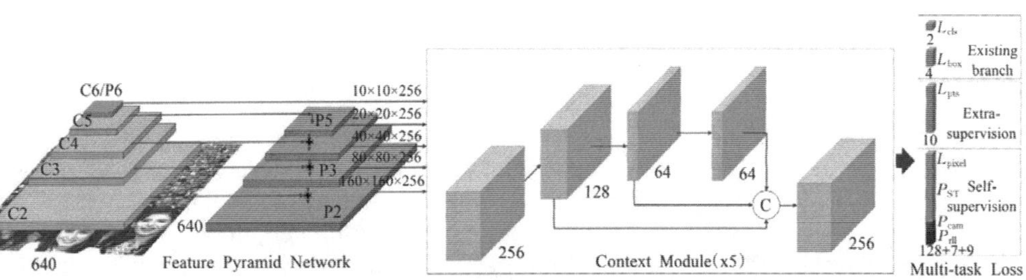


Figure.3 RetinaFace algorithm network structure

*3.4. Symbolic algorithm.*

Zhang Xiubao proposed a mask wearing recognition technology scheme combining attention learning to distinguish the mask wearing area of the face. This scheme consists of two technical modules: face detection with feature fusion and segmentation supervision and mask wearing recognition. The former realizes the function of accurately locating the face region in the image, and the latter uses attention learning to further analyze the face attributes and judge whether the face wears a mask on the basis of a single face region. The scheme adds spatial and channel attention learning mechanism on the basis of ResNet50 to enhance the learning of mask wearing recognition model on mask regional features [8].

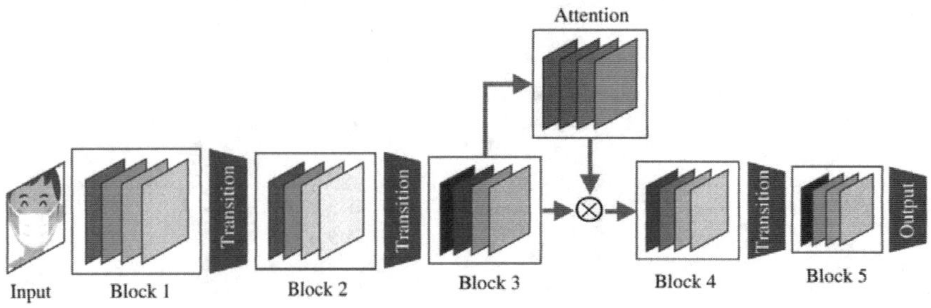


Figure.4 (Color online) Structure diagram of mask recognition model based on Resnet50

*3.5. Mask wearing detection algorithm based on improved SSD.*

Inspired by the FPN network structure, Li Yuyang and others introduced CA (coordinate attention) module to improve the learning and processing ability of the algorithm for detailed features, so that the algorithm can make more effective use of detailed feature information and improve the detection ability of mask wearing errors. By improving the feature extraction and loss function in the SSD target detection framework, they enhanced the feature expression ability of the model, alleviated the problem of category imbalance in the process of model training, and finally improved the detection accuracy of mask wearing state [9].

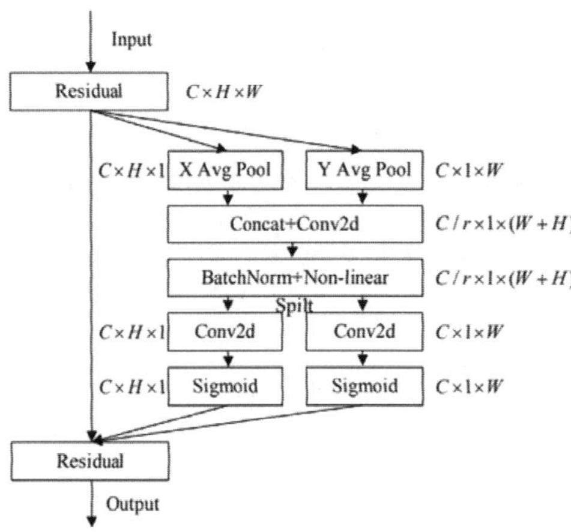


Figure.5 CA module

*3.6. Research on mask wearing detection method based on YOLOv5 enhancement model.*

In order to realize model compression and accelerated detection under multi-scale conditions, Peng Cheng and others proposed a lightweight algorithm based on improved YOLOv5. By replacing some of the feature extraction modules with lighter Ghost Bottle Neck and Shuffle Conv modules, a lighter mask wearing detection framework was realized, which alleviated the problem of computational occupation in the process of feature channel fusion and enhanced the ability of feature expression.

After experimental detection, the improved model has almost no loss in recognition accuracy, and the amount of model parameters and calculations have increased, which reduces the requirements of platform volume, power consumption, storage and computing power. Compared with the existing model, it has obvious advantages, which is more conducive to the deployment of the model in the mobile terminal with limited resources, making it easier to obtain the wearing supervision function of face mask [10].

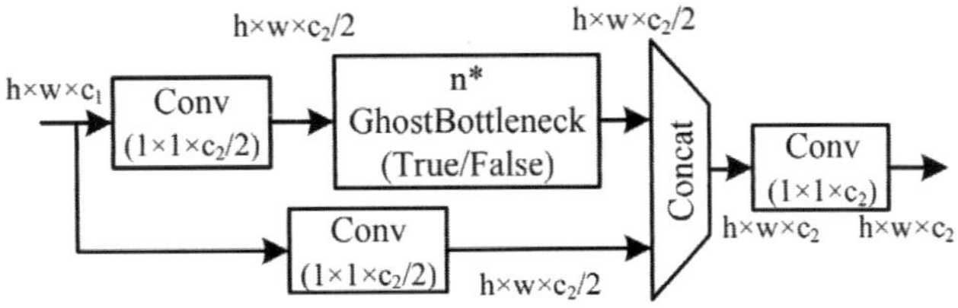


Figure.6 GhostBottleneckCSP module

### *3.7. Intelligent monitoring model of aggregated infectious risk.*

Chun Yutong organically combines gathering area detection, pedestrian detection and pedestrian mask wearing detection, and proposes a multi-functional and multi scene target detection framework, Cascade Attention R-CNN. Aiming at the problem that the target scale changes too much in the epidemic monitoring task, Cascade Attention R-CNN, a high-precision two-stage target detection algorithm, is selected as the basic detection framework. By adding a spatial attention mechanism to the cascaded candidate region classification regression network, we can highlight the important features in the candidate region features and suppress the noise features, so as to improve the expression ability of features and achieve higher detection accuracy. In addition, the detection results are fused with the risk level through the result fusion module to realize contactless and intelligent prevention, control and monitoring of aggregated infection [11].

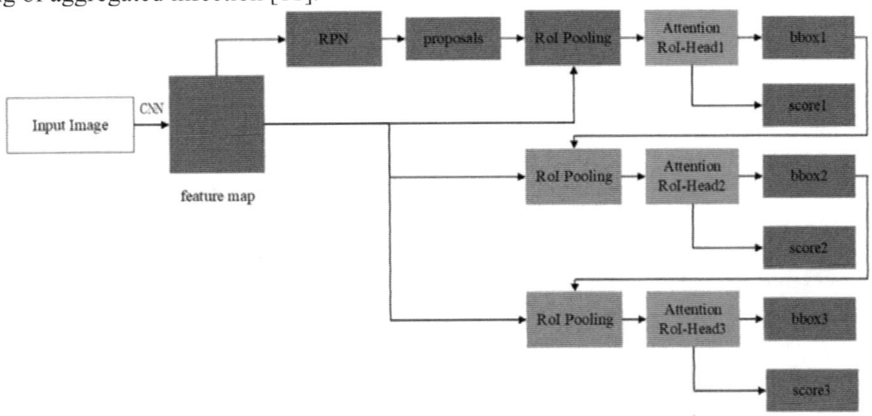


Figure.7 Structure of Cascade-Attention R-CNN

Chun Yutong use the mean average precision (mAP) as the evaluation index of the target detection module, which takes into account the accuracy rate and recall rate. The calculation formulas of P (Precision) and R (Recall) are as follows:

$$P = \frac{TP}{TP + FP} \times 100\% \tag{1}$$

$$R = \frac{TP}{TP + FN} \times 100\% \tag{2}$$

TP is the number of correctly identified targets, FP is the number of incorrectly identified targets, and FN is the number of missing targets. By selecting the values of recall rate and accuracy rate as abscissa and ordinate, the area under the P-R curve is the Average Precision (AP). Calculate AP for all categories and take the mean value as the mAP index. mAP is a commonly used evaluation index in target detection, which can reflect the overall performance of the algorithm.

*3.8. Application of infrared thermometer to COVID-19.*

As novel coronavirus pneumonia is needed in the infrared thermometer, the temperature measurement is very close to the patients. And the measurement accuracy is low, which leads to the low safety of body temperature measurement and the risk of erroneous judgement. [12] In order to improve the accuracy of infrared temperature measurement equipment in detecting human body temperature, Ren Jingqiu and other scholars designed an infrared temperature measurement gun system with long distance, accurate measurement and wide application. The system compensates a group of individual infrared dot matrix temperature readings collected by the sensor module of the infrared thermal imager for light perception and distance temperature, processes the collected temperature points at different distances by using the fast descending sort method, and compensates the temperature points at different distances through multiple linear function fitting, so as to improve the accuracy. After many experiments and debugging, the system can meet the accuracy requirements of human body temperature measurement and improve the accuracy of long-distance temperature measurement [13].

*3.9. Bispectral intelligent temperature detection and health big data management system.*

According to the characteristics of Chinese population density, high flow rate and easy spread of New Coronavirus pneumonia, Zhang Jieru and other scholars have established an intelligent temperature detection and health data management system integrating dual spectral imaging monitoring of visible and infrared light. The system realized the face recognition without contact rapid temperature detection and wearing masks, and completed the registration of personnel information quickly. The system has been deployed in many places and has passed the verification of effectiveness and reliability. The system can monitor and track the flow of personnel in real time, make statistical analysis on personnel information and big data of epidemic development in a multidimensional manner, model and predict the development dynamics of the epidemic, and improve the epidemic prevention and control strategy according to the analysis results, which is conducive to the accurate and efficient development of epidemic prevention and control [14].

*3.10. High precision infrared thermal imaging temperature measurement technology.*

Zhang Zhiqiang and other scholars established a matching model between the output gray value of the infrared temperature measurement system and the measured temperature of the measured object. They found a method to accurately calibrate the temperature measurement curve, and proposed the V-T curve applied to high-precision temperature measurement, as well as the V-TFPA, V-T lens barrel and V-T shutter curve applied to temperature drift compensation [15]. Through experimental verification, the ideal effect is achieved, and the temperature measurement accuracy is significantly improved. This system changes the infrared thermal imager from absolute temperature measurement to relative temperature measurement, which will solve the influence caused by noise drift to a great extent. The principle of relative temperature measurement is to use a reference body to compensate the temperature difference to the measured target by measuring the temperature of the reference body.

$$T_{r0}^{n} = \varepsilon_0 T_{00}^{n} + (1 - \varepsilon_0) T_{u0}^{n} \tag{3}$$

$$T_{rr}^{n} = \varepsilon_r T_{0r}^{n} + (1 - \varepsilon_r) T_{ur}^{n} \tag{4}$$

In the formula: $T_{r0}$、$T_{rr}$、$T_{00}$、$T_{0r}$、$\varepsilon_0$、$\varepsilon_r$ is the measured temperature, real temperature and emissivity of the target and reference respectively.

## 4. Development direction of epidemic surveillance system in the future.

COVID-19 will coexist with mankind for a long time. Therefore, before the development and implementation of virus specific drugs, the epidemic situation monitoring system will still play an important role in blocking the virus transmission chain in public places and clustered people. Exploring a set of epidemic surveillance system with higher accuracy and efficiency is still the main task at this stage, and the development research of such system will show the following characteristics:

The first is scalability. When designing the scheme, fully consider the scalability of the system to meet the continuous development of business and form an easy to manage and sustainable architecture. The more functions the system itself can carry, the better the scalability, the more it can meet the needs of future upgrading and development, and it is convenient to make modification and debugging according to the actual situation.

The second is the unity of advanced nature and practicality. The epidemic surveillance system needs to adopt advanced design ideas and methods, and use some current and technologically mature development tools to assist the construction. Its core purpose is to reduce power consumption, enhance resource utilization and continuously improve the advanced technology. Practicality is the necessary prerequisite for the advanced nature to put into practice. The system design needs to be fully investigated and analyzed, and from the actual needs of social health, safety and health, the software function module is developed and designed to make the system more in line with the usage habits and use needs of the base end. The system function design should fully consider the use scenario and simulate the problems in practical application, so as to ensure that the system can be truly used for health and safety monitoring when it is put into use.

The third is safety and reliability. In the era of information and data, the security of information needs to be paid great attention. The epidemic monitoring system has a large number of personnel flow information, and its security design needs to be strengthened.

The fourth is economic effectiveness. The application scenarios of epidemic surveillance system are very wide. Public transportation and public places all over China need to be widely distributed. The design scheme with low economic cost, strong technical realizability and high economic benefit will be the focus of development [16].

In the post epidemic era, the actual value of epidemic monitoring may be greatly reduced, health and safety monitoring and regular risk regulation will occupy a dominant position, and the market will continue to develop in the direction of upgrading of health and safety monitoring functions [17]. The epidemic prevention and control function will be transformed and upgraded and put into use in daily life.

## 5. Conclusion

The novel coronavirus pneumonia is still not yet over, and the surveillance system needs to be strengthened. As the epidemic situation enters the normalization stage, people's awareness of epidemic prevention continues to improve, consciously reduce going out and actively isolate themselves. At the same time, the epidemic prevention and control management is tired, and some personnel who despise the harmfulness of the epidemic also need the supervision and warning of external forces. The epidemic monitoring system will become an external supervision machine for epidemic prevention and control and play a key role in the stage of normalized epidemic prevention and control [18].

The epidemic monitoring system adds infrared temperature measurement function and mask detection function through target detection algorithms such as face recognition to coordinate the transmission of IOT communication technology. The intelligent and automatic monitoring system effectively reduces the exposure risk of “people” in the epidemic site, reduces the labor cost and avoids the formation of clustered people. The formation of this kind of system meets the current epidemic prevention requirements and has a wide range of practical application scenarios. The data storage and epidemic prevention alarm of local visualization platform and cloud platform can effectively reduce the risk of epidemic spread and strengthen epidemic prevention guarantee. The epidemic monitoring system is conducive to public health construction, ensuring the health and life safety of the people, serving the society with science and technology and promoting economic development [19].

The establishment of epidemic surveillance system in public places such as buses and high-speed railway stations has long-term development space. Its target detection algorithm based on neural network is widely used in pedestrian target detection, face detection, remote sensing image target detection, medical image detection and natural scene text detection.

## References

[1] Li Zhimeng, Sheng Fangfu. (2020) Influence of new coronavirus pneumonia on industry and consumption in China and its countermeasures. Jiangxi Social Science, 40(03):5-15.

[2] Wu Tingting, Zhu Angang. (2020) Novel coronavirus pneumonia's impact on China's economy and coping strategies. South China Finance, 2020(05):3-11.

[3] Liang Luhong, Ai Haizhou, Xu Guangyou, Zhang Ba. (2020) Overview of face detection research. Chinese Journal of Computers, 2002(05):449-458.

[4] Liu Li, Kuang Gangyao. (2009) Overview of image texture feature extraction methods. Journal of Image and Graphics, 2009,14(04):622-635.

[5] Wang, Yihao, Ding Hongwei, Li Bo, Yang Zhijun and Yang Jundong. (2020) Mask wearing detection algorithm based on improved YOLOv3 in complex scenes. Computer Engineering, 46(11):12-22.

[6] Cao Chengshuo, Yuan Jie. (2021) Mask-Wearing detection method based on YOLO-Mask. Laser & Optoelectronics Progress, 58(8):211-218.

[7] Niu Zuodong, Qin Tao, Li Handong, Chen Jinjun. (2020) Improved algorithm of RetinaFace for natural scene mask wear detection. Computer Engineering and Applications, 56(12):1-7.

[8] Zhang Xiubao, Lin Ziyuan, Tian Wanxin, Wang Yan, Shen Haifeng and Ye Jieping. (2020) Mask-wearing recognition in the wild. Scientia Sinica Informationis, 50(7):1110-1120.

[9] Li Yuyang, Shen Jiquan, Zhai Haixia and Feng Weihua. (2021) Mask wearing detection algorithm based on improved SSD. Computer Engineering, 1-9[2021-12-06]. https://doi.org/10.19678/j.issn.1000-3428.0062150.

[10] Peng Cheng, Zhang Qiaohong, Tang Chaohui and Gui Weihua. (2021) A face mask wearing detection method based on YOLOv5 enhancement model. Computer Engineering, 1-12[2021-12-06].https://doi.org/10.196 78/j.issn.1000-3428.0061502.

[11] Chun Yutong, Han Feiteng and He Mingke. (2022). Intelligent monitoring model of novel coronavirus pneumonia in the background of new crown pneumonia epidemic. Computer Engineering, 1-13[2022-03-23].DOI:10.19678/j.issn.1000-3428.0063195.

[12] Wang Hongjun, Bai Chengyu, Wang Jinghui, Yuan Zundong, Li Jinyuan and Gao Wei. Application of infrared thermometer to COVID-19. China Metrology,2020(05):17-20.DOI:10.16569/j.cnki.cn11-3720/t.2020.05.006.

[13] Ren Jingqiu, Qiu Pang, Zhang Lin and Fan Zongyuan. (2021) Study on temperature measurement compensation of human infrared temperature measuring gun. Electronic Measurement Technology,2021,44(16):62-67.DOI:10.19651/j.cnki.emt.2106950.

[14] Zhang Jieru, Su Feng, Yuan Peijiang, Wang Tianmiao, Tao Yining and Ning Dong. (2020) Bispectral intelligent temperature detection and health big data management system. Journal of Beijing University of Aeronautics and Astronautics,2020,46(09):1739-1746.DOI:10.13700/j.bh.1001-5965.2020.0063.

[15] Zhang Zhiqiang, Wang Ping, Yu Xudong, Zhang Jian and Li Ting. (2020) Research on high precision infrared thermal imaging temperature measurement technology. Chinese Journal of Scientific Instrument, 2020,41(05):10-18.DOI:10.19650/j.cnki.cjsi.J2006177.

[16] He Yi, Lu Yinhao, He Yongchao and Huang Xiaoyan. (2020) Construction and thinking of monitoring system for public health emergencies in Shanghai. Chinese Health Resources,2020,23(02):94-98.DOI:10.13688/j.cnki.chr.2020.19360.

[17] Li Deren, Shao Zhenfeng, Yu Wenfu, Zhu Xinyan, and Zhou Suhong. (2020) Public epidemic prevention and control services based on spatio-temporal location big data make cities smarter. Geomatics and Information Science of Wuhan University,2020,45(04):475-487+556.DOI:10.13203/j.whugis20200145.
[18] Sun Yongsheng, Jin Wei and Tang Yuchao. (2020) Application of covid-19 in prevention and control of new crown pneumonia. Science & Technology Review, 2020,38(04):39-49.
[19] An Guojun, Jia Fuwei. (2020) Analysis of COVID-19's impact on the economy and Countermeasures. Financial Theory & Practice, 2020(03):45-51.

# Research on the detection method of wheel defects of electromagnetic ultrasonic phased array trains

**Xiao-yu ZHANG** [1,a*], **Zhi-chun WANG** [1,b*]

[1]Inner Mongolia University of Science and Technology College of Information Engineering, Baotou, 014010,China

[a*]email:944636062@qq.com

[b*]email:wangzhichun0472@163.com

Corresponding author's e-mail:wangzhichun0472@163.com

**Abstract.** Train wheels are important moving parts of railway locomotives, with the development of China's railways, the reliability of its quality is of great significance to ensure the safe operation of trains and reduce the danger of train driving. However, the traditional electromagnetic ultrasonic transduction efficiency is low, and the detection is difficult. Based on the principle of electromagnetic ultrasonic phased array detection, this paper uses the finite element method to establish a two-dimensional simulation model of Comsol Multiphysics 16-channel electromagnetic ultrasonic phased array, and adopts the detection method based on B-scan imaging to process the electromagnetic ultrasonic phased array echo signal and identify the internal defects of train wheels.

## 1. Introduction

With the development of China's economy, China's railways have experienced 6 major speed-ups, China has become the country with the longest operating mileage in the world, and the importance of train wheel safety is self-evident [1-2]. Crack is a more typical damage in wheel failure, refers to the internal fatigue crack of the wheel, the crack originates from the inside of the rim, if found in time, it is likely to cause vicious traffic accidents such as train axle cutting and derailment, endangering the safety of the train [3].

At present, no small progress has been made in the detection and development of train wheel defects at home and abroad. As early as the 1990s, the Institute of Nondestructive Testing of the Franchoeffer Association in Germany successfully developed a set of wheel pedal flaw detection devices using electromagnetic ultrasonic transducers (EMAT) to realize the detection of running vehicle wheels [4]. In 2015, researchers in Italy developed a laser ultrasonic inspection technology for the inspection of train wheels, using air as a couplant to ensure that there is no contact between the probe and the workpiece under test, thereby improving the inspection efficiency [5]. In 2021, Benjamin of Germany proposed an axlebox acceleration (ABA)-based approach that combined velocity information obtained by gnavi satellite systems with inertial measurement units to detect wheel defects.

The modern detection technology for train wheels in China was developed late, and the conventional detection technology such as ultrasonic and magnetic powder was the mainstay in the early days [6]. In 2013, Cheng Hongzhao et al. used non-contact 3D scanner technology to reconstruct the wheel tread model to achieve the detection of wheel tread defects . In 2019, Zeng Yuanchen et al.

of Southwest Jiaotong University designed a vehicle wear model that combines physics and data to predict wheel damage through closed-loop alternation between wear and editing .

Based on the current situation of domestic and foreign research, although various countries have continuously proposed new train wheel detection methods in recent years, they are more in the detection method, and have not made good improvements to the drawbacks of conventional testing, and the detection effect of internal defects in wheels is poor. Therefore, this paper combines phased array and electromagnetic ultrasonic ultrasonic technology to propose an electromagnetic ultrasonic phased array B-scan imaging detection method, which can realize the deflection and focus of the wave by controlling the delay of each array element, and solve the problem of low transducer energy transfer efficiency, which has high research value for the detection of internal defects of the wheel.

## 2. Materials and Methods

Electromagnetic ultrasonic testing technology is different from the traditional piezoelectric ultrasonic detection technology, and the resulting beam is not the transmission of sound waves, but the vibration generated by the particles inside the specimen under the action of the external environment, the essence of the electromagnetic ultrasonic transducer is that the electrical energy at the electromagnetic ultrasonic transmitter is converted into the kinetic energy of the particles inside the workpiece.

Based on phased array technology, this paper develops electromagnetic ultrasonic phased array detection technology for wheel internal defects, and uses ultrasonic shear wave phased reflection method to complete the identification and positioning of wheel internal defects.Firstly, using the phased array technology, the two-dimensional simulation model of the Comsol Multiphysics 16-channel electromagnetic ultrasonic phased array is established by using the finite element method, and the focus of the ultrasonic beam is realized by the time-delay array element, which verifies the feasibility and necessity of the focusing of the electromagnetic ultrasonic phased array detection technology. In order to locate the wheel defects, the electromagnetic ultrasonic phased array defect echo signals are collected, and finally the characteristic values are extracted by filtering, denoising and detection, and the B-scan imaging of the wheel defects is completed.

## 3. EMAT simulation analysis

The application of Comsol simulation software to design the electromagnetic ultrasonic transducer model, with a circumference of 150mm, a thickness of 40mm of 60°circular arc steel as the object of the wheel to be measured, the steel using CL60 high carbon steel, ultrasonic transducer using four reverse polarity magnets juxtaposed permanent magnet group excitation, static magnetic field and the test piece perpendicular, excitation ultrasonic transverse wave. The lower 16 sets of coils are set for different delays, using time-sharing independent coils, the left 8 coil current direction vertical paper facing outward, the right 8 coil current direction vertical paper facing inward. The diameter of the permanent magnet and the diameter ratio of the transducer coil are 1.2, and the ultrasonic effect obtained by excitation is the best, and the two-dimensional detection simulation model is established by using Comsol simulation software as shown in Figure 1. ：

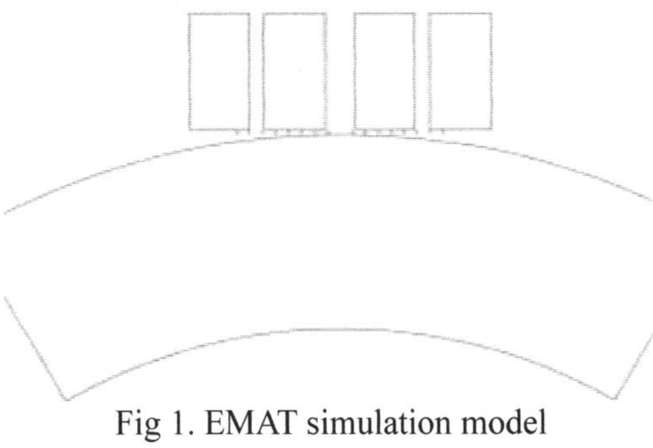

Fig 1. EMAT simulation model

In the electromagnetic ultrasonic transducer, the reverse polarity permanent magnet group provides a static magnetic field, and the distribution of the magnetic field and magnetic field lines of the permanent magnet is shown in Figure 2 .

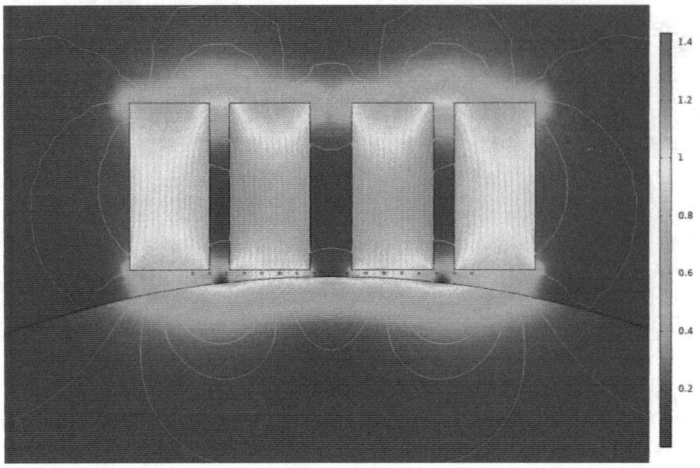


Fig 2. Static magnetic field distribution of permanent magnets

To excite ultrasonic transducers in an electromagnetic transducer, the model excitation coil input signal uses a burst signal, with the number of periodic waves set to 6 and the frequency as a 1MHz excitation signal.The beam focus point is set at 30mm below the wheel specimen, and according to the ultrasonic phased array deflection theory, the delay of each coil array element is set to achieve the effect of beam focusing and improve the signal strength.The focus simulation results of the EMAT are shown in Figure 3.

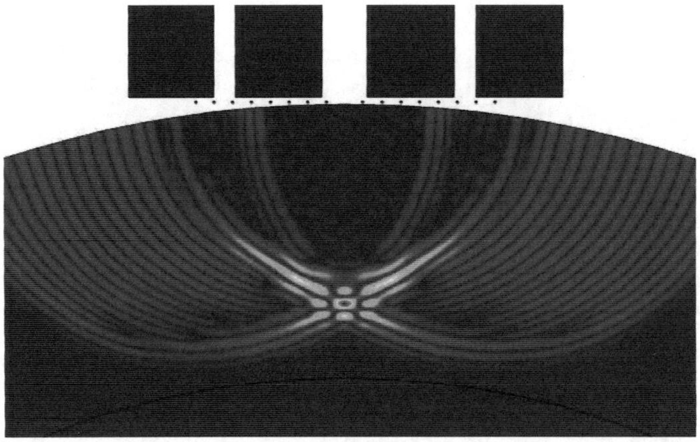

Fig 3. EMAT's soundstage focus map

Collect the echo signal before and after EMAT focusing for research, from Table 1 can be seen, after focusing, due to the setting of signal delay, the pulse excitation signal propagation after focusing is 2us slower than reaching the signal peak point when not focused, the amplitude of the echo signal increases, the maximum displacement of the detection point before focusing is 12mv, the maximum displacement of the monitoring point after focusing is 20mv, and the signal strength is increased by 66%, and the simulation verifies the necessity of the focused phased array.

Table 1. Formatting sections, subsections and subsubsections

| | Excitation signal peak/mv | Peak echo signal/mv | Peak delay time after focusing/us |
|---|---|---|---|
| Before focusing | 35 | 12 | 2 |
| After focusing | 29 | 20 | |

## 4. Processing and imaging of defective signals

The electromagnetic ultrasonic wheel defect detection imaging system designed in this paper includes: EMAT excitation unit, defect detection unit, EMAT signal processing and imaging three units.This is shown in Figure 4.In the EMAT excitation unit, the pulse signal transmitter generates an adjustable frequency excitation pulse signal loaded into the EMAT phased array, the wheel generates ultrasonic waves, and the echo signal is converted into an electrical signal by the array coil of the transceiver, and the signal is processed and the visual defect scanning is completed by the B-scan imaging program.

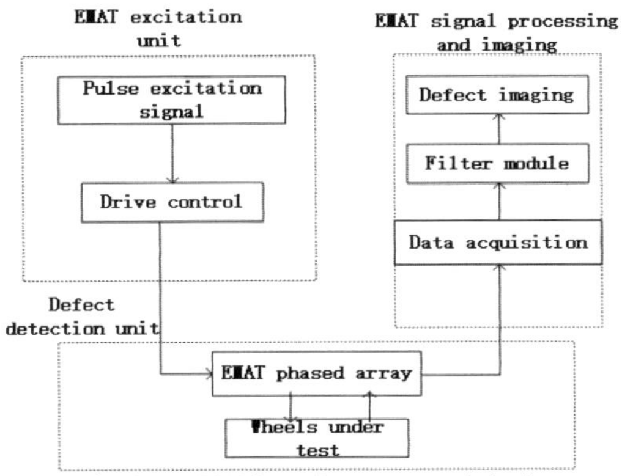


Fig 4. EMAT imaging protocol design

In the operation of the electromagnetic ultrasonic phased array defect detection system, due to the interference of the power supply and the external environment, the electromagnetic ultrasonic phased array receiving signal is flooded by random noise, the useful signal cannot be identified, and the signal-to-noise ratio is seriously reduced. Therefore, the wavelet denoising algorithm is used to de-denoise the noisy signal received by the coil array to obtain the defect echo signal that is easy to detect.

The defect signal of the electromagnetic ultrasonic phased array inside the original wheel, the noisy signal and the defect signal after wavelet denoising are as shown in Figure 5. After wavelet denoising, the signal-to-noise ratio of defective signals has been greatly improved,Conditions are provided for the subsequent extraction of defect signal features.

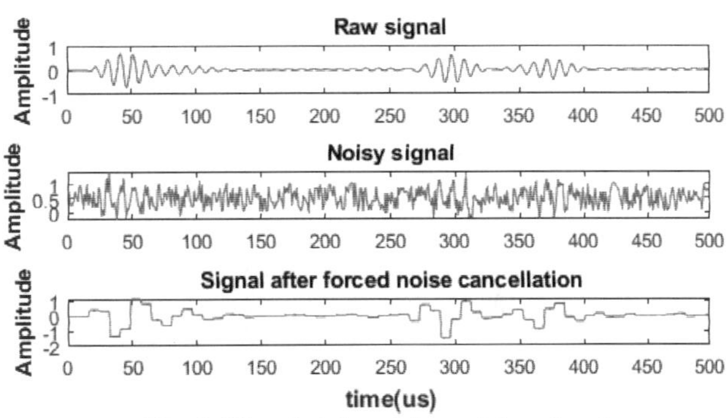


Fig 5. Wavelet denoising of the signal

Read the processed internal defect echo signal of the wheel,and use Hilbert transform for envelope-demodulation,B-scan imaging of defects at different depths.The result of the B-sweep with a defect depth of 30 mm is shown in Figure 6.

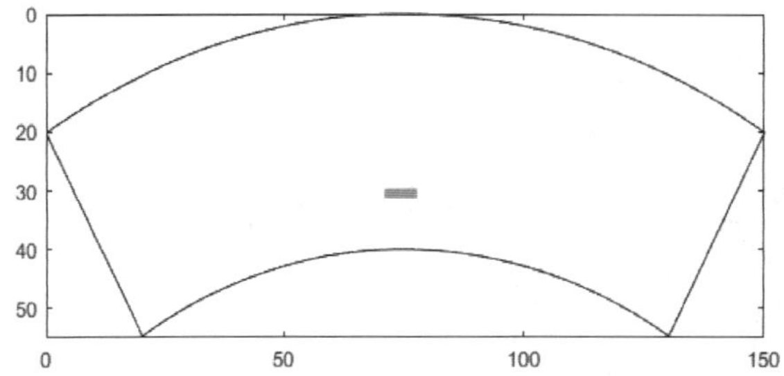


Fig 6. Schematic diagram of an imaging with a defect depth of 30 mm

Defective echo signals at different depths are processed and imaged.The experimental results are shown in Table 2 .It can be seen from the experimental calculation results that with the increase of the defect depth, the peak of the electromagnetic ultrasonic longitudinal defect echo signal inside the wheel decreases, and as the depth increases from 10mm to 35mm, the receiving peak of the normalized defect echo signal decreases to 88.3, which is reduced by 41.2%. When the defect is near the reference position, the defect detection error of the electromagnetic ultrasonic phased array is small, and the average error of defect detection imaging is 2.73% when the depth is in the range of 20mm-35mm, and the detection accuracy is high. When the defect depth is 10mm, the error of electromagnetic ultrasonic imaging detection defects is expanded to 12.1%. That is, the defect position when the internal defect of the wheel is close to the surface of the wheel, the error of the detection defect position is large.

Table 2. Experimental results at different depth defect locations

| Defect depth/mm | Normalize signal peaks | Defect imaging depth/mm | Standard error/% |
|---|---|---|---|
| 10 | 150.1 | 6.37 | 12.1 |
| 20 | 145.1 | 21.31 | 4.37 |
| 25 | 101.6 | 23.61 | 4.67 |
| 30 | 100 | 30 | 0 |
| 35 | 88.3 | 2.67 | 34.23 |

## 5 .Conclusion

This paper gives the overall scheme of internal defect detection of EMAT phased array wheels, completes the focusing of the EMAT phased array beam, collects the echo signals of defects of different depths, remove noise from it, and completes defect imaging by B-scan.Through experimental studies of multiple depth defects,It is found that the EMAT phased array imaging inspection system has an average error of 3.79% for the detection of defects inside the wheel.

## Acknowledgments

Supported by the National Natural Science Foundation of China (61463041); Project supported by science and technology plan of Inner Mongolia Autonomous Region (2020gg0157);Kundulun science and technology plan project of Baotou City, Inner Mongolia Autonomous Region (yf2020010)

Journal of Physics: Conference Series **2264** (2022) 012025 doi:10.1088/1742-6596/2264/1/012025

**Reference**

[1] Luo Chunxiao. (2008) China's railways open the era of high-speed-remembering the opening ceremony of the Beijing-Tianjin intercity railway. Railway knowledge,4:4-5.

[2] Tong Xi,Wang Xiao.(2017)A new paradigm for Made in China: high-speed rail.Communist,6:58-59.

[3] Hu Junhui.(2014)Analysis of common fault characteristics and causes of rim cracks in railway passenger cars.Shanghai railway technology,4;78-79.

[4] Kappes,W.,Kroning.M.(2001)Non-Destructive Testing of Wheel-Setsas a Contribution to Safety of Rail Traffic.In:CORENDE 2000 Congreso Regional de Ensayos No Destructivos y Estructurales.Germany.pp.1-12.

[5] Cavuto,A.,Martarelli, M.,Pandarese, G.(2015)Train wheel diagnostics by laser ultrasonics.Measurement,80:99.

[6] Pang Zhihui,Yu Runqiao,Hu Chen, Zhao Longchan.(2017)Application of micromagnetic detection in the detection of train wheelsets.Failure analysis and prevention,12(02):94-100.

# Test and Measurement Data Analysis of Fuel Consumption Measurement and Control System of Floating Ball Engine

**Dong Xu**

Automotive Technical College, Wuxi Vocational Institute of Commerce, Wuxi Jiangsu 214153, China

xd1228@126.com

***Abstract:*** Fuel consumption is an important index to evaluate the fuel economy of automobile engine, and it is one of the important measurement parameters in engine test research. This paper uses the engine eddy current test bench to carry out the repeatability verification test, measurement accuracy and accuracy verification test, and engine bench test of the designed and developed floating ball fuel consumption meter. The data obtained from the test are analyzed to verify the floating ball fuel consumption. performance indicators of the instrument.

## 1. Introduction

The test of the engine on the test bench is called a bench test. Various performance indicators, parameters and various characteristic curves of the engine are usually measured on the engine test bench according to the prescribed test methods. The test bench must ensure that the test conditions meet the standard requirements, and can quickly and accurately measure various working parameters of the engine [1-2].

The measurement of engine fuel consumption in this paper is carried out on the engine eddy current test bench. The measurement process is to start the engine, run the engine to the normal working temperature, use the dynamometer to load it to the limited conditions, make the engine work stably in the specified working condition, and then use the fuel consumption meter to measure the fuel consumption of the engine.

## 2. The connection of the oil circuit and the removal of air bubbles in the oil circuit

### *2.1. Oil connection*

At present, the most widely used in automobiles are EFI engines and diesel engines. EFI engines are often equipped with oil return lines. The amount of fuel supplied to the engine by the gasoline pump of this kind of engine is much higher than its actual demand, and the excess fuel will be returned to the fuel tank by the oil return line. The conventional fuel consumption meter cannot solve the problems of fuel return and high pressure fuel supply of the EFI engine, so it cannot be directly applied to the measurement of fuel consumption of the EFI engine. In order to measure the fuel consumption of EFI engines, special measuring instruments are often used. Although these special fuel consumption meters can meet the requirements of fuel consumption measurement of EFI engines, there are also some problems in the bench measurement of EFI engines. The fuel consumption meters for EFI engines are connected in series between the fuel pump and the engine, which will increase the pressure drop in the fuel circuit of the EFI engine. In addition, some special fuel consumption meters also need to install a pressure regulator when connecting with the fuel return pipe of the EFI engine. If the selected pressure regulator is not suitable, the actual fuel supply state of the engine will be changed, so that the EFI engine will continue to consume fuel. The working state during measurement is inconsistent with the working state when no measurement is performed, resulting in a large measurement error [3].

In the test of measuring the fuel consumption of the EFI engine on the bench, in order to use the conventional floating ball fuel consumption meter to measure without reducing the accuracy of the fuel consumption measurement of the engine, the connection method shown in Figure 1 can be used [4].

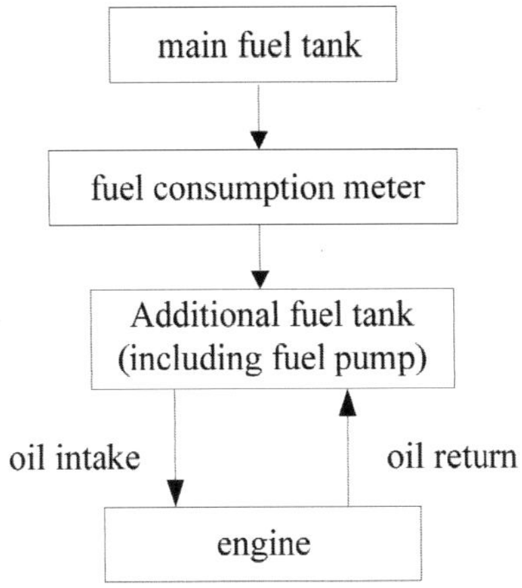


Fig.1 The connection of the fuel consumption meter in the oil circuit of the EFI engine

Add an additional fuel tank between the fuel consumption meter and the engine, install the fuel pump in the additional fuel tank, supply the engine with high-pressure fuel, and return the engine oil to the additional fuel tank, which is consistent with the original fuel supply state of the engine, avoiding the need for fuel supply. The adverse effect of the fuel consumption meter connected in the middle of the system on the engine fuel supply system.

The fuel in the main fuel tank flows to the additional fuel tank through the fuel consumption meter, and the fuel consumed by the engine is the fuel in the main fuel tank measured by the fuel consumption meter, which solves the problem of oil return and high-pressure fuel supply of the EFI engine, and avoids the fuel consumption. The fuel consumption measurement data is distorted due to the repeated measurement of the fuel consumption meter.

For diesel engines, the connection method shown in Figure 1 can also be used to measure fuel consumption.

With this connection method, the float type fuel consumption meter can be applied to the fuel consumption measurement of various types of engines, and the application range of the float type fuel consumption meter can be expanded.

*2.2. Removal of air bubbles in the oil circuit*

(1) Removal of Air Bubbles in Gasoline Engine Oil Circuit

Eliminating air bubbles in the oil circuit of a gasoline engine is a time-consuming task, especially when there is a blockage or leak in the pipeline, which will prevent the air bubbles from being completely exhausted. Once the air bubbles are generated, the impact on the fuel consumption test results will be very large. The fuel consumption meter will measure the volume occupied by the air bubbles as the fuel consumption, so that the measured data is higher than the actual number, which will cause the distortion of the measured value [5-6].

Air bubbles must be removed when measuring fuel consumption. Usually, the following methods can be used: "short-circuit" the pipeline from the fuel tank to the gasoline pump, install a new, well-sealed, non-clogging fuel pipe, and use an electric motor with stable performance. The gasoline pump and gasoline filter replace the corresponding parts of the original car, shorten the length of the oil pipe from the oil pump to the sensor, so that the resistance from the oil pump to the fuel consumption sensor is greatly reduced, thereby avoiding the adverse effect of air bubbles on the detection results [7-8].

(2) Removal of air bubbles in diesel engine oil circuit

After the fuel consumption meter is installed in the oil circuit of the diesel engine, the oil must be pumped with a manual pump to remove the air bubbles in the oil circuit with the pump oil pressure. One of the differences between it and the gasoline engine is that the gasoline engine can remove the air bubbles after starting, while Diesel engines must remove air bubbles in the oil circuit before starting; the second difference is that gasoline engines do not need to remove air bubbles when the fuel consumption meter is removed to restore the crude oil circuit, while the diesel engine still remains after the fuel consumption meter is removed to restore the crude oil circuit. Air bubbles just created in the oil circuit need to be removed [9].

## 3. Fuel consumption meter measurement repeatability verification test

### *3.1. How to determine the duplication of data*

The fuel consumption measurement data of the engine must meet the requirements of the following formula:

$$(B_{max} - B_{min})/B_A \leq R \quad (1)$$

where: $B_{max}$ is the maximum value in the measured data of fuel consumption per hour, $B_{min}$ is the minimum value in the measured data of fuel consumption per hour, $B_A$ is the arithmetic mean value of the measured data of fuel consumption per hour, $R$ is the proportionality coefficient, and its value is shown in Table 1.

The confidence interval $\Delta B_V$ (90% confidence level) of the test results is calculated as follows:

$$\Delta B_V = \pm \frac{0.031}{\sqrt{n}} B_A \quad (2)$$

The arithmetic mean of the measured data of fuel consumption per hour is calculated as follows:

$$B_A = \frac{B_1 + B_2 + \cdots + B_n}{n} = \frac{\sum_{i=1}^{n} B_i}{n} \quad (3)$$

where: $B_i$ is the hourly fuel consumption for each measurement, and n is the number of measurements.

Table.1 The value of the scale factor R

| Test count (n) | 2 | 3 | 4 | 5 | 10 |
|---|---|---|---|---|---|
| R | 0.053 | 0.063 | 0.069 | 0.073 | 0.085 |

### *3.2. Test content and test data*

The repeatability of the measurement data is an important indicator to verify the stability of the fuel consumption meter. The purpose of this test is to verify the measurement repeatability of the float type fuel consumption meter.

The test equipment includes a floating ball fuel consumption meter and a bench test simulation device.

The test method is to use the fuel pump to fill the fuel consumption meter with oil, install a throttle valve at the oil outlet of the fuel consumption meter, and adjust the throttle valve to control the fuel flow rate of the fuel consumption meter outlet, simulate the fuel consumption of the engine, and verify the fuel consumption meter. measurement repeatability. The density of the fuel used in the test is 0.75g/ml.

The test data are shown in Table 2.

Table.2 Fuel consumption meter repeatability verification test data

| test count | Measurement results (kg/h) | Throttle valve position |
|---|---|---|
| 1 | 13.86 | Position 1 |
| 2 | 13.88 | |
| 3 | 13.82 | |
| 4 | 13.83 | |
| 5 | 13.76 | |
| 6 | 13.71 | |
| 7 | 13.77 | |
| 8 | 13.76 | |
| 9 | 13.73 | |
| 10 | 13.75 | |
| 11 | 25.79 | Position 2 |
| 12 | 25.77 | |
| 13 | 25.73 | |
| 14 | 25.78 | |
| 15 | 25.62 | |
| 16 | 25.71 | |
| 17 | 25.72 | |
| 18 | 25.67 | |
| 19 | 25.82 | |
| 20 | 25.66 | |
| 21 | 39.83 | Position 3 |
| 22 | 39.84 | |
| 23 | 39.79 | |
| 24 | 39.95 | |
| 25 | 39.71 | |
| 26 | 39.85 | |
| 27 | 39.69 | |
| 28 | 39.82 | |
| 29 | 39.76 | |
| 30 | 39.74 | |

### *3.3. Test Conclusions*

Through the above data analysis, it can be seen that when the floating ball fuel consumption meter is doing the repeatability verification test, when the throttle valve is at position 1, the repeatability of the fuel consumption meter measurement data is 0.012, and the confidence interval is indicating that the measurement of the fuel consumption meter Repeatability is good.

When the throttle valve is in different positions, the repeatability and confidence interval of the measurement data are relatively close, indicating that the fuel consumption meter not only has good repeatability, but also has good stability.

## 4. Fuel consumption meter measurement accuracy and accuracy verification test

### *4.1. Verification method of measuring accuracy of fuel consumption meter*

The fuel consumption measured by the engine bench test shall be equal to the arithmetic mean of the fuel consumption of 3 consecutive measurements. If the difference between the limit value of fuel consumption and the average value after 3 tests is more than 5%, continue the test until a measurement accuracy of at least 5% is obtained. If the measurement accuracy does not reach 5% after 10 tests, it means that the measurement accuracy of the fuel consumption meter is unqualified [10].

The measurement accuracy is calculated by the following formula:

 2264 (2022) 012026 doi:10.1088/1742-6596/2264/1/012026 

$$D = K * \frac{S}{\sqrt{n}} * \frac{100}{B} \% \tag{4}$$

where: $D$ is the measurement accuracy of the fuel consumption meter, $S$ is the standard deviation of n times measured fuel consumption, $B$ is the hourly fuel consumption, and $n$ is the number of measurements, $K$ given in Table 3.

Table.3 The value of the coefficient K

| Number of measurements | 4 | 5 | 6 | 7 | 8 | 9 | 10 |
|---|---|---|---|---|---|---|---|
| K | 3.2 | 2.8 | 2.6 | 2.5 | 2.4 | 2.3 | 2.3 |

The standard deviation S is calculated by:

$$S = \sqrt{\frac{\sum_{i=1}^{n}(B_A - B_i)^2}{n-1}} \tag{5}$$

where: $B_A$ is the arithmetic mean of the measured data of fuel consumption per hour, $B_i$ is the hourly fuel consumption obtained in each measurement, and $n$ is the number of measurements.

### *4.2. Measurement accuracy verification method of fuel consumption meter*

In this paper, the comparison method is used to verify the measurement accuracy of the floating ball fuel consumption meter. The specific method is to connect the float type fuel consumption meter and the standard fuel consumption meter in series in the engine oil circuit, carry out the engine fuel consumption measurement test, compare the measurement data of the float type fuel consumption meter and the standard fuel consumption meter, calculate the relative error, and verify the floating ball The measurement accuracy of the fuel consumption meter.

The relative error is calculated by:

$$E = \left|\frac{\overline{B} - \overline{B_S}}{\overline{B_S}}\right| * 100\% \tag{6}$$

where:$E$ is the relative error of the measurement results of the floating ball fuel consumption meter, $\overline{B}$is the average value of the measurement results of the floating ball fuel consumption meter, and $\overline{B_S}$is the average value of the measurement results of the standard fuel consumption meter.

### *4.3. Test content and test data*

The purpose of this test is to verify the measurement precision and accuracy of the float type fuel consumption meter. The test method is to connect the floating ball fuel consumption meter and the FCMM-2 fuel consumption meter to the engine oil circuit, and use the dynamometer to simulate the engine working conditions. When the engine works at a fixed speed (1700r/min) and a fixed torque (35.0Nm) ) to measure the engine fuel consumption to verify the measurement accuracy of the fuel consumption meter. By comparing the measurement data of the float type fuel consumption meter and the FCMM-2 fuel consumption meter, the measurement accuracy of the float type fuel consumption meter is verified. The density of the fuel used in the test is 0.75g/ml.

The test data are shown in Tab 4.

Table.4 Fuel consumption meter measurement accuracy verification test data

| Number of measurements | Measurement results of float type fuel consumption meter (kg/h) | FCMM-2 fuel consumption meter measurement results (kg/h) | difference | Relative error(%) |
|---|---|---|---|---|
| 1 | 2.88 | 2.98 | 0.10 | 3.356 |
| 2 | 2.85 | 2.99 | 0.14 | 4.682 |
| 3 | 2.79 | 2.90 | 0.11 | 3.793 |

| | | | | |
|---|---|---|---|---|
| 4 | 2.84 | 2.95 | 0.11 | 3.728 |
| 5 | 2.90 | 2.99 | 0.09 | 3.041 |
| 6 | 2.94 | 2.96 | 0.12 | 4.054 |
| 7 | 2.80 | 2.91 | 0.11 | 3.780 |
| 8 | 2.85 | 2.98 | 0.13 | 4.682 |
| 9 | 2.81 | 2.90 | 0.09 | 4.362 |
| 10 | 2.80 | 2.93 | 0.13 | 4.437 |

### *4.4. Test Conclusions*

According to the above data analysis, after 10 tests, the measurement accuracy of the fuel consumption meter when the fuel consumption is the largest is 1.21%, and the measurement accuracy of the fuel consumption meter when the fuel consumption is the smallest is 1.28%, both less than 5%, indicating that the fuel consumption meter has The measurement accuracy is high.

The measurement results of the float type fuel consumption meter and the FCMM-2 fuel consumption meter are relatively close, and the relative error is only 3.493%, indicating that the measurement accuracy of the float type fuel consumption meter is high, reaching the level of similar products.

## 5. Engine bench test

### *5.1. Test content and test data*

The purpose of this test is to verify the floating ball by analyzing the relationship between the engine's fuel economy index and load (expressed as effective power $P_e$ , torque $T_{tq}$ or mean effective pressure $p_{me}$ etc.) at a certain speed, that is, the engine's load characteristics performance of the fuel consumption meter.

Test instruments include floating ball fuel consumption meter, FCMM-2 fuel consumption meter, gasoline engine and engine eddy current test bench, etc.

The test method is to connect the floating ball fuel consumption meter and FCMM-2 fuel consumption meter in series to the engine oil circuit, use the dynamometer to simulate the engine working conditions, make the engine work under different working conditions, measure the engine fuel consumption, and obtain the engine oil consumption. Load characteristic data, observe the working condition of the fuel consumption meter under different engine conditions, and draw the engine load characteristic curve, including the hourly fuel consumption curve and the effective fuel consumption rate curve. The performance of the float type fuel consumption meter is verified by analyzing and comparing the engine load characteristic curves. The engine speed in the test was 1200 r/min. The density of the fuel used in the test is 0.75g/ml.

The test data are shown in Table 5.

Table.5 Engine Load Characteristic Data

| Effective power (kW) | Float Type Fuel Consumption Meter | | FCMM-2 fuel consumption meter | |
|---|---|---|---|---|
| | Fuel consumption (kg/h) | Fuel consumption rate [g/(kW.h)] | Fuel consumption (kg/h) | Fuel consumption rate [g/(kW.h)] |
| 10.44 | 4.61 | 441.11 | 4.73 | 453.07 |
| 17.76 | 6.18 | 347.78 | 6.29 | 354.17 |
| 23.40 | 7.24 | 309.41 | 7.38 | 315.38 |
| 26.52 | 7.85 | 296.04 | 7.95 | 299.77 |
| 33.96 | 9.52 | 280.27 | 9.66 | 284.45 |
| 35.64 | 9.81 | 275.14 | 9.92 | 278.34 |

| | | | | |
|---|---|---|---|---|
| 37.92 | 10.34 | 272.77 | 10.46 | 275.84 |
| 41.04 | 11.33 | 276.14 | 11.47 | 279.48 |
| 43.52 | 12.45 | 286.08 | 12.56 | 287.22 |
| 46.68 | 14.24 | 305.12 | 14.39 | 308.27 |

*5.2. Test Conclusions*
According to the load characteristic data of the gasoline engine listed in Tab 5, the load characteristic curve of the engine is drawn as shown in Fig 2.

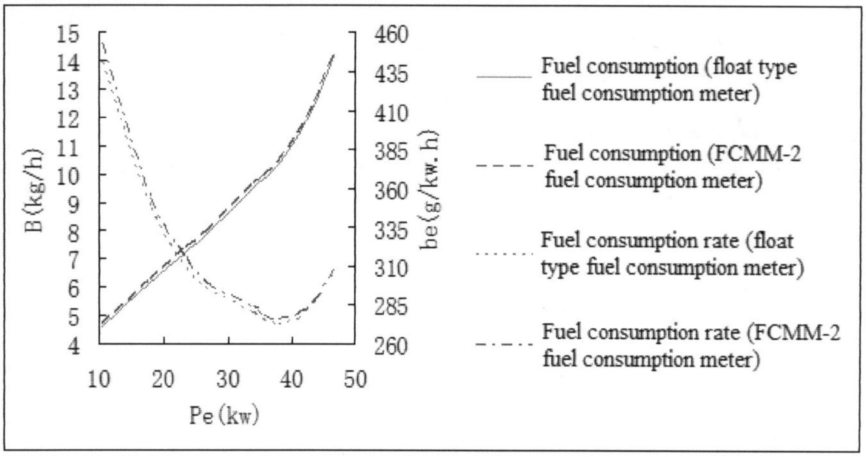


Fig.2 Engine load characteristic curve

It can be seen from the engine load characteristic curve in Figure 3 that when the engine speed is constant, its hourly fuel consumption B changes almost linearly with the change of load. When the engine is idling ($P_e = 0$), the fuel consumption rate $b_e$ is infinite. As the load increases, the fuel consumption rate curve decreases rapidly. When the load increases to a certain extent, the fuel consumption rate is the lowest. When the load continues to increase, the fuel consumption rate increases. The fuel consumption curve and the fuel consumption rate curve are in line with the load characteristics of the gasoline engine, and the curves are relatively smooth, indicating that the floating ball fuel consumption meter has good performance and high reliability.

## 6. Conclusion
Through repeatability verification test, measurement accuracy and accuracy verification test, engine bench test, and data analysis, comparing the load characteristic curves measured by the two fuel consumption meters, it can be seen that the floating ball fuel consumption meter and the FCMM-2 fuel consumption meter The changes of the measurement data are basically the same, the curves are nearly coincident, and the error of the measurement data is small, indicating that the measurement accuracy of the floating ball fuel consumption meter is high.

## References
[1] XU Zhao-kun. Principle of automobile engine [M]. Beijing: Tsinghua University Press, 2010,01..
[2] YAN Fu-wu. Principles of Automobile Engines 4th Edition [M]. Beijing: Machinery Industry Press, 2017,09.
[3] TANG Yi. Measurement and Research on Fuel Consumption of EFI Gasoline Engine[D]. Dalian: Dalian University of Technology, 2005: 23-24.
[4] YAO Chun-de, LI Yun-qiang. A simple method to accurately measure the fuel consumption of gasoline injection engines [J]. Internal Combustion Engine, 2004(3): 11-12.

[5] Vocational Qualification Center of the Ministry of Transport. Motor vehicle inspection and evaluation and application technology inspection and maintenance engineer [M]. Beijing: People's Communications Press, 2012.07.
[6] MAO Mao. Used car identification and evaluation [M]. Beijing: National Defense Industry Press, 2013,08.
[7] LI Cheng-gong, XU Guang-yong. Motor vehicle technical evaluation including testing personnel qualification examination must read [M]. Beijing: Golden Shield Press, 2008,05.
[8] ZHU Gui-ying, SUN Wei, GUO Hong-wei. Automobile Performance and Testing Technology [M]. Beijing: Beijing University of Posts and Telecommunications Press, 2012,01.
[9] ZHAO Pei-quan; SU Ling-lin et al. Automobile identification and evaluation [M]. Beijing: China Water Resources and Hydropower Press, 2016,04
[10] DU Dan-feng.Automobile Experimentation[M].Beijing:People's Communications Press, 2017,01.

IOP Publishing
**2264** (2022) 012027 doi:10.1088/1742-6596/2264/1/012027

# Research on eddy current testing method of mould molten steel

**Yanqing Liu[1,a], Zhichun Wang[2,b]**

[1]Inner Mongolia University of Science and Technology College of Information Engineering ,Baotou, Inner Mongolia,014010,China

[a]email:1937782220@qq.com

[b]email:wangzhichun0472@163.com

**Abstract.** A double eddy current sensor is proposed to measure the liquid level of molten steel in mold and the liquid level of protective slag molten pool. The finite element models of the sensor and the measured object are established by using the finite element COMSOL simulation software. According to the different conductivity of each material in the mold, the measured object is set with the composition layer structure, and the liquid level of molten steel and the liquid level of protective slag molten pool are obtained by using double eddy current sensor and double frequency measurement. The factors affecting the measured value are analyzed and the structure of the sensor is optimized to detect the liquid steel in a large distance within 50-160mm, and the liquid level of molten pool in the mold is measured by high-frequency coil.

## 1. Introduction

In the process of continuous casting production, the mold plays an important role in the continuous caster. The control of molten steel level and molten pool thickness of protective slag is not only related to improving product quality, but also closely related to reducing safety accidents. Molten steel level detection and molten pool thickness detection are very important. At present, foreign imported equipment is mostly used in the production line, and there is no detection equipment for the molten pool thickness of protective slag in the mold at home and abroad. This paper presents a double eddy current sensor for measuring the liquid level of molten steel in the mold and the liquid level of protective slag molten pool.

Liquid steel level detection methods mainly include isotope method, infrared method, ultrasonic method, thermocouple method and eddy current method [1]. Eddy current method has become the preferred equipment for the production line because of its simple structure, convenient installation, low cost, high temperature state, workpiece narrow area detection, non-contact and other advantages. At present, the electromagnetic induction sensor produced by vuhz A.S company in the Czech Republic is mainly used in the production line. Eddy current testing was first proposed by French scientist Francois Arago in 1842. In the 1950s-1960s, scientists Libby odd Schmidt successively verified and improved the theory of Foerster, developed eddy current testing instruments with good performance, and developed new testing technologies such as multi frequency eddy current testing, far-field eddy current testing and pulse eddy current testing [2]. Huang pingjie [3], Chen Peihua, Cai Qinghua, etc. of Zhejiang University in China have carried out a large number of theoretical and Experimental Research on thickness detection of multilayer conductive structures. In 2020, Wang Baochao first developed a 150 mm large range eddy current sensor [4] and applied it to the deformation detection equipment in the pipeline with a diameter of 1219 mm, It provides valuable results for measuring large range (50mm-160mm) molten steel level and small range molten pool thickness (2mm-20mm) by electromagnetic method.

## 2. Principle of eddy current testing

The probe of mutual inductance eddy current sensor is composed of excitation coil and detection coil. Its working principle is based on the principle of electromagnetic induction. When the excitation coil is connected with AC signal, alternating current I will be generated in the coil, because the electromagnetic induction phenomenon will produce eddy current on the measured object. At the same time, the eddy current will also produce an opposite magnetic field to affect the sensor coil. When the position of the measured object and the sensor changes, the induced magnetic field will also change. Due to the change of the induced magnetic field, the induced voltage of the coil will also change.

In the process of continuous casting, protective slag needs to be added into the mold, so that the mold forms a three-layer structure from top to bottom, namely powder slag, melt and molten steel, as shown in Figure 1.

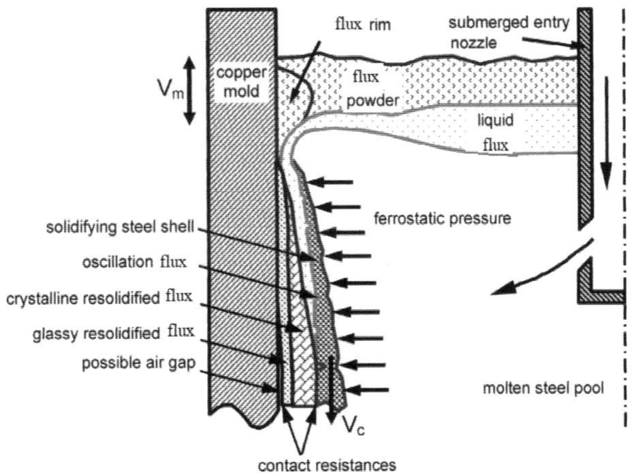


Fig. 1 Schematic diagram of crystallizer structure

For this multilayer structure, the impedance change of the coil can be expressed as a function of multiple parameters such as the characteristics of the tested material, coil parameters, excitation signal and so on:

$$Z=f(r1,r2,L1,L2,L3,\sigma 1,\sigma 2,\ldots,\sigma m,\mu 1,\mu 2,\ldots,\mu m,H1,H2,\ldots,H m,\omega) \quad (1)$$

In formula (1), parameters of probe coil: R1, R2 - inner diameter and outer diameter of probe coil, L1 - probe length; L2 lift off (calculated from the data measured by the millimeter wave sensor and the installation position of the millimeter wave sensor and eddy current sensor), the distance between the L3 sensor and the measured object; Material conductivity: $\sigma$ 1, $\sigma$ 2,..., $\sigma$ m; Material permeability: $\mu$ 1, $\mu$ 2,..., $\mu$ m. H1, H2,... H M-1, h m - thickness of each layer; $\omega$: Excitation angular frequency. When the parameters of the sensor and the properties of the measured object are determined, the impedance Z can be obtained from equation (1) is only related to the size of L3, and the position of molten steel level can be obtained through analysis. For the distance measurement in the mold, two positions need to be measured: the surface of liquid steel, which changes between 50-160mm from the mold mouth. When the liquid steel is stable in the current period, the liquid steel level is about 50mm from the mold mouth. Production experience points out that when the liquid steel is in normal production, the fluctuation range of the liquid level needs to be controlled at about ± 10mm. When the liquid steel is stable, the upper surface of the protective slag molten pool is measured at this time.

## 3. Sensor simulation analysis

The test object of this subject is the liquid steel level in the mold. As shown in Figure 1, the substances in the mold are composed of protective slag, molten protective slag and liquid steel level. Protective slag is a key material necessary for the continuous casting production process. Its main component is a powdery substance based on the low melting point and low viscosity of cao-sio2-al2o3 silicate system, with a melting point of about 1300-1500 ℃ [5], When it is placed on the molten steel, part of the protective slag will melt due to high temperature, forming a molten layer of about 6-15mm. These two substances will have a certain impact on liquid steel detection, so the parameters of the two substances need to be set in the simulation.

In this paper, the finite element simulation software COMSOL multiphysics is used to establish the simulation model of molten steel level detection. The specific parameters are shown in Table 1 below.

Table 1.Parameters of simulation model for liquid steel measurement

| Name | Parameter | Name | Parameter |
|---|---|---|---|
| Excitation coil radius (mm) | 60 | Detection coil radius (mm) | 60 |
| Inner diameter of excitation coil (mm) | 80 | Inner diameter of detection coil (mm) | 80 |
| Number of turns of excitation coil (n) | 400 | Number of turns of detection coil (n) | 4500 |
| Slag thickness (mm) | 3 | Weld pool thickness (mm) | 1 |

The setting of sensor diameter follows the following rules: the larger the coil diameter, the larger the measuring range. When the coil diameter is small, the measuring range is small but the sensitivity is relatively high. In the simulation design, the above factors need to be comprehensively considered to obtain the best size. For the selection of this subject, because the size of the mold is large, the distance between the molten steel level and the top of the mold is far (50mm-160mm), and the distance between the molten pool layer of the protective slag is 2mm-20mm. Considering the requirements of the measured distance, the large coil is used as the simulation model, and the coil size is simulated with COMSOL finite element simulation software to analyze the change of the coil from 20mm to 60mm, the radius changes by 5mm each time, and the detection results are normalized, Get the simulation data diagram as shown in the figure. According to the simulation figure 2 (a), when the radius of the coil is 60mm, the rising curve is smoother than the measured value of the coil with other radius, which is conducive to detection.

Considering the influence of the inner diameter of the coil on the detection results, under the condition that the simulation parameters remain unchanged, change the inner ring radius of the excitation and detection coil from 35mm to 50mm in steps of 5mm, and measure the distance change of the liquid steel surface. As shown in Figure 2 (b), when the radius of the inner ring of the coil changes between 30-50mm, the induced voltage changes basically in the total range. When the radius of the inner ring is 45mm, the curve drawn by the induced voltage is better than that measured by other radii. Therefore, 45mm is selected as the size of the inner ring radius of the coil.

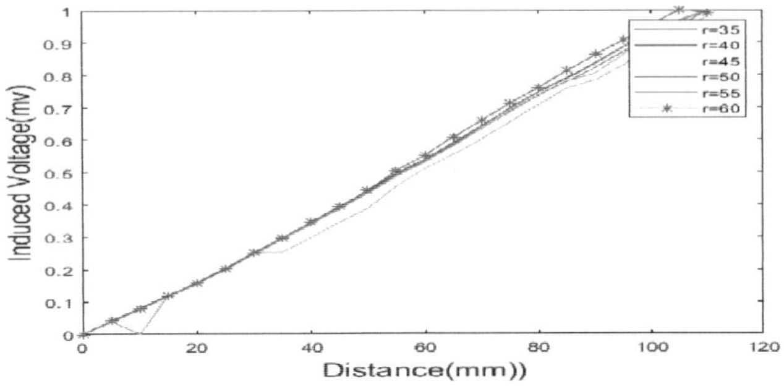


(a) Outer diameter change

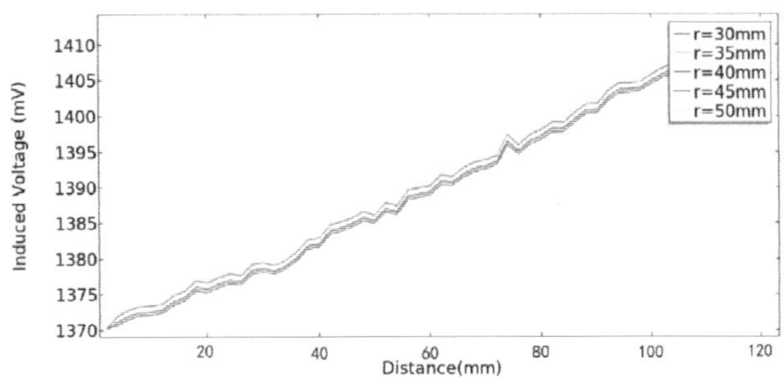


(b) Inner diameter change

Fig. 2 Effect of coil size

The mold is located in a bad environment and there is more electromagnetic field radiation around it. Therefore, how to choose the shape and material of the shielding structure is particularly important. In the simulation, the shape of the shielding structure adopts three forms: full closure, vertical opening and opening. The simulation results are shown in Figure 3.

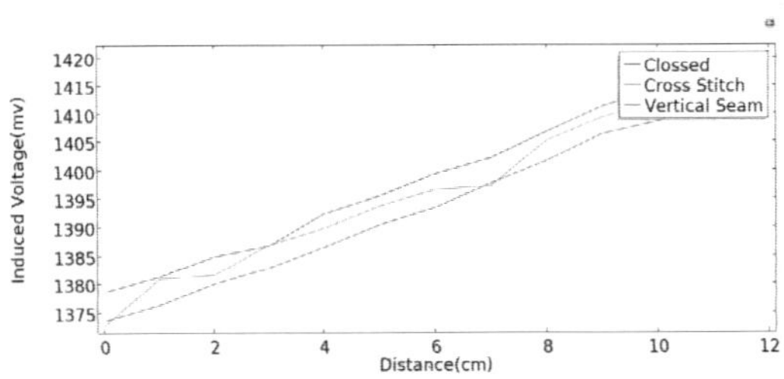


(a) Induced voltage diagram

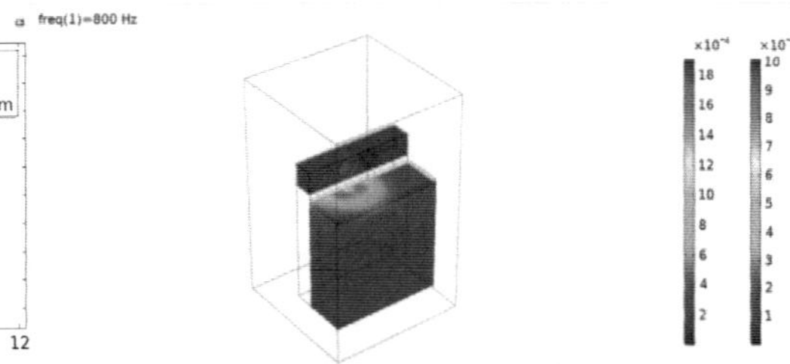

(b) Closed magnetic field

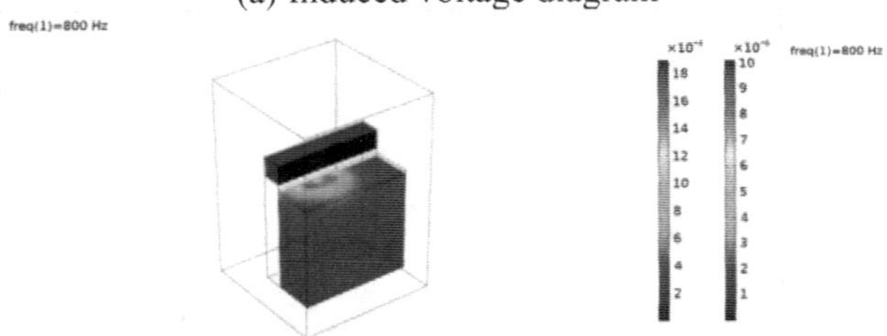

(b) Transverse magnetic field

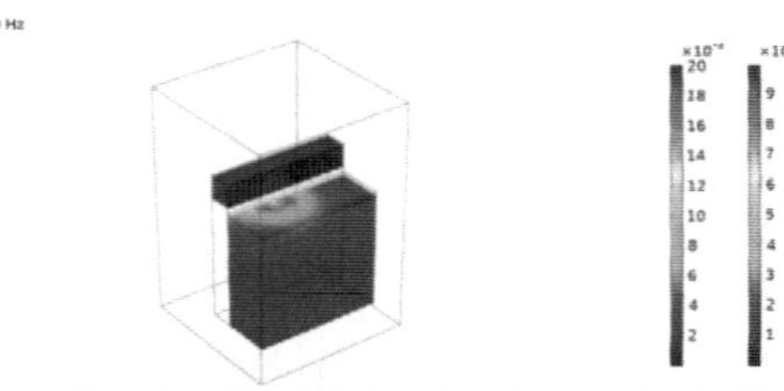

(d) Vertical magnetic field

Fig. 3 Shielding structure

The molten pool has weak conductivity. In the process of measuring the molten steel, the molten pool always exists above the molten steel. Therefore, when measuring the molten steel, it is necessary to analyze the influence of the molten pool on the measurement. When the molten steel is at a stable position, i.e. 50mm, simulate the molten steel with or without molten pool respectively to obtain the induced voltage curve as shown in Figure 4.The conductivity of the molten pool is far lower than that of the liquid steel. The eddy current formed by the low-frequency eddy current on the molten pool can be ignored compared with the eddy current formed on the liquid steel. At the same time, the analysis of the induction curve also shows that the influence of the molten pool on the liquid steel detection can be ignored.

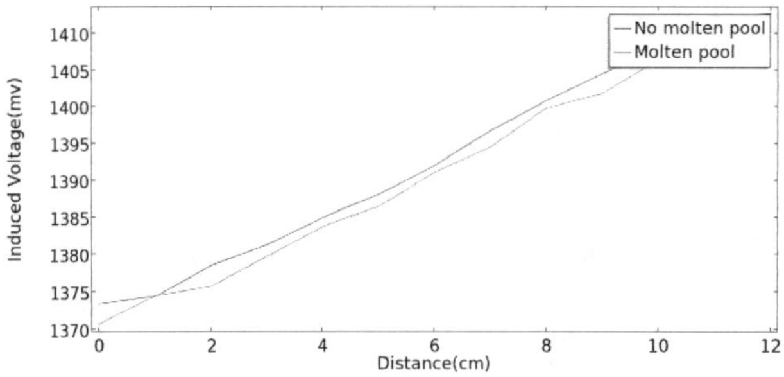


Fig. 4 Influence of molten pool on liquid steel detection

In order to improve the quality of continuous casting slab,a layer of protective slag will be added into the mold during continuous casting production, which plays the functions of thermal insulation, absorbing impurities, improving heat transfer and promoting lubrication. It plays a very important and irreplaceable role in continuous casting quality and the smooth operation of continuous casting process. At present, the main component of protective slag used in China is silicate.The protective slag will form a molten layer of about 6-15mm after contacting the high-temperature liquid steel. Therefore, the weak conductivity of the molten layer will have a certain impact on the detection of liquid steel. Therefore, accurate detection of the thickness of the powder slag molten layer can not only realize the accurate separation of the powder slag layer and the powder slag molten layer, but also realize the stratification between the powder slag molten layer and the surface of liquid steel, which is conducive to the detection of each layer. The simulation model is established through simulation, and the specific parameters are shown in Table 2 below:

Table 2.Simulation measurement of molten pool parameters

| Name | Parameter | Name | Parameter |
|---|---|---|---|
| Excitation coil radius (mm) | 30 | Detection coil radius (mm) | 30 |
| Number of turns of excitation coil (n) | 100 | Number of turns of detection coil (n) | 1000 |
| Excitation frequency (MHz) | 5 | Excitation current (A) | 0.5 |

In the simulation model, the parameters of the fixed liquid steel detection coil and the molten layer detection coil remain unchanged. 10V voltage and 0.5A current are used as excitation signals respectively to measure the change of the molten layer within 1-20mm, and the curve of induced voltage and excitation form is obtained, as shown in Fig. 5.

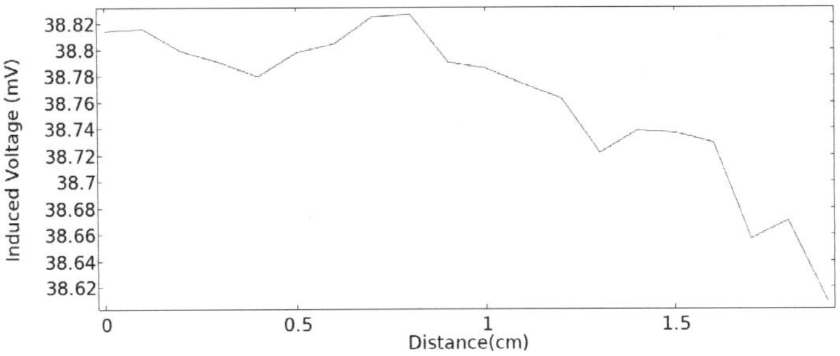


(a) Voltage excitation

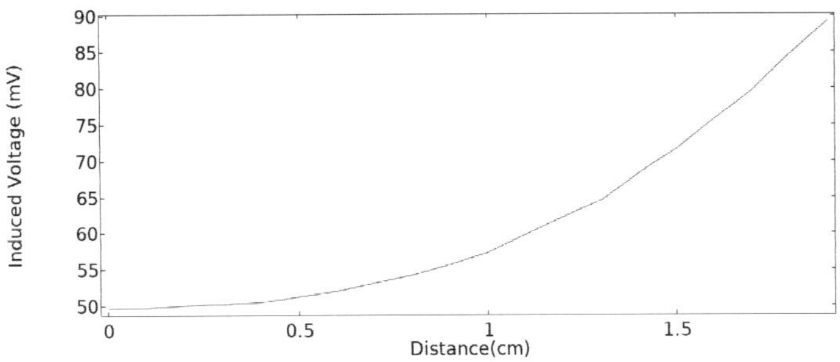


(b) Current excitation

Fig. 5 Incentive form

It can be seen from Figure 5 above that for the measurement of slag layer, the measured effect of current excitation is obviously higher than that of voltage excitation. The measured result of current excitation detection coil is linear for the change of slag layer within 1-20mm, which is more sensitive to the change of slag layer. The use of current excitation can improve the detection accuracy and facilitate observation. Therefore, current is used as the excitation source.

In the mold, because the liquid steel surface will change in 0-160mm, the liquid steel level is at the position of 50 ± 10mm when it is in production. Set the same parameters as the simulation in Fig. 5 (b), and add the liquid steel layer into the simulation model to check the influence of liquid steel on the molten pool. The voltage induction curve as shown in Figure 6 is obtained.

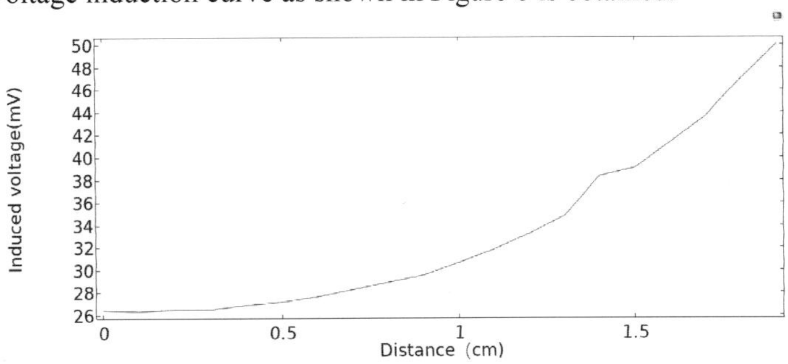


Fig. 6 Effect of molten steel on measurement of molten pool change

By comparing Fig. 5 (b) with FIG. 6, it can be seen that when the molten layer changes in the same range, the voltage amplitude changes from 40mv to 20mV. Through this phenomenon, it can be seen that the existence and change of liquid steel will have a certain impact on the measurement of molten layer, but the measured result is linear. Although the existence and change of liquid steel will have an impact, the measurement of the change of molten slag layer can be realized finally.

 **2264** (2022) 012027 doi:10.1088/1742-6596/2264/1/012027 

## 4. Experimental verification

In order to verify the accuracy of the simulation results, an experimental platform is built for verification. The experimental items include the sensor probe and the measured conductor. From the above, the measurement of the liquid steel distance can be obtained. As for the distance from the liquid steel to the sensor coil, the relationship between the distance from the liquid steel to the sensor and the induced voltage is obtained by using the least square method:

$$U = 0.0004l^2 - 0.47445l + 1590.9 \tag{2}$$

Put in the experimental data, and the inversion results are shown in table 3

Table 3.Liquid steel distance inversion

| Liquid steel distance(mm) | Inversion distance(mm) | relative error/% |
|---|---|---|
| 55 | 55.57 | -1.036 |
| 74 | 73.16 | 1.135 |
| 86 | 87.06 | -1.233 |
| 108 | 108.30 | -0.278 |
| 115 | 116.66 | -1.443 |
| 139 | 142.00 | -2.158 |
| 145 | 145.27 | -0.186 |
| 160 | 159.47 | -0.031 |

The measurement of molten pool can be obtained from the above. The detection of its distance is closely related to the position of molten steel and the thickness of molten pool itself. The relationship between the three is established by using multiple regression model, and the following relationship is obtained:

$$U_1 = 0.4454l_1 + 13.0292l_2 - 0.6940 \tag{3}$$

$U_1$ : molten pool induced voltage, $l_1$ : liquid steel distance(calculated by formula 2), $l_2$ : molten pool distance;Substituting the experimental data, the distance of molten pool is inversed, and table 4 is obtained.

Table 4.Molten pool distance inversion

| Molten pool location（mm） | Inversion distance(mm） | relative error/% |
|---|---|---|
| 5 | 6.048 | 4.96 |
| 15 | 14.53 | -3.13 |
| 13 | 12.95 | -0.38 |
| 19 | 18.65 | -1.84 |
| 2 | 1.58 | -6 |

The experimental results show that the experimental results are basically consistent with the simulation results, which verifies the accuracy of the simulation data.

## 5. Conclusion

Based on the principle of eddy current testing, the sensor coil is trial produced to test the large range of liquid steel and the small range of protective slag molten pool. The simulation model of mold eddy current testing is established by using COMSOL finite element simulation software. Through the simulation coil and optimizing various parameters of the coil, the change of liquid steel in the range of 50-160mm in the mold is measured, and the conductivity of each layer in the mold is greatly different, Through high-frequency signal excitation, the change of slag layer in the range of 6-20mm is detected, and the distance between molten steel is detected.

**Acknowledgments**
Supported by the National Natural Science Foundation of China (61463041); Project supported by science and technology plan of Inner Mongolia Autonomous Region (2020gg0157);Kundulun science and technology plan project of Baotou City, Inner Mongolia Autonomous Region (yf2020010)

**Reference**
[1]He Lamei, Hu Yan. Analysis on automatic control technology and characteristics of molten steel level in continuous casting mold [a]. Proceedings of 2007 National slab continuous casting mold technology seminar [C], Shanghai, 2010.254-257
[2]Bowler J R, Theodoulidis T P. Eddy cueernts induced in a conductingrod of finite length by a coaxial encircling coil.J. Phys D: Appl.Phys. 2005, 38(16):2861-2868
[3]Huang Pingjie, Peng Xiedan, Zhao Shuhao, Zhang Guangxin, Zhang Hongjian Composite detection method of pulsed eddy current and ultrasonic for defects of flat conductive structure [J] Control and decision making, 2019,34(04):743-750
[4]Wang Baochao. Development of Pipeline Deformation Detection Equipment Based on Eddy Current Sensing Technology [J]. Manufacturing Automation, 2020, 42(06): 141-145.
[5]Wang Zhen, Shu Qifeng, Zhou Guozhi. Study on Crystallization Kinetics of CaO-SiO2 (CaO / SiO2 = 1) - tio2-10 mass fraction B2O3 glass slag by differential thermal analysis [J]. Isij international 2015,55 (4): 709-716.

# Evaluation of measurement uncertainty of the orthogonal deviation angle of the target mirror based on AMCM

**Qiangxian Huang [a], Xiaoqian Guo [b], Hongli Li [c*], Zuyang Zhang [d], Zhiwen Chen [e]**

Anhui Province Key Laboratory of Measuring Theory and Precision Instrument, School of Instrument Science and Optoelectronics Engineering, Hefei University of Technology, Hefei , Anhui, 230009, China

[a]email: huangqx@hfut.edu.cn, [b]email: guoxq0617 @163.com, [c]email: hlli@hfut.edu.cn, [d]email: zhangzuyang11@163.com, [e]email: czw001314@gmail.com

[*]Corresponding author's e-mail: hlli@hfut.edu.cn

**Abstract.** The combination of the three-dimensional (3D) target mirror and laser interferometers can achieve 3D precision measurement, and the orthogonality between the target mirror has a significant impact on the accuracy of the measurement system. In order to ensure the accuracy of the measurement system, it is necessary to measure the orthogonal deviation angle of the 3D target mirror and evaluate the uncertainty to verify whether the machining accuracy of the target mirror meets the requirements. In this paper, to test the orthogonality of the X-Y reflection planes of the 3D target mirror, which is glued by three plane mirrors, the evaluation methods of measurement uncertainty of the orthogonal deviation angle $\theta_{xy}$ are studied. The measurement model of uncertainty evaluation is built based on the quantitative characteristics analysis method, and the evaluation results of measurement uncertainty are obtained based on the adaptive Monte Carlo method (AMCM). Moreover, the evaluation results are compared with those evaluation results obtained respectively based on the traditional assessment method of measurement uncertainty (called as GUM method) and Monte Carlo method (MCM). Through the comparison of the evaluation results, it can be concluded that the best estimate of the orthogonal deviation angle $\theta_{xy}$ is about -3.09", the corresponding standard uncertainty is about 0.69", and the machining accuracy of the X-Y reflection planes of the target mirror meets the requirements. Moreover, the AMCM evaluation results is reliable, and AMCM is more convenient and efficient than GUM and MCM methods. The research work provides valuable reference for solving practical measurement evaluation problems.

## 1. Introduction

Based on the principle of interferometry and using laser as length reference, laser interferometer can accurately measure the position accuracy and geometric accuracy of NC equipment, and it is often used in high precision measurement systems. In the design and development of the micro/nano coordinate measuring machine (CMM), the measurement framework is used to realize 3D precision measurement, so it is an important part of micro/nano measuring machine. The measurement frame of the CMM is usually formed by three laser interferometers arranged in the X, Y and Z measurement directions and a 3D target mirror that can reflect the laser beam [1]. The 3D target mirror is made up of three plane mirrors bonded together. In the actual measurement of micro/nano measuring machine, the orthogonality between any two planes of the target mirror has a significant influence on the precision of the measuring system. To guarantee the precision of the measuring system, it must be test whether the machining accuracy of the target mirror meets the requirements.

Measurement uncertainty is a non-negative parameter that characterizes the dispersion of measured values, and is an important index to evaluate the quality of measurement results [2]. Measurement uncertainty is required for modern product inspection and laboratory accreditation [3]. Therefore, in order to test whether the machining accuracy of the target mirror meets the accuracy requirements of the measurement system, it is necessary to measure the orthogonal deviation Angle of the 3D target mirror and evaluate its uncertainty. In this paper, aiming at the demand of the 3D target mirror machining accuracy inspection, the measurement uncertainty of the orthogonal deviation Angle between the reflection planes of the target mirror is evaluated. The measurement model of uncertainty evaluation is determined through analyzing and modeling based on the quantitative characteristics analysis of the measurement system, and the evaluation results of measurement uncertainty are obtained based on the adaptive Monte Carlo method (AMCM), a new evaluation method of measurement uncertainty [4,5]. Finally, by comparing the evaluation results of AMCM with those of GUM and MCM methods, the reliability and efficiency of AMCM evaluation results are verified, and the inspection conclusion of whether the machining accuracy of the target mirror meets the accuracy requirements of the measurement system is obtained.

## 2. Measurement of the orthogonal deviation angle of target mirror

In this paper, it is necessary to inspect whether the machining accuracy of the orthogonal deviation angles between any two planes of the 3D target mirror are within 3". In order to get the test conclusion, the orthogonal deviation angles between any two planes of the 3D target mirror must be measured, then the measurement uncertainties can be evaluated. The orthogonal deviation angle $\theta_{xy}$ between X and Y reflection planes will be taken as an example in the following.

In the study, the Nikon 6B autocollimator and the DP300 precision rotary worktable of RPI company have been used to measure the orthogonal deviation angle $\theta_{xy}$. The measurement experiment is shown in Figure 1. The measuring accuracy of autocollimator can reach ±0.5" in the 5' measuring range. The rotating precision of rotary table can reach ±1", and the resolution is 0.2". Before measuring, the rotation accuracy of rotary table is checked.

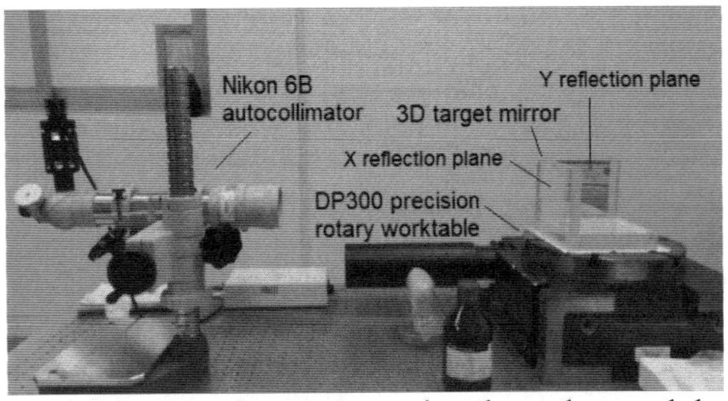


Figure 1. Schematic diagram of the experiment measuring the orthogonal deviation angle of the target mirror.

As shown in Figure 1, install the target mirror, adjust the autocollimator to align the scribed line, rotate the turntable 90° anticlockwise, then rotate the turntable 90° clockwise or 360 ° directly, and observe the reading of the autocollimator. The accuracy of the turntable meets the requirements because the reading is within ±1". Place the target mirror at the initial position as shown in Figure 1. Face the autocollimator to the X reflection plane, adjust the autocollimator to the position where the reading is 0, and then control the turntable to rotate 90° anticlockwise. After the rotation is completed, record the reading of the autocollimator. This data is the measured orthogonal deviation angle $\theta_{xy}$

between X and Y reflection planes. After reading, control the turntable to rotate to the initial position, repeat the above measurement steps, and repeat the experiment until 10 groups of repeated measurements are completed. In order to determine the reproducibility error of the measurement results, change the measurement environment or measurement personnel, 5 groups of reproducibility measurements are performed on $\theta_{xy}$, and 10 measurements are also carried out for each group. The measured data are shown in Table 1 and Table 2.

Table 1. Measurement data of the orthogonal deviation angle $\theta_{xy}$ of the target mirror.

| Number | 1 | 2 | 3 | 4 | 5 | 6 | 7 | 8 | 9 | 10 | Best estimate |
|---|---|---|---|---|---|---|---|---|---|---|---|
| $\theta_{xy}$ | -2.5″ | -3″ | -3.5″ | -2.5″ | -3″ | -3″ | -3.5″ | -3.8″ | -2.8″ | -3.3″ | -3.09″ |

Table 2. Reproducibility measurement data of the orthogonal deviation angle $\theta_{xy}$ of the target mirror.

| Group | Number | 1 | 2 | 3 | 4 | 5 | 6 | 7 | 8 | 9 | 10 |
|---|---|---|---|---|---|---|---|---|---|---|---|
| 1 | $\theta_{xy}$ | -2.5″ | -3″ | -3.5″ | -2.5″ | -3″ | -3″ | -3.5″ | -3.8″ | -2.8″ | -3.3″ |
| 2 | $\theta_{xy}$ | -3.5″ | -3″ | -2.6″ | -3″ | -4″ | -4″ | -3.5″ | -4″ | -3.3″ | -3.7″ |
| 3 | $\theta_{xy}$ | -3.5″ | -3.5″ | -3.2″ | -3.6″ | -3.5″ | -3″ | -3″ | -3.8″ | -3.5″ | -3.7″ |
| 4 | $\theta_{xy}$ | -4.3″ | -3.7″ | -3.8″ | -4.2″ | -3.5″ | -3.2″ | -3.4″ | -3″ | -3.3″ | -2.5″ |
| 5 | $\theta_{xy}$ | -3.7″ | -3.5″ | -3″ | -3″ | -3.3″ | -3.4″ | -3.2″ | -3.5″ | -3.7″ | -3″ |

Usually, the measurement result is composed of the best estimate and the measurement uncertainty. According to the measured data, the best estimate of the orthogonal deviation angle $\theta_{xy}$ of the target mirror is -3.09″. In order to give accurate measurement result and inspect whether the machining accuracy of the 3D target mirror meets the requirements, the measurement uncertainty of $\theta_{xy}$ needs to be evaluated.

## 3. Analysis and modeling of measurement uncertainty

In practical measurement, there are many sources of uncertainty in the measurement system. It is necessary to analyse those main sources of measurement uncertainty according to the actual measurement situation. Generally, there are two different analysis methods: error traceability analysis method and quantitative characteristics analysis method of measurement system. In this paper, the main sources of measurement uncertainty are determined according to the quantitative characteristics analysis method.

The quantitative characteristics analysis method is to analyze the main sources of uncertainty from the quantitative characteristics of the measured results. Generally, the quantitative characteristics indexes that can be applied to the evaluation of measurement uncertainty include offset, linearity, stability, resolution, repeatability and reproducibility.

When measuring the orthogonal deviation angle $\theta_{xy}$ of the target mirror, through the analysis of the above six quantitative characteristics indexes, it is concluded that the main uncertainty sources affecting the measurement results include the system indication error, measurement repeatability, measurement reproducibility, and the change of ambient temperature. In the case of uniform expansion in the target mirror, it can be considered that the expansion will not change the angle between the reflection planes. Therefore, the influence of the change of ambient temperature on the measurement results can be ignored in the actual evaluation.

Thus, the corresponding measurement model for uncertainty evaluation can be established based on the black box modeling method [6]:

$$\theta_{xy} = \theta_0 + \delta_E + \delta_{RT} + \delta_{RD} \tag{1}$$

Where $\theta_0$ is the best estimate of the measurement result ( $\theta_0 = -3.09''$ ), $\delta_E$ represents the system indication error, $\delta_{RT}$ represents measurement repeatability, $\delta_{RD}$ represents measurement reproducibility. These are all input quantities.

Formula (1) will be the application model of AMCM and MCM evaluation methods.

If the corresponding standard uncertainty components of $\delta_E$, $\delta_{RT}$ and $\delta_{RD}$ are independent of each other, the synthetic standard uncertainty expressed in the traditional GUM evaluation method can be obtained:

$$u_C = \sqrt{u_E^2 + u_{RT}^2 + u_{RD}^2} \tag{2}$$

## 4. Uncertainty evaluation by AMCM

Adaptive Monte Carlo method is an upgrade of Monte Carlo method. In the execution process of AMCM, the number of Monte Carlo experiments is continuously increased until the required results reach a statistical stability [7], which can reduce the number of experiments and in turn reduce the cost of simulation experiments. Based on the measurement model in formula (1), the measurement uncertainty of the orthogonal deviation angle $\theta_{xy}$ of the target mirror is evaluated by AMCM method. The assessment process of AMCM is shown in Figure 2.

In the measurement model in formula (1), the distribution of each input can be determined that the system indication error $\delta_E$ follows uniform distribution $R(-1.118'',1.118'')$, repeatability error $\delta_{RT}$ follows normal distribution $N\left(0,(0.137'')^2\right)$, and reproducibility error $\delta_{RD}$ follows normal distribution $N\left(0,(0.191'')^2\right)$. The confidence probability is 95% ( namely P=95% ) .

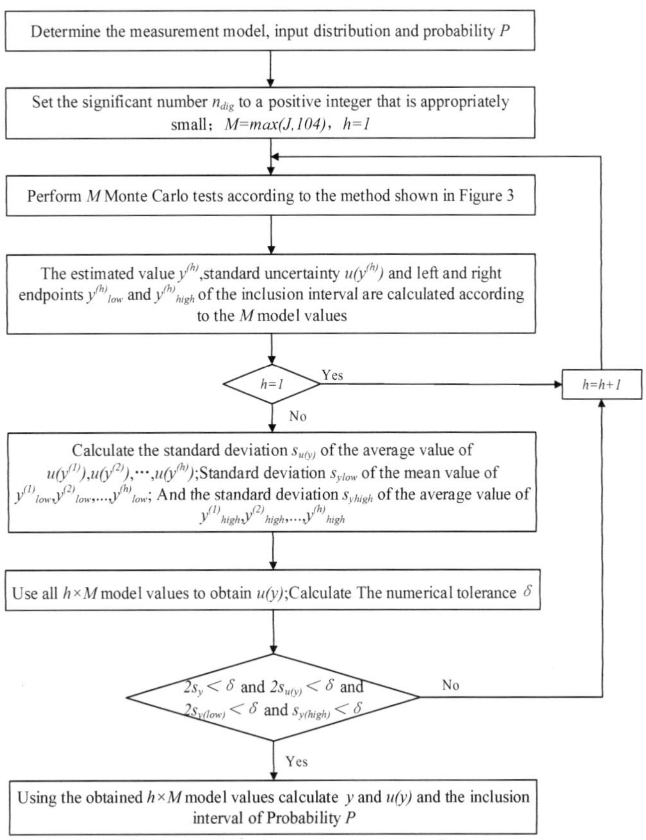


Figure 2. Assessment process of AMCM.

The evaluation results obtained by AMCM are shown in Table 3, and the number of experiments is $23\times10^4$.

Table 3. Evaluation results of AMCM.

| Angle | Best estimate | Standard uncertainty | Inclusive interval |
|---|---|---|---|
| $\theta_{xy}$ | -3.090″ | 0. 687″ | [-4.294″, -1.884″] |

## 5. Comparison of the results of different evaluation methods

In order to verify the reliability and evaluation efficiency of AMCM method, the above evaluation results are compared with those of GUM method and MCM method.

Based on the measurement uncertainty evaluation model shown in formula (2), GUM method is used to evaluate the measurement uncertainty of the orthogonal deviation angle $\theta_{xy}$, and the assessment process of GUM method is shown in Figure 3 [2].

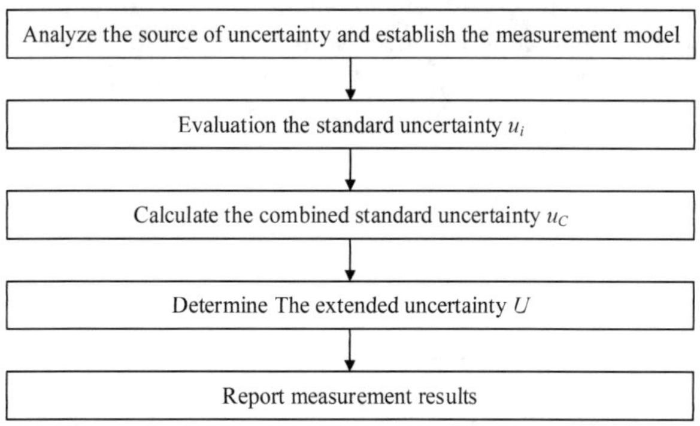


Figure 3. Assessment process of GUM method.

Based on the uncertainty evaluation measurement model shown in formula (1), MCM method is used to evaluate the measurement uncertainty of the orthogonal deviation angle $\theta_{xy}$, and the assessment process of MCM [7] as shown in Figure 4.

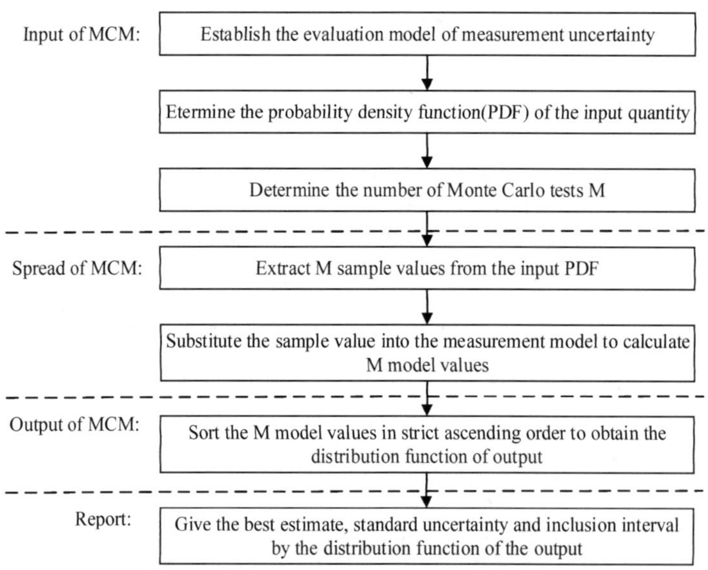


Figure 4. Assessment process of MCM.

The GUM method and the MCM method are respectively used to evaluate the measurement uncertainty of the orthogonal deviation angle A of the target mirror. The comparison results with AMCM method are shown in Table 4. The PDF histogram corresponding to the evaluation results of the three methods is shown in Figure 5.

Table 4. Comparison of results of three evaluation methods ( P=95% ) .

| Angle | Method | Number | Best estimate | Standard uncertainty | Inclusive interval |
|---|---|---|---|---|---|
| $\theta_{xy}$ | GUM | — | -3.090″ | 0.686″ | [-4.434″, -1.744″] |
| | MCM | $10^6$ | -3.089″ | 0.687″ | [-4.297″, -1.883″] |
| | AMCM | $23\times10^4$ | -3.090″ | 0.687″ | [-4.294″, -1.884″] |

It can be seen from Table 4 that when the measurement model is the same, the evaluation results of measurement uncertainty obtained by GUM, MCM and AMCM methods are basically consistent. It has proved that the evaluation results of AMCM method are accurate and reliable, and the evaluation efficiency of AMCM is better than that of MCM evaluation method. The evaluation results show that the machining accuracy of the orthogonal deviation angle of the 3D target mirror basically meets the requirements.

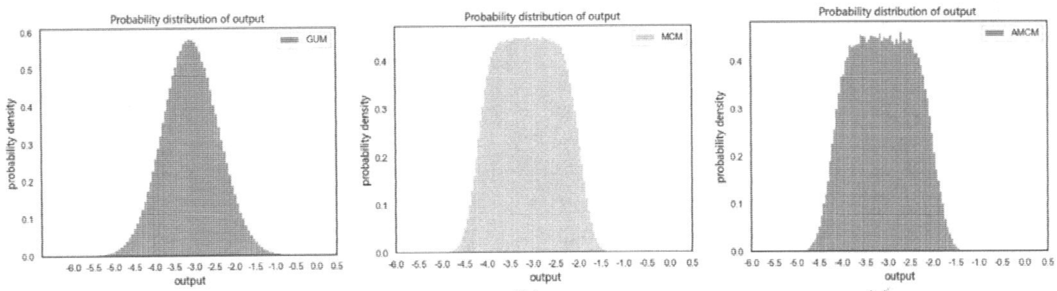

(a) (b) (c)

Figure 5. Corresponding PDF histograms of the evaluation results of the three methods.

At the same time, it can be seen from Figure 5 that MCM and AMCM methods are more in line with the actual situation through a large number of experimental sampling evaluations. Relatively speaking, the output obtained by GUM method still shows that the output conforms to t-distribution type, which does not conform to the actual situation, so GUM method has limitations.

## 6. Conclusion

In summary, the measurement uncertainty evaluation of orthogonal deviation Angle on X-Y plane of the 3D target mirror has been studied, and the specific evaluation process and evaluation results of AMCM have been given. Compared the evaluation results of AMCM with those of GUM and MCM methods, the reliable conclusions can be obtained:

(1) The best estimated value of the deviation angle $\theta_{xy}$ of the target mirror is about -3.09 ", and its standard uncertainty is about 0.69 ". The machining accuracy of the X-Y planes of the target mirror basically meets the accuracy requirements of the measurement system.

(2) Compared with the traditional GUM method, AMCM has more reasonable evaluation results, and it reduces the test number compared with MCM method making the evaluation process more efficient.

The research work can not only test whether the machined target mirror meet the accuracy requirements of the measurement system, judging the machined target mirror whether can be used as an important part of the measurement framework of the micro/nano CMM, but also provide the valuable suggestions about the choice of the uncertainty evaluation method, which will be benefit for spreading the applications of measurement uncertainty.

In the future, on the basis of this research, that how to use error correction to reduce the measurement influence in the actual measurement of micro-nano CMM caused by the orthogonal deviation angle between the target mirror will be studied to improve the measurement accuracy of the measurement system.

**Acknowledgments**

This paper is one of the phased achievements of the national key R & D Program "Basic theory and technology of cross scale micro-nano three-coordinates measuring " (SQ2019YFB200056).

**References**

[1] E Manske, T Fröhlich, R Füßl.(2020) Progress of nanopositioning and nanomeasuring machines for cross-scale measurement with sub-nanometre precision. Measurement Science and Technology, 31:1-8.
[2] GB/T 27418-2017: Evaluation and Expression of Uncertainty in Measurement [S].
[3] CNAS-CL01-G003 2021: Requirements for Measurement Uncertainty.
[4] Guan J Y, Wang Z Y, Li SH. (2021) Monte Carlo assessment of measurement uncertainty for handheld laser scanners. Electronic Measurement Technology.,44:72-76.
[5] ZHANG X J, CHENG Y B, WU J, et al. (2020) Evaluation of measurement uncertainty of gauge block calibration by Adaptive Monte Carlo method. Electronic Measurement Technology, 43(20): 84-88.
[6] GB/T 18779.2-2004: Geometrical Product Specifications(GPS)--Inspection by measurement of workpieses and measuring equipment--Part 2:Guide to the estimation of uncertainty in GPS measurement, in calibration of measuring equipment and in product verification [S].
[7] GB/T 27419-2018: Guide to the evaluation and expression of uncertainty in measurement—Supplement 1: Propagation of distributions using a Monte Carlo method [S].

# A Simultaneous Localization And Mapping Technology Based On Fusion of Radar And Camera

**Junyao Xiao[1], Dejun Ying[1*], Zijian Wu[1] and Zezheng Fang[1]**

[1] School of Mechanical Engineering, Nanjing University of Science and Technology, Nanjing, Jiangsu, 210000, China

[*]Corresponding author's e-mail: xiaojunyao@njust.edu.cn

**Abstract.** In order to solve the problem of incomplete and inaccurate environment perception in the process of simultaneous localization and mapping for autonomous vehicle, this paper proposes a SLAM system for autonomous vehicle by fusing LiDAR and depth camera, installing a relatively cheap single-line LiDAR, depth camera and odometer on the robot, designing a SLAM framework, and using the localization data of LiDAR to assist visual map building for a loosely coupled fusion. It is demonstrated that the method makes the environmental information for building the map richer and more accurate compared with the single sensor mapping. It also improves the perception ability of the driverless trolley for the environment, and provides a better basis for the path planning.

## 1. Introduction

In China, machinery manufacturing has always been the pillar industry of national economic development. Intelligent manufacturing is the main direction of "Made in China 2025", and research on technologies such as intelligent vehicles and robots has become a hot spot. The above research requires a key technology: Simultaneous Localization and Mapping (SLAM). SLAM refers to a subject equipped with a specific sensor, which senses the surrounding environment and pose information through its own sensors in a completely unknown environment to locate itself, builds an incremental map of the environment, and finally completes its own positioning and navigation.

Presently, SLAM is mainly divided into laser SLAM and visual SLAM. Laser SLAM has a small sensing range, serious lack of environmental information, and high cost of better radar. Although visual SLAM can obtain rich environmental features, it is easily affected by the environment and has a large amount of data processing. Therefore, SLAM based on sensor fusion such as laser and vision has become a major research and development trend in the future. This paper proposes a SLAM solution that integrates 2D lidar and RGB-D cameras, and builds an experimental platform for mobile robot, and conducts mapping experiments.

## 2. Sensor performance and SLAM algorithm

### 2.1. Sensor performance

The laser sensor used in this paper is the SLAMTEC RPLIDAR S2. Its measurement radius is 30m, sampling frequency is up to 32k. The vision sensor used in this paper is the ORBBEC Astra Pro depth camera. Its operating range is 0.6-8 m, with a horizontal viewing angle of 66.1°, a vertical viewing angle of 40.2°, and an operating accuracy of ±3 mm at a range of 1 m. Figure 1 shows the two sensors.

Figure 1. SLAMTEC RPLIDAR S2 and ORBBEC Astra Pro.

### 2.2. 2D mapping based on Gmapping

The Gmapping algorithm is an open-source SLAM algorithm based on the RBPF (Rao-Blackwellized Particle Filters) particle filtering framework, which improves the proposed distribution mainly on the RBPF. The RBPF constructs a joint distribution of maps as Eq. 1.

$$p\left(x_{1:t}, m \mid z_{1:t}, u_{1:t-1}\right) = p\left(m \mid x_{1:t}, z_{1:t}\right) \cdot p\left(x_{1:t} \mid z_{1:t}, u_{1:t-1}\right) \tag{1}$$

Where $x_{1:t}$ is the location data; m is the environment data; $u_{1:t}$ is the control data; $z_{1:t}$ is the observation data. To improve the proposed distribution, Gmapping uses the most recent observation, so the proposed distribution becomes Eq. 2. The formula for the weight of the particle becomes Eq. 3.

$$p\left(x_t \mid m_{t-1}^{(i)}, x_{t-1}^{(i)}, z_t, u_{t-1}\right) = \frac{p\left(z_t \mid m_{t-1}^{(i)}, x_t\right) p\left(x_t \mid x_{t-1}^{(i)}, u_{t-1}\right)}{p\left(z_t \mid m_{t-1}^{(i)}, x_{t-1}^{(i)}, u_{t-1}\right)} \tag{2}$$

$$\omega_t^{(i)} = \varphi_{t-1}^{(i)} \cdot \int p\left(z_t \mid x\right) p\left(x \mid x_{t-1}^{(i)}, u_{t-1}\right) dx \tag{3}$$

To obtain the improved proposal distribution, the particles are collected from the motion model in the first step, and these particles are weighted in the second step to select the best particles, and then the improved proposal distribution is simulated using these weighted particles. K values are collected near the peak and a Gaussian function is used to construct the proposed distribution as in Eq. 4 and Eq. 5.

$$\mu_t^{(i)} = \frac{1}{\eta^{(i)}} \cdot \sum_{j=1}^{K} p\left(z_t \mid m_{t-1}^{(i)}, x_j\right) \cdot p\left(x_j \mid x_{t-1}^{(i)}, u_{t-1}\right) \cdot \left(x_j - \mu_t^{(i)}\right)\left(x_j - \mu_t^{(i)}\right)^{\mathrm{T}} \tag{4}$$

$$\sum_t (i) = \frac{1}{\eta^{(i)}} \cdot \sum_{j=1}^{K} p\left(z_t \mid m_{t-1}^{(i)}, x_j\right) \cdot p\left(x_j \mid x_{t-1}^{(i)}, u_{t-1}\right) \cdot \left(x_j - \mu_t^{(i)}\right)\left(x_j - \mu_t^{(i)}\right)^{\mathrm{T}} \tag{5}$$

With the simulated proposal distribution, the positional information of the robot at the next moment can be sampled. With the pose information, a local map can be constructed to generate a 2D map.

### 2.3. 3D mapping based on RTABMap

RTABMap (Real-Time Appearance-Based Mapping) is the mainstream RGB-D SLAM method. The basic process of RGB-D SLAM method is that the camera can acquire a certain number of frames of RGB images and Depth images at the same time per second, and after greying out the RGB color images, the feature extraction algorithm is used to extract the feature points of the grayscale images before and after the two frames. Assuming that there are two image feature point sets before and after the frame, a series of 3D spatial coordinates corresponding to the feature points in the 2D image are calculated using Eq. 6 in combination with the camera model parameters and the depth image. where u and v denote the coordinates of a pixel in the grayscale image, dep$(u,v)$ is the depth value corresponding to the image coordinate point $(u,v)$, $f_x$, $f_y$, $c_x$ and $c_y$ are the built-in parameters of the camera, and $x$, $y$, and $z$ denote the 3D spatial position corresponding to the image plane coordinate point $(u,v)$.

$$z = \frac{\mathrm{dep}(u,v)}{s}$$
$$x = \frac{(u - c_x)\cdot z}{f_x} \tag{6}$$
$$y = \frac{(u - c_y)\cdot z}{f_y}$$

Note that the set of 3D points corresponding to the set of feature points $I_1$ and $I_2$ of the before and after two frames are $F_1$ and $F_2$: $P = p_1, p_2, \ldots, p_N \in F_1$ , $Q = q_1, q_2, \ldots, q_M \in F_2$.

$p$ and $q$ are both points of the current frame point set. Based on the correctly matched pixel pairs between the previous and the next frame images, the RANSAC algorithm is used to perform the alignment between the 3D coordinates P and Q of the key points of the two frames to find the minimized set of interior point matches to estimate the initial relative positional transformation of the mobile robot. Subsequent steps such as optimization of relative poses using ICP algorithm are also performed, which are not described here.

Figure 2 shows the RTABMap system architecture. The camera acquires RGB images, depth images and camera parameters to the visual odometer node, extracts image features and performs feature matching, calculates the initial relative position using the RANSAC method, and then calculates the optimal relative position of the robot using the ICP algorithm; the environmental map construction node RTABMap acquires RGB images, depth images, odometer and camera information to perform closed-loop detection and optimize the trajectory to construct a 3D point cloud map.

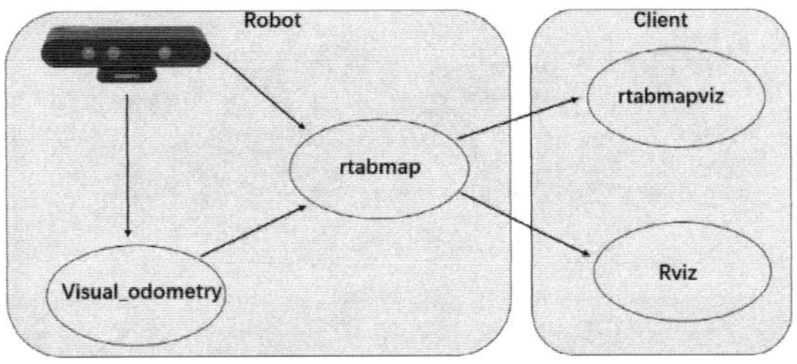


Figure 2. RTABMap system architecture diagram.

## 3. Radar and camera fusion mapping

Figure 3 shows the fusion scheme in this paper.

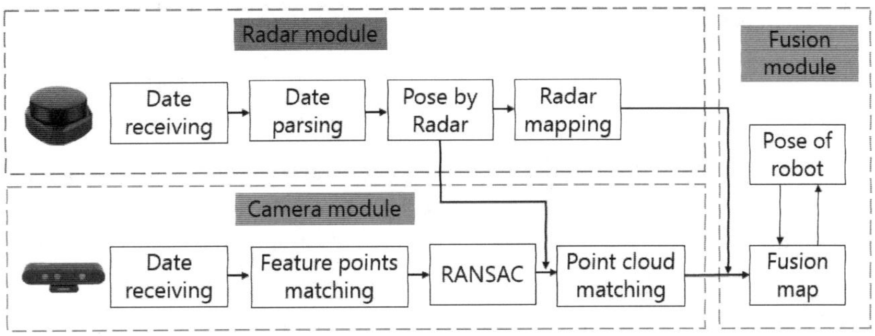


Figure 3. Workflow of multi-sensor fusion.

In this paper, a loosely coupled method is used to perform pose fusion of laser and vision sensors. Both vision and laser sensors are independent to a certain extent, and they can be fused after motion estimation is completed independently. At the same time, binocular vision is an independent and complete high-precision mileage method, and there is no scale uncertainty like monocular vision. The loosely coupled algorithm cannot destroy integrity and facilitate its independent optimization.

Specifically, the mobile robot uses the positioning information provided by the lidar and the odometer, and the camera provides the point cloud. The system stitches the point cloud information according to the laser positioning information, performs loosely coupled pose fusion, and outputs a fusion map.

## 4. Experiment

The experimental platform built in this paper is a mobile robot, as Figure 4. The robot is equipped with RPLIDAR S2, Astar Pro and ROS main controller with Jetson Nano TX1. The bottom main controller is based on STM32F103RCT6 driver board and the chassis of the robot is Ackerman chassis with independent suspension. The PC side is equipped with the ROS melodic operating system running under the Linux (ubuntul8.04) system. The experiment is carried out under outdoor conditions and the results were visualized by Rviz of ROS melodic. The real scene diagram is shown in Figure 5.

Figure 4. Ackerman mobile robot.

Figure 5. Real experimental scene.

The result of mapping by Astar Pro is shown in Figure 6. It can be seen from this method that the error of the robot trajectory obtained by this method is large, which increases the difficulty of loop detection and leads to point cloud splicing. The effect is not ideal, there are errors in the positions of objects such as pits and cones on the ground, and the point cloud map of the real scene cannot be well reflected. In addition, environmental information at the glass door is also missing.

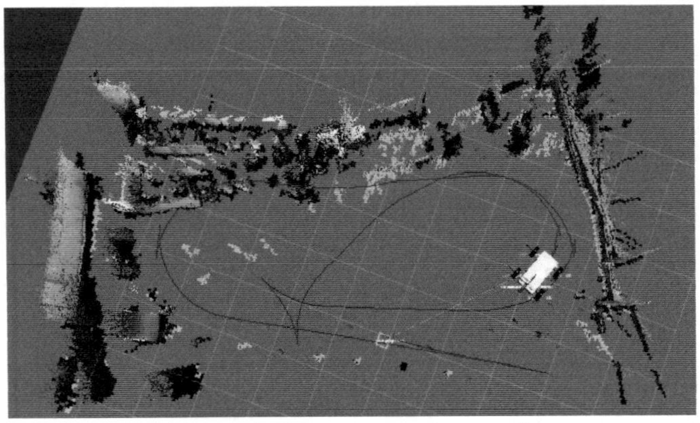

Figure 6. Camera mapping result.

The result of mapping by RPLIDAR S2 is shown in Figure 7. It can be seen that this method cannot detect objects below the radar installation level, such as the cone barrel and the small pit on the ground. Failure to output good environmental information for the subsequent path planning of the robot will affect the safety and stability of the robot situation.

Figure 7. Radar mapping result.

After using the fusion algorithm in this paper, the result of mapping is shown in Figure 8. It can be clearly seen that the positional accuracy of each point cloud is high, and the point cloud stitching effect is excellent. The locations of point clouds are also very accurate when facing challenging scenes like cone barrels, ground pits, and glass doors.

It can be seen From Figure 9 that the visualization effect of the details of the point cloud image is very ideal, which is due to the effective loop closure detection by laser-assisted vision, which optimizes the mapping effect. This experiment optimized the completeness and accuracy of the map, and laid a good foundation for the subsequent path planning of the robot.

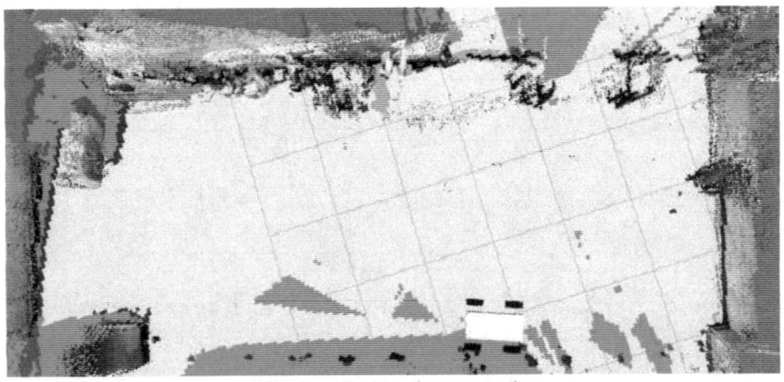

Figure 8. Fusion result.

Figure 9. Partial view of fusion result.

## 5. Conclusion

The SLAM system by fusing LiDAR and Depth Camera built in this paper enables richer and more accurate environmental elements of the map. It greatly improves the fit with the actual scene while controlling the sensor cost. At the same time, there are some shortcomings. On the one hand, the fusion algorithm increases a certain amount of calculation, and has high requirements for the computing power platform. If a master controller with a large performance gap from Jetson TX1 is used, the real-time performance of SLAM will be affected due to operational problems; on the other hand, this paper The experimental scene is a closed-loop small scene. The positioning accuracy of the robot is relatively high and the pose is relatively accurate. If the large-scale scene with a huge displacement is using for experiment, the algorithm needs to be further studied.

## References

[1] Cadena, C., Carlone, L., Carrillo, H. (2016) Future of Simultaneous Localization and Mapping: Towards the Robust Perception Age. IEEE Trans on Robotics., 32: 1309-1332.

[2] Jin, Q.Y., Liu, Y.G., Man, Y.C. (2019) Visual SLAM with RGB-D Cameras. In: Chinese Control Conference. Guangzhou. pp. 4072-4077.

[3] Serafin, J., Grisetti, G. (2015) Dense Normal Based Point Cloud Registration. In: International Conference on Intelligent Robots & Systems Systems. Hamburg. pp. 742-749.

[4] Wei, S.F., Pang, F., Liu, Z.B. (2020) Survey of LiDAR-based SLAM algorithm. Application Research of Computers., 37: 327-332.

[5] Zhou, Y.Q., Mao, X.S. (2021) Simultaneous distance and speed measurement method for automotive lidar. Optoelectronics Technology., 41: 153-158.

[6] He, K., Zhang, X., Ren, S., Sun, J. (2016) Deep residual learning for image recognition. In: 29th IEEE Conference on Computer Vision and Pattern Recognition. New Jersey. pp. 770–778.

[7] Liu, W., Anguelov, D., Erhan, D. (2016) SSD: Single Shot MultiBox Detector. In: European Conference on Computer Vision. Amsterdam. pp. 21-37.

[8] Rashinkar, P., Krushnasamy, V.S. (2017) An overview of data fusion techniques. In: 2017 International Conference on Innovative Mechanisms for Industry Applications. Bengaluru. pp. 694-697.

[9] Fung, M.L., Chen, M.Z.Q., Chen, Y.H. (2017) Sensor fusion: A review of methods and applications. In:Control & Decision Conference. New York. pp. 165-178.

[10] Zhang, Y.X., Feng, K.P. (2014) Research and Performance Analysis on Pedestrian Detection Based on Optimized Hog and SVM. Applied Mechanics & Materials., 511: 247-250.

# Coordinated Control of Path Following and Yaw Stability for Distributed Drive Autonomous Vehicle

**Zhikun Liu, Dejun Yin* and Ruiming Guo**

School of Mechanical Engineering, Nanjing University of Science and Technology, Nanjing, Jiangsu, 210000, China

*Corresponding author's e-mail: yin@njust.edu.cn

**Abstract.** This paper researches a coordinated control approach of path following and yaw motion stability for distributed drive autonomous electric vehicle. In the upper-level controller, a linear time-varying model predictive control algorithm is used to obtain the front wheel steering angle to guarantee the vehicle to drive along the planned path. Meanwhile, the sliding mode control theory is aimed to calculate the external yaw moment to ensure the vehicle yaw stability. In the low-level controller, this research proposes the particle swarm optimization algorithm to execute the torque distribution control. The Matlab/ Simulink simulation results indicate that the coordinated control strategy can not only effectively improve the accuracy of path following, but also guarantee yaw motion stability of the autonomous vehicles.

## 1. Introduction

Autonomous ground vehicle (AGV) is a high-tech fusion which comprises environmental information perception, target path planning, vehicle motion decision-making and chassis control [1]. Path following is one of the key technologies to realize intelligentization of the automobiles and is closely related to the active security and ride comfort. At present, the main popular path following control methods include PID, optimal preview control, LQR, MPC, etc.

Model predictive control was first applied in the industrial field and is a feedback control method based on the principle of optimization [2]. It can predict the output variables of the system in the future according to the kinematics or dynamic model of the system, and can also consider the dynamic characteristic constraints of each actuator in the system as well as the vehicle state constraints [3].

At present, MPC has been widely used in vehicle active safety and path following control. Ren *et al.* proposed a nonlinear MPC control method, which achieved effective trajectory tracking results with high-speed and on low-adhesion roads [4]. Liu designed a nonlinear MPC controller, which realized the collision avoidance function of the autonomous vehicle [5]. Yakub and Lee designed MPC controllers based on front wheel steering and conducted path following verification on low-adhesion roads [6]. Guo proposed a path tracking control strategy with yaw motion stabilization, and established a controller considering yaw rate, sideslip angle and Lyapunov stability condition [7].

By analyzing the above reference literatures, we will learn that lots of literatures mainly concentrate on improving precision of path following, but neglect the vehicle yaw motion stability in the control process, which causes the AGVs have bad tracking performance under extreme conditions (high vehicle speed or low adhesion road) [8]. Therefore, a coordinated control strategy is adopted to decrease the tracking deviation while guaranteeing yaw motion stability of the AGVs under such extreme condition.

The structure of this article is as follows: Part 2 presents the vehicle dynamic model, part 3 shows the coordinated control strategy of path tracking and torque distribution strategy based on particle swarm optimization (PSO). Part 4 is the analysis and comparison of the simulation results, and then the conclusions are summarized in part 5.

## 2. Vehicle Dynamics Model and LTV MPC Algorithm

Path following is essentially the control of the vehicle motion state. To reduce the process complexity and maintain the real-time performance of the LTV MPC controller, it is necessary to simplify constraints from suspension, tires and road surfaces in the modeling. Based on this, this paper adopts the monorail model with three DOF of longitudinal, lateral and yaw motion, considering the vehicle dynamics model and tyre model, the non-linear dynamic equations are given as follows:

$$\begin{aligned} m(\dot{u} - vr) &= 2F_{lf}\cos\delta_f - 2F_{cf}\sin\delta_f + 2F_{lr} \\ m(\dot{v} + ur) &= 2F_{lf}\sin\delta_f + 2F_{cf}\cos\delta_f + 2F_{cr} \\ I_z\dot{r} &= 2a\left(F_{lf}\sin\delta_f + F_{cf}\cos\delta_f\right) - 2bF_{cr} \end{aligned} \tag{1}$$

The lateral force can be expressed as:

$$F_{cf} = C_{\alpha f}\left(\delta_{wf} - \frac{v+ar}{u}\right) \quad F_{cr} = C_{\alpha r}\left(\frac{br-v}{u}\right) \tag{2}$$

This paper chooses $\xi = [v, r, \varphi, u, X, Y]^T$ as the state variables of the path following system, $u = [\delta]$ as the input.

In order to improve the calculation efficiency to satisfy the real-time demand of the controller, this paper linearises the non-linear system equations by adopting the first-order expansion of Taylor series.

$$\dot{\xi} = f(\xi_r, u_r) + \frac{\partial f}{\partial \xi}(\xi - \xi_r) + \frac{\partial f}{\partial u}(u - u_r) \tag{3}$$

$$\Delta\dot{\xi} = A(t)\cdot\Delta\xi + B(t)\cdot\Delta u \tag{4}$$

The first-order difference quotient method is used to obtain discrete state expression (5):

$$\begin{aligned} \Delta\xi(k+1) &= A(k)\Delta\xi(k) + B(k)\Delta u(k) \\ A(k) &= I + TA(t),\ B(k) = TB(t) \end{aligned} \tag{5}$$

To effectively constrain the increment of the control amount of the vehicle, the discrete linear state space model established in (5) needs to be transformed as (6).

$$\xi(k,t) = \begin{bmatrix} \Delta\xi(k,t) \\ u(k-1,t) \end{bmatrix} \tag{6}$$

The state variables and system output vectors in the prediction horizon can be calculated as (7).

$$\begin{aligned} \xi(t+N_p|t) &= A_t^{N_p}\xi(t|t) + A_t^{N_p-1}B_t\Delta u(t|t) + \cdots \\ &+ A_t^{N_p-N_c-1}B_t\Delta u(t+N_c|t) \\ \eta(t+N_p|t) &= C_tA_t^{N_p}\xi(t|t) + C_tA_t^{N_p-1}B_t\Delta u(t|t) + \cdots \\ &+ C_tA_t^{N_p-N_c-1}B_t\Delta u(t+N_c|t) \end{aligned} \tag{7}$$

The output matrix is represented as (8).

$$Y(t) = \vartheta_t\xi(t|t) + \theta_t\Delta U(t) \tag{8}$$

## 3. Coordinated Controller Design

The coordinated control method for path following and yaw motion stability is presented in this part, and the comprehensive control system architecture diagram is showed in Fig. 1. The MPC controller is used to decrease the lateral deviation and heading angle error in the path tracking, while keeping the desired sideslip angle and yaw rate to sustain the yaw motion stability. In the low-level controller, this part proposes torque distribution strategy based on PSO thePMBDA2021ory in order to fulfill the upper-level control requirement by applying the extra motor torque to each in-wheel motor of the AGVs.

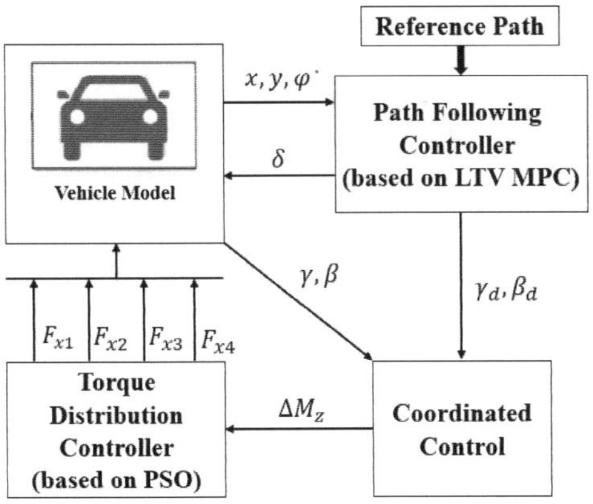

Figure 1. hierarchical control system.

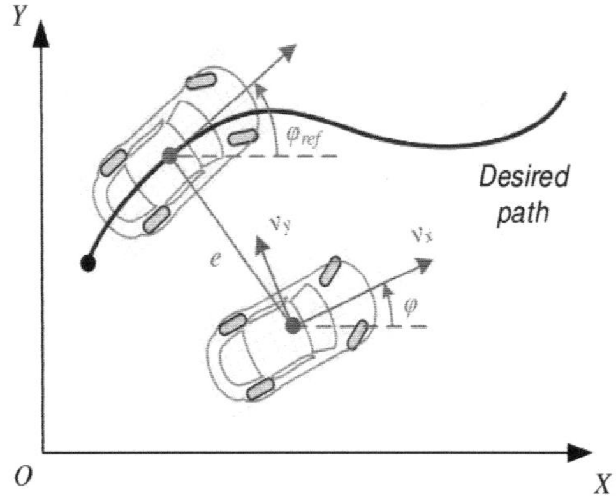

Figure 2. Path following model.

### 3.1. Path Following Controller

This research takes advantage of the model predictive control method, which possesses better ability to predict future trajectories and process multi-objective constraints, meanwhile it possesses robustness to sensor noise and variation of vehicle state parameters.

The optimal solution calculated from LTV-MPC controller aims to obtain the best control input, that is front wheel steering angle. The path following model is shown in Fig. 2. The cost function of path following can be defined as (9).

$$J(\xi(t), u(t-1), \Delta u(t)) = \sum_{j=1}^{N_p} \left\| \eta(t+j|t) - \eta_{ref}(t+j|t) \right\|_Q^2 + \sum_{i=1}^{N_c-1} (\|\Delta u(t+i|t)\|_R^2 + \|u(t+i)\|_S^2) + \rho\varepsilon^2 \quad (9)$$

We transform the cost function to the quadratic form and then solve the quadratic programming (QP) problem on the purpose of calculating the optimal solution vector. We will finally obtain the standard quadratic form of the cost function as (10).

$$J(\xi(t), u(t-1), \Delta U(t)) = [\Delta U(t)^T \ \ \varepsilon]^T H_t [\Delta U(t)^T \ \ \varepsilon] + G_t [\Delta U(t)^T \ \ 0] + P_t \quad (10)$$

Equation (10) is calculated for each control cycle to get the sequence of control increments in the control horizon:

$$\Delta U_t = [\Delta u_t \ \ \Delta u_{t+1} \ \ \cdots \ \ \Delta u_{t+N_c-1}]^T \quad (11)$$

The first increment $\Delta u_t$ of the function solution will be added to the control variable $u$(t-1) from last iteration process:

$$u(t) = u(t-1) + \Delta u_t$$

### 3.2. Yaw Motion Stability Control

Constraints on system variables of output and control increments have been added in the previous section. In addition, to improve the yaw stability of the AGVs during path following and the adaptability to the road environment, it is also necessary to add vehicle dynamics constraints. Thus we need to constrain the state parameters related to vehicle handling stability. The yaw rate and the sideslip angle have a great relationship with trajectory retention and vehicle stability. Assume that $\gamma_d$ and $\beta_d$ are respectively the desired yaw rate ($\gamma$) and sideslip angle ($\beta$), when the tire is in a linear area, $\gamma_d$ and $\beta_d$ will be expressed as (12).

$$\gamma_d = \frac{1}{1-\frac{mu^2}{2l^2}\frac{l_fC_f-l_rC_r}{C_fC_r}}\frac{u}{l}\delta \qquad \beta_d = \frac{1-\frac{mu^2}{2l}\frac{l_f}{l_rC_r}}{1-\frac{mu^2}{2l^2}\frac{l_fC_f-l_rC_r}{C_fC_r}}\frac{l_r}{l}\delta \quad (12)$$

This research adopts the sliding mode control (SMC) to obtain desired yaw moment. The sliding surface to which the parameters converge and switching control law are defined as

$$S = a(\gamma - \gamma_d) + b(\beta - \beta_d) \qquad \dot{S} = -k_1 sgn(S) - k_2 S \tag{13}$$

We assume the AGV is steering steadily for that steering angle $\delta$ can be obtained from the LTV MPC controller. In addition, the target external yaw moment produced by the longitudinal tire force of four wheels can be calculated as follows.

$$\Delta M_z = I_z\left(\dot{\gamma_d} + \frac{1}{a}\left(-k_1 sgn(S) - k_2 S - b(\dot{\beta} - \dot{\beta_d})\right)\right) - l_f(F_{y1}+F_{y2})\cos\delta + l_r(F_{y3} + F_{y4}) - \frac{d_f}{2}(F_{y1}-F_{y2})\sin\delta \tag{14}$$

In Eq. (19), a=1, b= -1, $k_1$=0.01, $k_2$=50.

The torque distribution controller aims to track the desired yaw moment $\Delta M_z$ by adjusting the motor torque from each wheel. According to the optimal control theory, this research adopts particle swarm optimization (PSO) algorithm to achieve torque distribution.

The optimal torque distribution control strategy proposed in this paper is described as

$$\min J = \sum_{i=1}^{4} C_i \frac{M_{xi}^2}{0.001 + M_{yi}^2} \qquad C_i = 1 - \frac{\Delta F_{zi}}{F_{zi\_c}}$$

$$s.t. \quad M_{zdesire} = \textstyle\sum_{i=1}^{4} (M_{xi} + M_{yi}) \qquad |F_{xi}| \le \frac{T_{imax}}{R_w} \qquad |F_{xi}| \le \mu F_{zi} \tag{15}$$

$F_{zi\_c}$ is the static load of each wheel, $\Delta F_{zi} = F_{zi} - F_{zi\_c}$ , $M_{xi}$ and $M_{yi}$ are respectively the yaw moment produced by longitudinal and lateral force of each tyre. $T_{imax}$ is the maximum value of torque in each motor. Equation (15) indicates the cost function of vehicle yaw stability. Vehicle should generate sufficient lateral force to maintain its handling stability when steering extremely.

In order to reduce the amount of calculation, it is vital to simplify the model, and at the same time try to satisfy the tire force ellipse model. Therefore, the longitudinal force is used to represent the lateral force, which is expressed as (16).

$$F_{yi} = \mathrm{sgn}\,(\delta) \cdot \sqrt{(0.8 \cdot \mu \cdot F_{zi})^2 - F_{xi}^2} \tag{16}$$

The desired total longitudinal force $F_{desire}$ and yaw moment $M_{zdesire}$ satisfy the following equations:

$$(F_{x1} + F_{x2}) \cos\delta + F_{x3} + F_{x4} = F_{desire}$$

$$(F_{x1} + F_{x2}) l_f \sin\delta + (F_{x2} - F_{x1}) \frac{d_f}{2} \cos\delta + \frac{d_r}{2}(F_{x4} - F_{x3}) = M_{zdesire} \tag{17}$$

Substituting (17) into (15), the cost function is expressed as:

$$J = J_1 + J_2 + J_3 + J_4 \tag{18}$$

$$J = \frac{\left(l_f \sin\delta - \frac{d_f}{2}\cos\delta\right)^2}{\left(l_f \cos\delta + \frac{d_f}{2}\sin\delta\right)^2} \cdot \frac{F_{x1}^2}{(\mu s F_{z1})^2 - F_{x1}^2} + \frac{\left(l_f \sin\delta + \frac{d_f}{2}\cos\delta\right)^2}{\left(l_f \cos\delta - \frac{d_f}{2}\sin\delta\right)^2} \cdot \frac{F_{x2}^2}{(\mu s F_{z2})^2 - F_{x2}^2}$$
$$+ \frac{(\frac{d_f}{2})^2}{l_r^2} \cdot \frac{F_{x3}^2}{(\mu s F_{z3})^2 - F_{x3}^2} + \frac{(\frac{d_r}{2})^2}{l_r^2} \cdot \frac{F_{x4}^2}{(\mu s F_{z4})^2 - F_{x4}^2}$$

Particle swarm optimization (PSO) algorithm aims to find the optimal solution vector through cooperation and information sharing among individual particles in the swarm. It has the characteristics of less calculation time, relatively high allocation accuracy and less adjusted parameters [9]. Therefore, it possesses high efficiency for real-time control and is suitable for improving the efficiency of torque distribution strategy in this paper.

According to PSO algorithm, the particle will find the individual optimal solution and the global optimal solution for each iteration, and it will update its speed and position through the following formulas:

$$V_{id}^{k+1} = \omega V_{id}^k + c_1 r_1 (P_{id}^k - x_{id}^k) + c_2 r_2 (P_{gd}^k - x_{id}^k) \qquad x_{id}^{k+1} = x_{id}^k + V_{id}^{k+1}$$

In this research, the search position range of particle is [-1600, 1600], the range of motion speed is [-500, 500], the acceleration constant ($c_1$ , $c_2$) is 2, the maximum number of iterations is 30 and the number of particles is 10. And the fitness function of this strategy is F, expressed as (19).

$$F = J = J_1 + J_2 + J_3 + J_4 \tag{19}$$

## 4. Simulation Results

This research adopts the emergency obstacle avoidance condition on the purpose of verifying the effectiveness of the coordinated controller. The vehicle speed is 80 km/h, and the road friction coefficient μ is 0.5. The simulation results are showed in Fig .3 and Fig .4.

Fig. 3 presents the path following comparison, lateral deviation and heading angle error results, which show the path following capability of the AGV. As presented in Fig. 3(a) and (b), the vehicle under the method of autonomous steering without yaw moment control has a large lateral position error, while adopting the coordinated controller, the error is maintained within a smaller range (0.5 m). This conclusion also applies to the comparison of vehicle heading angle deviation in Fig. 3(c). The absolute value of the heading angle deviation with the MPC method reaches 3.5 deg, which is larger than the coordinated control.

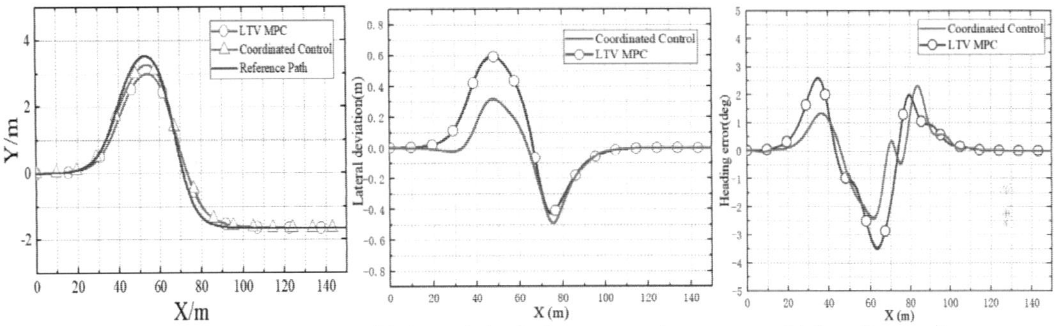


(a) path comparison (b) lateral deviation comparison (c) heading angle error
Figure 3. Path following simulation results.

Fig. 4 are the results of yaw rate comparison and sideslip angle comparison for the AGV, which demonstrates the benefit of the coordinated control method on vehicle yaw motion stability. In Fig. 4(a), the yaw rate of MPC method clearly has larger fluctuation which will cause a bad effect for driving comfort, on the contrary, the yaw rate of coordinated control method changes smoothly, which means the vehicle is steering stably. Fig. 4(b) shows the side slip angle comparison, the simulation result indicates that yaw stability control can obtain the minimum sideslip angle which is able to guarantee the vehicle to drive along the planned track stably. Besides, when AGV is under bad driving condition (higher vehicle speed and lower road adhesion coefficient), where it is easier to lose stability, coordinated controller is beneficial for maintaining better yaw motion stability. Though both of them do not exceed the vehicle stable boundary (about 6 deg), however, the maximum value of the sideslip angle is always less than 2 deg and changes smoother with coordinated control method, which is easier for the driver to handle.

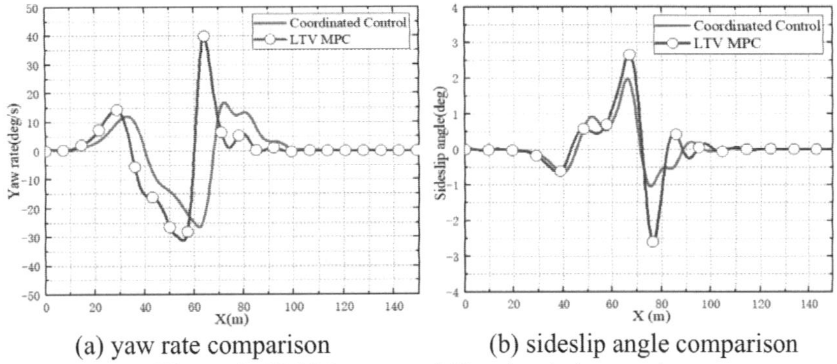


(a) yaw rate comparison (b) sideslip angle comparison
Figure 4. Vehicle yaw stability simulation results.

## 5. Conclusion

This paper proposed a coordinated control method to achieve path following and yaw motion stability for distributed drive autonomous electric vehicle. The LTV model predictive control method is used to reduce the lateral deviation and heading angel error according to the target path, meanwhile the SMC theory is used to generate the control demand of extra yaw moment. Then the torque distribution controller based on PSO algorithm is aimed to optimally distribute torque to four in-wheel motors to maintain the yaw motion stability. The results obtained in the simulation process show that the maximum lateral deviation has been reduced by 0.3 m and side slip angle remains within 2 deg under the proposed coordinated control approach. It can effectively improve the precision of path following as well as guarantee yaw stability of the autonomous vehicle. In the next step, this research will conduct actual vehicle experiment to illustrate the effectiveness of the coordinated control method.

**Acknowledgments**

This work is supported by Natural Science Foundation of Jiangsu Province (BK20201307).

## References

[1] Singh, S., Saini, B. S. (2021). Autonomous cars: Recent developments, challenges, and possible solutions. In IOP Conference Series: Materials Science and Engineering (Vol. 1022, No. 1, p. 012028). IOP Publishing.

[2] Hu, J., Xiong, S., Zha, J., Fu, C. (2020). Lane detection and trajectory tracking control of autonomous vehicle based on model predictive control. International journal of automotive technology, 21(2), 285-295.

[3] Chu, D., Li, H., Zhao, C., Zhou, T. (2022). Trajectory Tracking of Autonomous Vehicle Based on Model Predictive Control with PID Feedback. IEEE Transactions on Intelligent Transportation Systems.

[4] Ren, B., Chen, H., Zhao, H., Yuan, L. (2016). MPC-based yaw stability control in in-wheel-motored EV via active front steering and motor torque distribution. Mechatronics, 38, 103-114.

[5] Liu, J., Jayakumar, P., Stein, J. L., Ersal, T. (2017). Combined speed and steering control in high-speed autonomous ground vehicles for obstacle avoidance using model predictive control. IEEE Transactions on Vehicular Technology, 66(10), 8746-8763.

[6] Lee, J. Y., Yi, K. S. (2015). MPC based steering control using a probabilistic prediction of surrounding vehicles for automated driving. Journal of Institute of Control, Robotics and Systems, 21(3), 199-209.

[7] Guo, N., Zhang, X., Zou, Y., Lenzo, B., Zhang, T. (2020). A computationally efficient path-following control strategy of autonomous electric vehicles with yaw motion stabilization. IEEE Transactions on Transportation Electrification, 6(2), 728-739.

[8] Zhou, W., Fu, J., Yan, H., Du, X., Wang, Y., Zhou, H. (2021). Event-triggered approximate optimal path-following control for unmanned surface vehicles with state constraints. IEEE Transactions on Neural Networks and Learning Systems.

[9] Ma, H., Chu, L., Guo, J., Wang, J., Guo, C. (2020). Cooperative adaptive cruise control strategy optimization for electric vehicles based on SA-PSO with model predictive control. IEEE Access, 8, 225745-225756.

# Bias error in measurement of complex flow regime with ultrasonic flowmeter

**Zhili Wang[1], Qiutong Chen[2], Danxun Li[1*] and Shangwu Liu[1]**

[1] State Key Laboratory of Hydroscience and Engineering, Tsinghua University, Beijing 100084, China

[2] School of Civil Engineering, Tianjin University, Tianjin, 300354, China

[*]Corresponding author's e-mail: lidx@tsinghua.edu.cn

**Abstract.** Ultrasonic flowmeters are widely used for flow measurement. We proposed a theoretical model to quantify the measurement error caused by the deviation of ultrasonic propagation trajectory in complex flow regimes. The three-dimensional ultrasonic propagation model is based on the assumption of ray acoustics. The results show that the trajectory offset of ultrasonic waves and velocity measurement errors in a three-dimensional non-uniform flow field is of the first order of the Mach number. Unlike gas flow, the Mach number of open channel flow is small, and the velocity measurement error of open channel ultrasonic flowmeter can be safely ignored.

## 1. Introduction

Flow measurement is important to industrial and agricultural production as well as people's daily life [1]. Thus, it attracts the attention of a vast spectrum of scholars and engineers [2]. In water conservancy projects, for instance, flow measurement and control constitute two core tasks for practitioners, with the first laying the basis for the second [3].

Currently available flow measurement methods include differential pressure flowmeters [4], rotameters [5], electromagnetic flowmeters [6], vortex flowmeters [7], and ultrasonic flowmeters. Thanks to its minimal intrusion to the flow to be measured and its immunity to the temperature and conductivity of the fluid, the ultrasonic approach has found wide application in irrigation areas [8].

For ultrasonic flowmeters, the measured flow rate represents the line averaged value in uniform flows. In non-uniform flow fields, however, the nonlinear trajectory of the ultrasonic wave may affect the measurement accuracy [10-13]. For instance, in measuring a two-dimensional non-swirling flow field the ultrasonic gas flowmeter was found to produce an error in the order of the square of the Mach number [14-15].

While the performance of ultrasonic gas flowmeters has been extensively examined by previous studies, the ultrasonic flowmeters for measuring open channel flows have drawn far less attention. In addition, in Franchini's research [14], the vertical flow velocity was ignored to reduce the three-dimensional flow field into a two-dimensional one. In practical cases of open channel flow measurement, such as the right-angle water intake scenario, the flow is typically spiral. Simplifying the flow to two-dimensional flow will induce unquantified errors [16].

In this study, we present analytical results of the bias error in ultrasonic transit time velocity measurement of non-uniform three-dimensional open channel flows.

## 2. Method

### *2.1. The principle of the ultrasonic time difference method*

The ultrasonic time difference method is to measure the flow velocity by analyzing the influence of the flow field on the propagation of the ultrasonic signal. The measurement principle is shown in Figure 1. A pair of transducers that transmit and receive ultrasonic waves define a measurement sound path. Due to the different propagation speeds of ultrasonic waves in the flow field, the transit time of the downstream propagation is different from that of the countercurrent propagation. The transit time of ultrasonic propagation between transducers is as follows:

$$t_+ = \frac{l}{c+u} + \tau_+, \quad t_- = \frac{l}{c-u} + \tau_- \tag{1}$$

in which the subscript + and - represent the downstream and upstream propagation of ultrasonic waves, respectively, $l$ represents the length of the measurement sound path, $t$ represents the transit time of the ultrasonic waves propagating between the transducers, and $\tau$ represents the propagation time delay of the ultrasonic waves caused by the transducers.

In the formula, the subscript + represents the upstream propagation of ultrasonic waves, - represents the upstream propagation of ultrasonic waves, $l$ represents the length of the measurement sound path, $t$ represents the transit time of the ultrasonic waves propagating between the transducers. $\tau$ represents the propagation time delay of the ultrasonic waves caused by the transducers, circuits, etc.

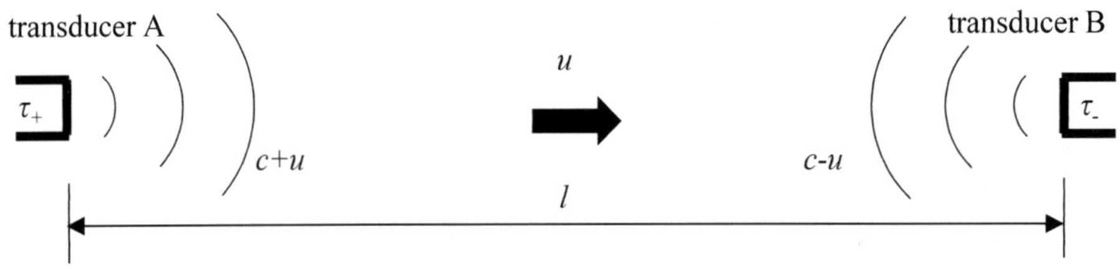


Figure 1. The principle of the ultrasonic time difference method for velocity measurement

Before application in the real world, the ultrasonic transit-time flowmeter requires a zero-flow inspection and on-site calibration [17]. Therefore, the time delay term, $\tau$, generally exerts so little effect on the flow measurement that it can be ignored. The measured velocity along the sound path can be converted into the flow velocity according to the channel angle between the channel and the flow direction, which is convenient for subsequent calculation of the overcurrent section flow rate.

After obtaining the transit time of the ultrasonic wave propagating in the co-current and counter-current between the transducers, one can solve the flow velocity of the flow field according to equation (1), such that,

$$u = \frac{l}{2}\left(\frac{1}{t_+} - \frac{1}{t_-}\right) \tag{2}$$

### *2.2. Ray Acoustics Assumption*

The propagation process of ultrasound can be described by the following wave equation [18]:

$$\nabla^2 p - \frac{1}{c(r)^2}\frac{\partial^2 p}{\partial t^2} = 0 \tag{3}$$

where $p$ represents the sound pressure and $c$ represents the speed of sound.

The velocity of open channel flow is relatively small, far less than the propagation speed of ultrasonic waves in water. The transit time of ultrasonic waves between transducers is of millisecond degree, and the time difference itself is of the order of nanosecond degree. As the open channel flow field and ultrasonic sound field can hardly change in such a short time, the flow field is considered to remain constant within one measurement time [19].

Ray acoustics is a practical method for estimating the propagation of high-frequency sound waves, especially for the problem of sound wave propagation trajectories in large spaces. It is assumed that a certain area around each point within the sound field space constitutes a plane wave that propagates along a ray. The peak value of each plane wave forms the front of the wave. If the amplitude and propagation direction of the sound wave exhibits little change with the wavelength, the characteristics of the sound wave can be ignored [18].

For most cases, the velocity of the flow field to be measured is much smaller than the speed of sound. When the energy carried by the ultrasonic wave is much smaller than the mechanical energy of the flow field, the influence of the sound wave on the flow can be ignored and only the influence of the flowing medium on the sound wave propagation is considered. The operating frequency of the ultrasonic flowmeter currently used in open channel flow measurement is about $4\times10^4$ Hz, and the wavelength is about 37.5 mm, much smaller than the length of the propagation path, which is in line with the assumption of the ray acoustic method. Therefore, the ray acoustic method can be used to analyze the propagation trajectory of ultrasonic waves in a three-dimensional non-uniform flow field.

### 2.3. Ultrasonic Propagation Trajectory Equation

The propagation speed of ultrasonic waves is related to various factors, such as temperature, sand content, and salinity of the flow. These factors generally change slowly with time and space, so no abrupt change will occur in the speed of sound traveling in the flow [19]. In one measurement, the time is extremely short and the space is relatively limited, thus the wave speed is considered constant.

Figure 2 shows downstream propagation of ultrasonic waves from transducer $P_1$ to $P_2$. The time-of-flight spans the interval when $P_1$ starts to emit ultrasonic waves until $P_2$ detects the transmitted ultrasonic waves, i.e.,

$$t_+ = \int_{x_1}^{x_2} \frac{dx}{u_r(x,y,z)} \tag{4}$$

in which $u_r(x,y,z)$ is the component of the actual propagation velocity of the ultrasonic wave along the x-axis.

According to Fermat's principle, the propagation trajectory of ultrasonic waves in a non-uniform flow field is the path that takes the shortest time for ultrasonic waves to pass through the flow field. Therefore, the ultrasonic wave propagation trajectory in the non-uniform flow field corresponds to a minimum value of $t_+$.

To solve the trajectory, the space curve equation can be represented by a parametric equation, with $x$ as the parameter, $y$ and $z$ as the dependent variables. Defining a function $L = L(x;y,z;y',z') = 1/u_r(x,y,y',z,z')$, one can abbreviate Equation (4) to the following form:

$$t_+ = \int_{x_1}^{x_2} L(x;y,z;y',z')dx \tag{5}$$

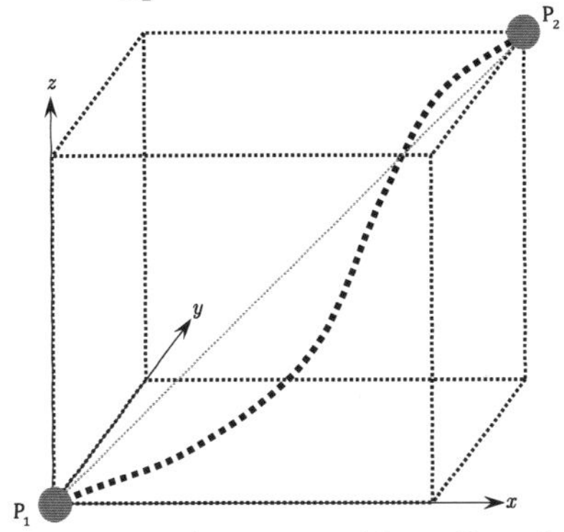


Figure 2. Ultrasonic propagation trajectory in a non-uniform flow field (the black line represents the actual propagation path, and the red line is the no-offset path)

The minimum value of the function $t_+$ is associated with the trajectory such that $\delta t_+ = 0$,

$$\begin{cases} \frac{\partial L}{\partial y} - \frac{d}{dx}\left(\frac{\partial L}{\partial y\prime}\right) = 0 \\ \frac{\partial L}{\partial z} - \frac{d}{dx}\left(\frac{\partial L}{\partial z\prime}\right) = 0 \end{cases} \tag{6}$$

## 3. Result

*3.1. Offset of Ultrasonic Propagation Trajectory in Three-Dimensional Non-Uniform Flow Field*

Boundary conditions can be obtained from Figure 2, namely, $y(x_1) = y_1$, $y(x_2) = y_2$, $z(x_1) = z_1$, $z(x_2) = z_2$.

In the natural coordinate system, the transit time is a function of the trajectory offset. Let $T_+ = c \times t_+/l$, and the nondimensional transit time is as follows,

$$T_+ = \int_{-1/2}^{1/2} \frac{1}{UM_r + \frac{1}{\sqrt{1+Z_r'^2}}\frac{1}{\sqrt{1+Y_r'^2}}} dX = \int_{-1/2}^{1/2} F_+ dX \tag{7}$$

in which the transducer connection line is the no-offset trajectory of ultrasonic propagation, $Y_r$ and $Z_r$ represents the offset of the propagation trajectory along the *y*-axis and the *z*-axis in the natural coordinate system, respectively. The dimensionless integrand $F_+$ is as follows,

$$F_+ = \frac{1}{UM_r + \frac{1}{\sqrt{1+Z_r'^2}}\frac{1}{\sqrt{1+Y_r'^2}}} \tag{8}$$

in which $M_r = v_0/c$ is the Mach number of the flow field.

It's simple for the derivation process to define the *x*-axis as the connecting line of transducers. The positive direction points to the transducer B along with transducer A (see Figure 1), and the coordinates of the two transducers A and B are (-1/2,0,0) and (1/2,0,0).

The ultrasonic propagation trajectory can be determined by replacing the function $L$ in equation (6) with the dimensionless integrand function in equation (8). The Mach number $M_r$ of open channel flows is about one-thousandth and the radius of curvature of the ultrasonic propagation trajectory is positively related to the Mach number of the flow field. Therefore, both ultrasonic propagation's vertical and lateral offsets and their derivatives are small high-order quantities. Expanding equation (8) through Taylor series at the origin yields the following:

$$F_+ = 1 - UM_r + \frac{Z_r'^2}{2} + \frac{Y_r'^2}{2} - UM_r Z_r'^2 - UM_r Y_r'^2 + O\left(M_r{}^2\right) \tag{9}$$

For the convenience of derivation, rewrite $F_+$ as a function of the ultrasonic trajectory offset in the fixed reference frame, $F_+ = F_+(Z_+, Z_+', Y_+, Y_+', X)$. After writing $Y_r{}'$and $Z_r{}'$as functions about $Y_+'$ and $Z_+'$ respectively and ignoring the higher-order terms of $M_r$ in the formula, one obtains the following:

$$\begin{aligned} Z_r' &\approx Z_+'(1+M_r U) - M_r W \\ Y_r' &\approx Y_+'(1+M_r U) - M_r V \end{aligned} \tag{10}$$

Substituting equation (10) into equation (9), one can get the expression of $F_+$:

$$F_+ = 1 - UM_r + \frac{Z_+'^2}{2} + Z_+'^2 M_r U - Z_+' M_r W + \frac{Y_+'^2}{2} + Y_+'^2 M_r U - Y_+' M_r V \tag{11}$$

Substituting $F_+$ in equation (11) for the function $L$ in equation (6) yields the following equation:

$$\begin{aligned} Z_+'' + M_r(U_z - W_x) &= 0 \\ Y_+'' + M_r\left(U_y - V_x\right) &= 0 \end{aligned} \tag{12}$$

The equations satisfied by $Z_{+1}$ and $Y_{+1}$ can be obtained:

$$\begin{aligned} Z_{+1}'' + (U_{Z0} - W_{X0}) &= 0 \\ Y_{+1}'' + (U_{Y0} - V_{X0}) &= 0 \end{aligned} \tag{13}$$

In the same way, the transit time of ultrasonic wave propagation in countercurrent can be analyzed.

The dimensionless transit time of countercurrent propagating ultrasonic waves in a three-dimensional flow field is given by:

$$T_- = \int_{-1/2}^{1/2} \frac{1}{UM_r - \frac{1}{\sqrt{1+Z_r'^2}} \frac{1}{\sqrt{1+Y_r'^2}}} dX = \int_{-1/2}^{1/2} F_- dX \tag{14}$$

The overall dimensionless integrand function of countercurrent propagation is denoted as $F_-$:

$$F_- = -(1 + UM_r + \frac{Z_-'^2}{2} + Z_-'^2 M_r U - Z_-' M_r W + \frac{Y_-'^2}{2} + Y_-'^2 M_r U - Y_-' M_r V) \tag{15}$$

Similarly, the equations satisfied by $Z_{-1}$ and $Y_{-1}$ can be obtained:

$$Z''_{-1} + (U_{Z0} - W_{X0}) = 0$$
$$Y''_{-1} + (U_{Y0} - V_{X0}) = 0 \tag{16}$$

The differential equations (13) and (16) can be solved by combining the boundary conditions. After ignoring the higher-order terms of $M_r$, the ultrasonic propagation trajectory equations in the two directions are obtained as follows:

$$\begin{cases} Z_+ = Z_1 M_r \\ Z_- = -Z_1 M_r \end{cases} \tag{17}$$

$$\begin{cases} Y_+ = Y_1 M_r \\ Y_- = -Y_1 M_r \end{cases} \tag{18}$$

### 3.2. *Actual velocity by ultrasonic time difference method*

It can be seen from equations (17) and (18) that the lateral and vertical offsets of the ultrasonic propagation trajectory are Mach orders of magnitude in the flow field. Due to the small Mach number in open channel flows, the higher-order terms of the Mach number are ignored in the subsequent derivation process. The functions $F_+$ and -$F_-$ can be rewritten as:

$$F_+ = 1 - UM_r - K_+ M_r - F_- = 1 + UM_r + K_- M_r \tag{19}$$

where $K_+$ and $K_-$ are functions of vertical and lateral offset derivatives:

$$K_+ = Z_+' W + Y_+' V K_- = Z_-' W + Y_-' V \tag{20}$$

Substituting the obtained ultrasonic wave propagation trajectory equation (19) into equations (7) and (14), respectively, one can solve the transit time of ultrasonic waves between transducers:

$$T_+ = 1 - \bar{U} M_r - I_{K_+} M_r$$
$$T_- = 1 + \bar{U} M_r - I_{K_-} M_r \tag{21}$$

Among them, $\bar{U}$ is the line average velocity of the measurement sound path, the theoretical measurement value of the ultrasonic transit time method under the condition of uniform flow.

$I_{K_+}$ and $I_{K_-}$ are the additional dimensionless flow velocity caused by the offset of the trajectory due to the ultrasonic wave propagating in the non-uniform flow field in upstream and downstream directions, respectively.

$$I_{K_+} = \int_{-1/2}^{1/2} (Z_+' W + Y_+' V) dX \tag{22}$$

$$I_{K_-} = \int_{-1/2}^{1/2} (Z_-' W + Y_-' V) dX \tag{23}$$

Substituting equation (21) into equation (2) and performing dimensionless processing on the result, one can obtain the dimensionless flow velocity value:

$$U_M = \bar{U} + \frac{I_{K_+} - I_{K_-}}{2} \tag{24}$$

It is noticeable that the final measured flow velocity equals the sum of the average flow velocity $\bar{U}$ of the transducer connection and the additional velocity term. The additional term is the bias error induced by the non-uniform flow field, the magnitude of which is the Mach order of magnitude of the flow field.

### *3.3. Calculation example of velocity measurement error*

Following is the calculation of the common trapezoidal section channel in the irrigation area. The transverse velocity profile of the trapezoidal section is $u = u_m(0.11\ln(2x/B) + 1.02)$, with an assumption that the actual flow velocity is 0.5, 1, 1.5, 2, 3 m/s, respectively. The actual flow rate is calculated by applying equation (45), and the calculation results are shown in Table 1.

Table 1. Results of the velocity error

| Theoretical velocity (m·s$^{-1}$) | Additional velocity (m·s$^{-1}$) | Bias error (%) |
|---|---|---|
| 0.5 | 0.000192 | 0.04 |
| 1 | 0.000540 | 0.05 |
| 1.5 | 0.002152 | 0.14 |
| 2 | 0.003759 | 0.19 |
| 3 | 0.006207 | 0.21 |

The results show that the velocity error of the ultrasonic transit time method applied to the typical section of the open channel is less than 5‰, which is much smaller than the error caused by the installation of the ultrasonic transducers. The velocity error caused by the offset of the ultrasonic propagation trajectory can be ignored in hydraulic engineering.

## 4. Discussion

Bias error in measurement of non-uniform flows with the ultrasonic transit time method has long been identified. As mentioned earlier, Franchini [14] used a two-dimensional model and reported $U_M = \bar{U}(1 + C_2 M_r^2)$, indicating an error in the order of squared Mach number. Here we have found that $U_M = \bar{U} + \left(I_{K_+} - I_{K_-}\right) / 2$, quantifying an error in the order of Mach, which is larger than Franchini's result.

There are two main reasons for the difference between the two models. Franchini used a two-dimensional model, ignoring the effect of vertical flow velocity on ultrasonic propagation, while we used a three-dimensional model such that the influence of the velocity in three directions at any point in the flow field on the ultrasonic propagation trajectory is considered. The second reason is the propagation medium for model establishment. The propagation medium studied by Franchini is gas, and the Mach number of the flow field is relatively large, in the order of 0.1, and the square term of the Mach number cannot be directly ignored. The propagation medium studied in this paper is water, and the Mach number of the flow field is very small, in the order of 0.001, and the higher-order terms exceeding the second order can be safely ignored.

We take two typical flow velocity distributions in Franchini's paper as an example to show the comparison of the two models. The first is a one-dimensional flow directional shear flow. The flow velocity is linearly distributed $U = 2Z + 1$, the lateral and the vertical velocity are zero, and the theoretical velocity is unity. The measured flow velocity of Franchini's model is 1.000008 m/s. Our model's additional flow velocity term is zero, and the measured flow velocity is just the theoretical flow velocity.

The second is the flow with uniform velocity and linear distribution in the lateral direction, i.e., $U = 1, V = 2X + 3$, and the theoretical flow rate is also unity. The flow velocity measurement value of Franchini's model is 0.9991 m/s. Since the vertical flow velocity is zero and the flow velocity is constant, the ultrasonic propagation trajectory is shifted in the vertical and lateral directions. Therefore, the measured flow velocity reported by our model is 1.0002 m/s.

For the second flow case, the results of the two models are quite different. The results of Franchini's model are smaller than the theoretical values, and the results in this paper are larger than the theoretical values. The main reason is that the former is a two-dimensional model and cannot capture the offset of the three-dimensional ultrasonic propagation trajectory. The flow velocity of the flow field contributes to the vertical and lateral offset of the ultrasonic propagation trajectory, and the two-dimensional model cannot solve the influence of the vertical offset of the propagation trajectory on the flow velocity measurement. In hydraulic engineering, due to the influence of the channel wall and hydraulic structures such as gates, the flow field is mostly three-dimensional and non-uniform. Therefore, the application of 3D models is recommended in the study of related flow fields in open channels.

## 5. Conclusion

Based on the assumption of ray acoustics, the propagation trajectory equation of ultrasonic waves in a three-dimensional non-uniform flow field is proposed, and the measurement results of the flow velocity in complex flow regimes by the ultrasonic transit time method are quantitatively given.

The following conclusions are drawn:

(1) The model in this paper considers the effect of vertical flow velocity on the ultrasonic offset, which is more accurate than the traditional two-dimensional model and is suitable for describing open channel flow.

(2) The lateral and vertical offsets of ultrasonic waves propagating in a three-dimensional inhomogeneous flow are the first-order quantities of the Mach number of the flow field. The propagation trajectory of ultrasonic waves in general open channel flow can be considered as straight-line propagation.

(3) The flow velocity measured by the ultrasonic flowmeter in the three-dimensional non-uniform flow field is slightly larger than the average flow velocity. The error is of the first-order of the Mach number of the flow field. The ultrasonic open channel flowmeter's flow velocity measurement accuracy is so high that the error can be ignored safely.

## Acknowledgments

This work was supported by the key program of Tsinghua-Ningxia Yinchuan joint institute of Internet of waters on digital water governance (SKL-IOW-2020TC2007). We express our deepest gratitude to the editor and reviewers, who have provided valuable insights for the revision of this paper. The contributions mentioned above are gratefully acknowledged.

## References

[1] Sun, Y., Tao, Z., Zheng, D. (2018) New Analysis Scheme of Flow-Acoustic Coupling for Gas Ultrasonic Flowmeter with Vortex near the Transducer. Sensors, 18(4): 1151.
[2] Mousavi, S.F., Hashemabadi, S.H., Jamali, J. (2020) Calculation of geometric flow profile correction factor for ultrasonic flow meter using semi-3D simulation technique, Ultrasonics, 106: 106165.
[3] Ankur, K., Aniruddha, D.G., Avinash, M.B. (2022) CFD Simulations of Conical Central Baffle Flumes, J. Irrig. Drain. Eng., 148: 06021014.
[4] Fang, L.D., Han, B.B., Lv, X.H., et al. (2021) Structural Optimization and Characteristics of a Novel Single-Ended Pressure Vessel with Inside and Outside Tubes. Measurement, 186: 109711.
[5] Lata, A., Mandal, N., Maurya, P., et al. (2018) Development of a Smart Rotameter with Intelligent Temperature Compensation. In: 12th International Conference on Sensing Technology (ICST). Limerick. pp. 303-308.
[6] Gupta, H., Arumuru, V., Jha, R. (2021) Industrial Fluid Flow Measurement Using Optical Fiber Sensors: A Review. IEEE Sens. J., 21(6): 7130-7144.
[7] Vershinin, V.E., Polkovnikov, F.I. (2019) Numerical Investigation of the Influence of Bluff Body's Shape on Flow Characteristics in Vortex Flowmeters. Autom. Remote Control, 80(7): 1368-1373.
[8] Lynnworth, L.C., Yi, L. (2006) Ultrasonic flowmeters: half-century progress report, 1955-2005. Ultrasonics, 44(supp-S): e1371-e1378.
[9] Lynnworth, L.C. (1989) Ultrasonic Measurements for Process Control: Theory, Techniques, Applications. Academic Press, New York.
[10] Koechner, H., Melling, A. (2000) Numerical simulation of ultrasonic flowmeters. Acustica, 86(1):

39-48.

[11] Iooss, B., Lhuillier, C., Jeanneau, H. (2002) Numerical simulation of transit-time ultrasonic flowmeters: uncertainties due to flow profile and fluid turbulence. Ultrasonics, 40(9): 1009-1015.

[12] Li, Y.Z., Wu, J.T., Hu, K.M. (2013) Numerical simulating nonlinear effects of ultrasonic propagation on high-speed ultrasonic gas flow measurement. Appl. Math. Inf. Sci., 7(5): 1963-1967.

[13] Zheng, D., Mei, J., Wang, M. (2016) Improvement of gas ultrasonic flowmeter measurement non-linearity based on ray tracing method. IET Sci. Meas. Technol., 10(6): 602-606.

[14] Franchini, S., Sanz-Andrés, A., Cuerva, A. (2007) Effect of the pulse trajectory on ultrasonic fluid velocity measurement. Exp. Fluids, 43: 969–978.

[15] Zheng, D., Lv, S.H., Mao, Y. (2019) Effect mechanism of non-ideal flow field on acoustic field in gas ultrasonic flowmeter[J]. IET Sci. Meas. Technol., 13(4): 469-477.

[16] Neary, V.S., Sotiropoulos, F., Odgaard, A.J. (1999) Three-Dimensional Numerical Model of Lateral-Intake Inflows. J. Hydraul. Eng., 125(2): 126-140.

[17] Van Willigen, D.M.V., Van Neer, P.L.M.J.V., Massaad, J., et al. (2021) An Algorithm to Minimize the Zero-Flow Error in Transit-Time Ultrasonic Flow Meters[J]. IEEE Trans. Instrum. Meas., 70: 1-9.

[18] Morse, P.M., Ingard, K.U. (1968) Theoretical Acoustics. McGraw Hill Book Company, New York.

[19] Yao, Y.X., Lu, Y. (2006) Application of Acoustic Time Transit Flowmeter in Discharge Measurement for Open Channel. Automation in Water Rescources and Hydrology, 1: 1-5.(in Chinese)

# A Study on Parallel Test Approach for the Flash Burn-in Experiment in Laboratory

**Biao Wang [1, 2*], Hao Li [1, 3], Zijie Zhang [4]**

[1]School of Instrument Science and Opto-electronics Engineering, Hefei University of Technology, Hefei, 230009, Anhui, China

[2]Anhui Province Key Laboratory of Measuring Theory and Precision Instrument, School of Instrument Science and Optoelectronics Engineering, Hefei University of Technology, Hefei 230009, Anhui, China

[3] Gigadevice Semiconductor (Beijing) Inc., Beijing, China

[4]School of Internet and Communication, Anhui Technical College Of Mechanical and Electrical Engineering, Wuhu 241000, China

* Corresponding author's e-mail: wangbiao@hfut.edu.cn

**Abstract:** With the increasing number of lab burn-in samples, this paper proposes a parallel test approach for the problem that the single-device test board cannot meet the requirements of the JEDEC standard for the large number of samples with high efficiency. The designed FPGA control board independently separates 4 control SCAN signals, different SCAN chips perform read operations in sequence, and complete the operation in parallel. The experimental results show that, compared with the single-device test board, the MT test board can test 80 chips at the same time, which meets the requirements of the JEDEC standard, and reduces the requirement for hardware resources in the burn-in test, and improves test efficiency.

## 1. Introduction

Flash memory[1-4] has the characteristics of low manufacturing cost, high storage density, and data can still be retained after power off, so thousands of different types of chips are involved in the fields of robots, drones, and space exploration. Each chip must go through the stages of analog design, logic verification, layout, and mass production testing before tape-out. So the chip will undergo strict wafer level[5] and package level tests during the manufacturing process. In order to clear all obstacles in the stage of mass production testing, the burn-in experiment[6-8] in the lab is required to detect all the parameter of the chip, and find the defects of the product. Although the existing single-device test board in the lab can meet the daily burn-in test, the number of samples in the aging test is also increasing, and the resources provided by the single-device test board are limited, so it is easy to reach the test bottleneck. Thus, in this paper, we design a multi-DUT test scheme, which adopts the idea of parallel testing, and drives 80 products at the same time through the FPGA[9-10] control board connected by gold fingers, and writes test code based on Delphi environment, realizes the product failure model is automatically counted and analyzed, so the test program effectively improves the efficiency of experimental evaluation and reduces the investment of hardware resources.

## 2. Memory burn-in experimental process

In order to adapt flashs to various application scenarios, we need to verify and evaluate them at different temperatures. For example, the evaluation experiment of PE cycling durability should cover the environment of -40℃~125℃. This experiment mainly measures the endurance of the PE cycles of the samples. During the test, all possible failure mechanisms in each cycle area should be checked, including chip programming disturbance, gate disturbance, programming-erasing capability, etc. And we need to test the data anti-interference and anti-loss ability of the cell at different temperatures.

Different types of burn-in experiments require specific temperature conditions and control information. For example, the PE cycle experiment uses a chamber to provide temperature environment and electrical signal of test-board, and the sample performs the PE cycle and the reading operation under corresponding temperature condition. The experiment can perform the PE cycle operation on multiple blocks or slices starting from the starting address. In the existing single-device test scheme, 20 NOR chips are placed on the single-device test board (Fig 1), and according to the bias voltage condition of the chip, the output voltage of the control board (Fig 2) is detected by using a multimeter in a jumper manner. In order to ensure that the electrical connection and bias signal of the experimental sample are accurate, it is necessary to verify at least 4 positions on the test board. In addition, we download the FPGA program to the control board and configure the experimental control information on the counting chip. After completing the preparatory work of the experiment, we will insert the single-device test board and the control board into the connector of the chamber at the same time to carry out the burn-in experiment. After the samples are tested, the parameter test and cycle erasure verification of the sample chip are completed by the ATE tester. According to JEDEC standard, each burn-in test requires 80 or more test samples. If the single-device test board test method is used, at least 4 single-device test board resources are required for each burn-in experiment project, which will occupy more experimental hardware resources, and increase the workload of engineers.

Figure 1 device test board

Figure 2 The output voltage of control board

## 3. Improvement of lab burn-in scheme

In order to realize that a single test board meets the requirements of the JEDEC for the number of samples, this paper develops an 80DUTS MT parallel test system based on the idea of I/O resource sharing. The system frame diagram is shown in Fig 3. The test code is developed on the PC based on the Delphi and downloaded to the control board through the network cable. The power module supplies working voltage for the control board, and the function of the LDO module is to stabilize the voltage signal and protect the load from short circuit. The FPGA chip decodes the code in the PC, and then according to the instructions stored in the SDRAM, the graphics generator generates the operation logic sequence to drive the chip on the test board. The buffer memory is used to temporarily store the data read out from the chip, and the data is used to compare with the expected pattern of the master PC. The LED module is used to display the working status of each link of the system.

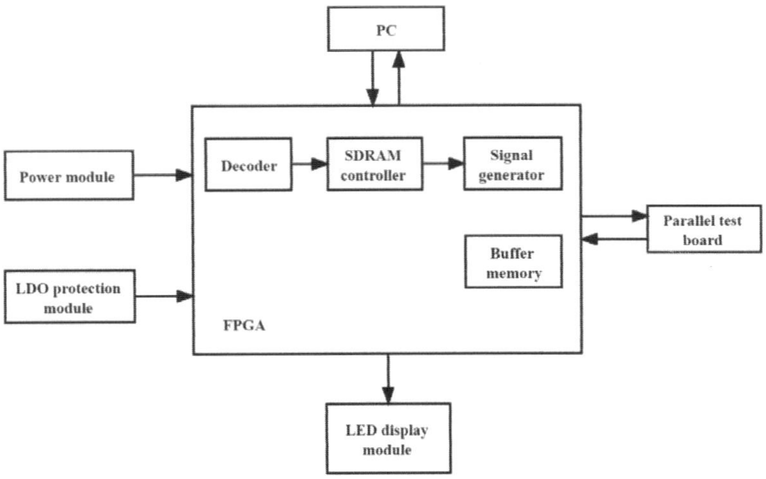


Figure 3 Test system frame diagram

When the chip receives the PE signal sent by the MT system, the internal finite state machine of the chip will decode and control the external circuit to generate excitation, and the internal charge pump will generate the voltage signal for PE operation, and the signal is applied to the gate, source, drain and substrate of the memory cell. In addition, the PE operation is completed through hot electron injection and FN tunneling, and there is no need to read out the internal data of the chip for comparison. Therefore, multiple chips can share the signal channel according to the signal excitation. As shown in Fig 4, each SCAN signal selects a column of chips to perform the PE operation. When the operation of this group of chips is completed, the SCAN signal of the next group is enabled to continue its operation. However, the read operation of the chip needs to pull out the internal data for comparison, and judge the PASS and FAIL through the slave computer system. As shown in Fig 5, the #CS enable signal of chips in the same column are connected to different signal channels of the MT system, and the output pins of chips in different columns and at the same position are connected to the same I/O channel of the MT system. Therefore, each chip realizes the independence of the data output through the SCAN method. If multiple chips share a data channel, fighting will occur when outputting data at the same time, which will lead to inability to accurately judge the output results of each chip.

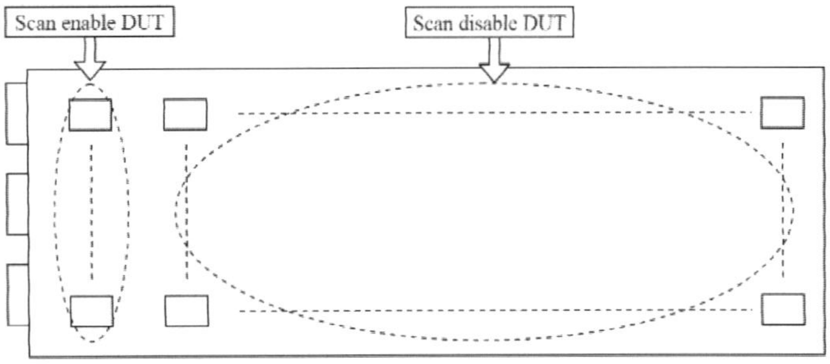


Figure 4 SCAN diagram of erase-program operation

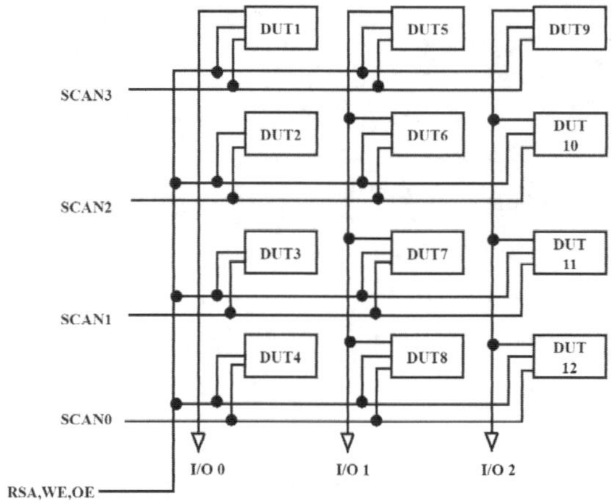


Figure 5 I/O diagram for read operation

Based on the erasing and reading operation characteristics of the chip, the actual layout of the chips on the parallel test board implemented in this paper is shown in Fig 6. 80 chips are arranged in 8 columns, and each SCAN can access two columns of chips. As shown in Fig 7, the MT hardware system includes a #CS signal controller and 20 chip controllers, and the #CS signal controller is used to control the #CS chip select enable pins of 4 groups of chips, and each chip controller can control 4 chips between different groups. Therefore, due to this signal generation structure, when sending an erase-program operation command to the chip, all 4 #CS chip select enable signal channels can be selected, that is, 80 chips can be erased and programed at the same time. When reading data from the inside of the chip, the test program is set to select one #CS chip select enable signal channel at a time, that is, 20 chips for read operation, and then after the read operation of this group is completed, jump to the next group of chips to perform the read operation, and the internal data of the 80 chips can be read out after 4 cycles.

Figure 6 Diagram of the chip layout

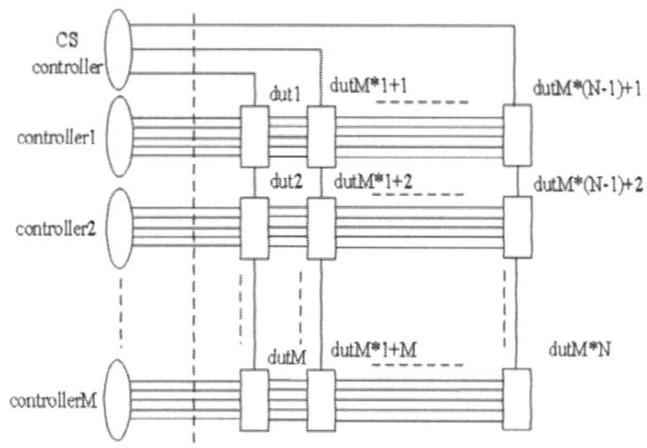


Figure 7 Chip Select Distribution

## 4. Validation of Parallel Test Schemes

During the debugging of the MT system, it is found that the higher the voltage, the more obvious the overshoot of the signal, but the chip will be damaged when the phenomenon is serious. In order to obtain better system signal stability, an appropriate matching resistor is added at the end of the system signal to perform impedance matching with the parasitic load generated by the long wiring of the system. Then we choose the best solution according to the debugging results, that is, a group of 360 ohm resistors are connected in series with the signal pins of the first and second groups of chips, and a group of 15 ohm resistors are connected in series with the output pins. The signal pins of the third and fourth groups of chips are connected in series with a set of 300 ohm resistors, and the output pins are connected in series with a set of 15 ohm resistors. The signal waveforms of the CLK, CS, and IO pins of the chip at a frequency of 10MHZ are shown in Fig 8. The driving capability of the MT hardware system is limited, the impedance will be mismatched and the signal will be distorted.

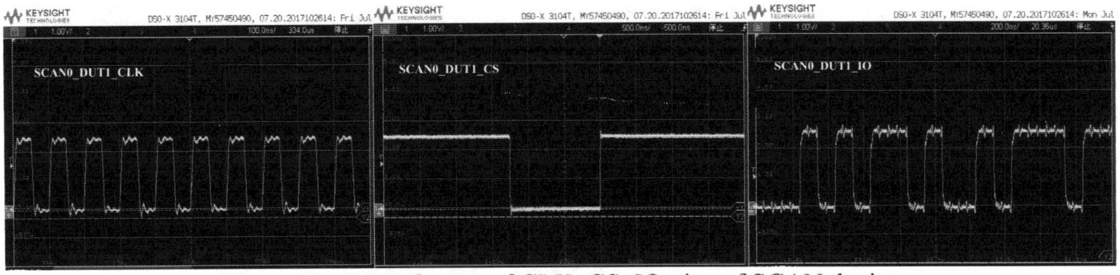


Figure 8 Wave format of CLK, CS, IO pins of SCAN device

After the MT test system is built, in order to verify the logic accuracy of the force and sense signals, we use an oscilloscope to select the toggle waveform of the load board chip pin signal (Fig 9), and through a large number of data verification and result analysis, the system is continuously adjusted and optimized. In addition, in the process of making the load board, it is necessary to make reasonable wiring to reduce the signal interference between chips, and adjust the resistance and capacitance of the load board traces to ensure the suppression and stability of the signal on the chip side, so as to meet the minimum response speed of 10HMZ.

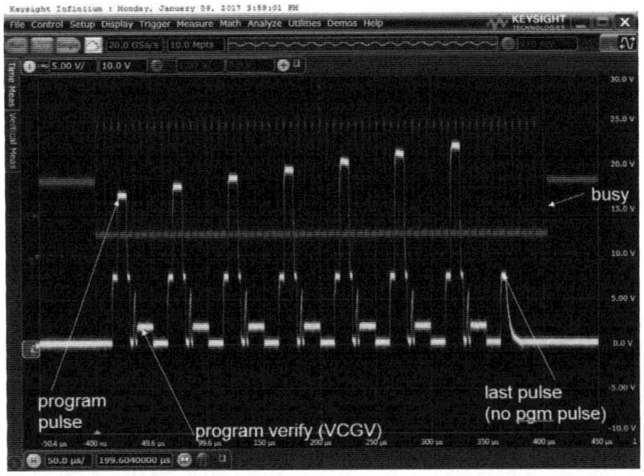


Figure 9 Verification of chip program operation

## 5. Results and Analysis of burn-in Experiments

In the same batch of P25Q32H chips, we selected 20 chips respectively and used the MT parallel measurement system and the single-device test board to conduct a comparison experiment of working life at 125 °C. This experiment adopts the diagonal vector mode, and uses T5830ES and Magnum 2 testers to test chip parameters respectively. In order to meet the requirements of JEDEC standards and customer needs, we test the data parameters of the experimental chip within 1000h (partial data are shown in Fig. 10 and Fig. 11), and analyze the performance and data retention ability of the chip after the 1000h high temperature life test.

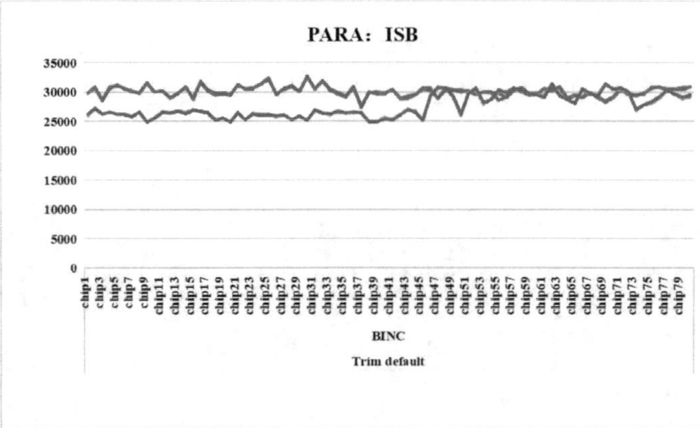


Figure 10 ISB current of the chip

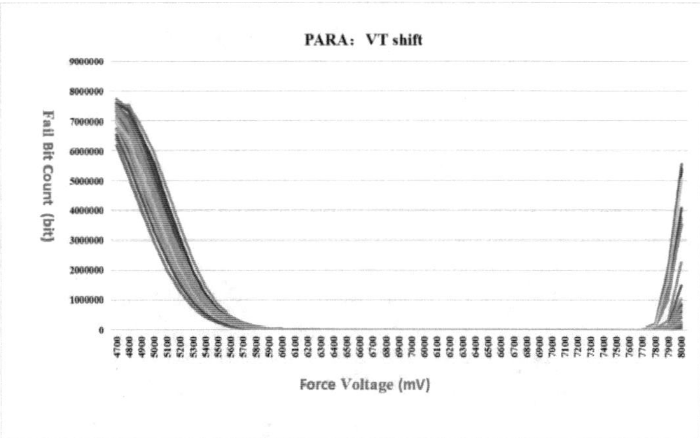


Figure 11 VT shift of the cell

The checkerboard vector mode requires that the data of two adjacent memory cells are physically opposite. We use the MT test system and the single-device test board to perform the XM25LU128C full-chip erase-program cycle test experiments at 125 °C. According to the chip technical manual, we develop the test code on the ATE, write the 01010101 vector data to the chip after the high- temperature erasing experiment to maintain the checkerboard pattern, and receive the initial threshold voltage data of the chip memory cell in the test mode. Chip1-5 in Fig 12 is the programing time curve of the chip on the single device test board, and chip 6-10 is the programing time curve of the corresponding MT test system board. Chip1-5 in Fig 13 is the erasing time curve of the chip on the single-device test board, and chip 6-10 is the erasing operation time curve corresponding to the MT test system board. From these two figures, we can see that the program and erase time of the two test systems is not much different. The single-device test board has relatively many chip failure units, and its cycling erase-program effect is strong. As shown in Fig 14, there are 2 steps at the boundary between the two groups of experimental strong and weak memory cells, which is caused by the hardware structure of the MT test system. Each chip is independent of each other in single-device testing. However, when the parallel test judges whether an erasing cycle is completed, the same group of chips will have a certain signal delay, so the erasing effect of the MT test system is weak.

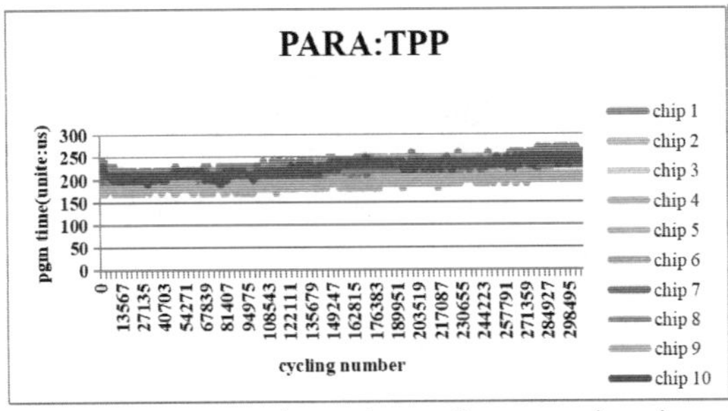


Figure 12 Comparison chart of programing time

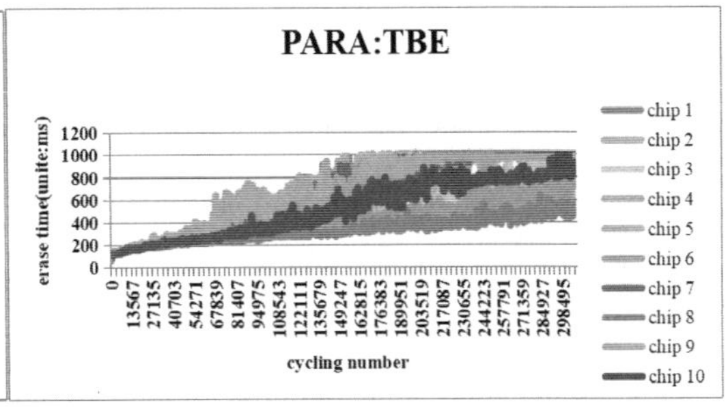


Figure13 Comparison chart of erasing time

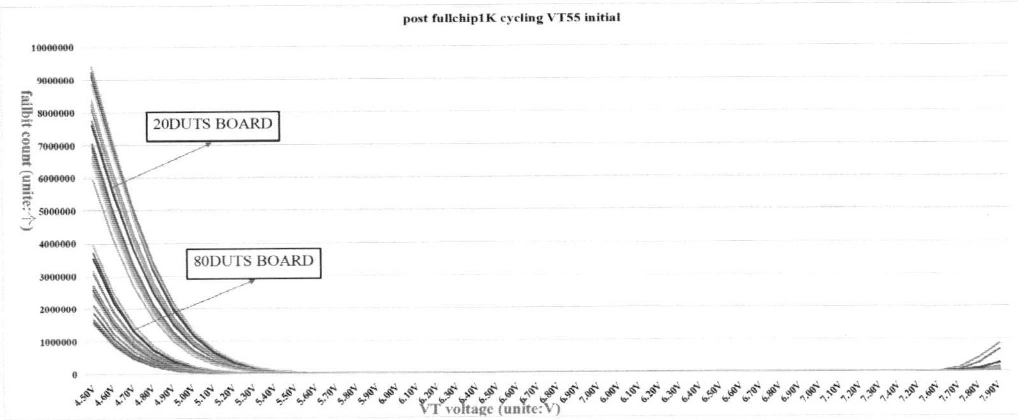


Figure 14 Variation of chip threshold voltage on different test boards

Table 1 FAIL statistics of wide-voltage FLASH products (pcs)

| Sample Test | fc_ersfail | fc_pgmfail | rd_ersfail | rd_pgmfail | maxpv | Maxev |
|---|---|---|---|---|---|---|
| Single Device Test Board | 51 | 33 | 15 | 11 | 7 | 3 |
| MT Test System | 12 | 8 | 6 | 4 | 3 | 0 |

Table 2 FAIL statistics of vehicle FLASH products (pcs)

| Sample Test | fc_ersfail | fc_pgmfail | rd_ersfail | rd_pgmfail | maxpv | Maxev |
|---|---|---|---|---|---|---|
| Single Device Test Board | 27 | 26 | 9 | 13 | 1 | 4 |
| MT Test System | 6 | 8 | 4 | 3 | 1 | 0 |

Table 1 and Table 2 are the FAIL type statistics obtained from the head-to-tail strong stress tests performed on 40 single-device test boards and 10 MT test system boards with 800 chips of different products respectively. Comparing the two test methods in general, it can be seen that the MT test system outperforms the single-device test system in most failure scenarios. The failed chip needs to be cross-analyzed on the ATE to determine the specific cause of the chip failure. When the chip's internal circuit or the manufacturing process defect of the chip foundry causes the failure, it needs to be fed back to the chip design engineer or process engineer for revision of the design process. In addition, the failed chip will be archived and sealed, and subsequent experiments will not be carried out. When the failure is caused by the hardware connection, pulse fluctuation and other problems of the experimental system, the failed chip is considered to be false. We need to reconnect the hardware resources and set the signal generator to continue the next experiment. The MT test system uses fewer load boards and experimental chambers, which reduces the preparatory work for experiments such as downloading FPGA programs, checking the voltage of experimental samples, experimental failure analysis, and collecting experimental sample data. Therefore, the MT test system reduces the chip failure rate, reduces the analysis and verification time, and large sample experiments will not be delayed either.

## 6. Conclusions

At present, in the aspect of mass sample testing, the traditional burn-in approach has shortcomings such as long time consumption and high hardware resource occupation. Therefore, this paper creates a MT parallel test system, which can reduce the hardware resources and risks of the experiment. Therefore, the MT test system can save the labor and time cost of testing, and has certain reference significance for the evaluation of mass burn-in test of different devices such as DRAM, NAND and NOR.

## Acknowledgments

This study was supported by the Natural Science Foundation of Anhui Province for Higher Education Institutions (Grant No. KJ2021A1574) and National College Students' innovation and entrepreneurship training program (Grant No. 202010359054).

## References

[1] Lee W H, Park C K, Kim K, (2006) Temperature dependence of endurance characteristics in NOR flash memory cells. IEEE International Reliability Physics Symposium Proceedings. San Jose. 701-702.
[2] Lee S, Lee Y T, Han W K, et al. (2004) A 3.3 V 4 Gb four-level NAND flash memory with 90 nm CMOS technology. IEEE International Solid-State Circuits Conference. San Francisco. 52-513.
[3] Kumar M P, Rao G J, Vani P K, et al. (2019) An Efficient Flash Memory Devices. IEEE International Conference on Intelligent Systems and Green Technology.Visakhapatnam. 49-493.
[4] Xiaodong H, Hui L, Ming S, et al. (2017) Analysis and research on failure of a certain nor FLASH for aerospace use. Prognostics and System Health Management Conference (PHM-Harbin). Harbin. 1-4.
[5] Chou C Y, Hung T Y, Huang C J, et al.(2010) Development of empirical equations for metal trace failure prediction of wafer level package under board level drop test. IEEE transactions on advanced packaging, 33(3): 681-689.
[6] Bahukudumbi S. (2010) Wafer-level testing and test during burn-in for integrated circuits. J.Dissertations & Theses - Gradworks .
[7] Bai B, Chen S, Wang W, et al. (2011) Thermal management of integrated circuits in burn-in environment. The Proceedings of 2011 9th International Conference on Reliability, Maintainability and Safety. Guiyang. 1092-1095.
[8] Wang Y, Zhang H, Wang W, et al. (2019) Investigation on the erase time of Nor Flash. 2019 2nd International Conference on Information Systems and Computer Aided Education. Dalian. 347-349.
[9] Fu N, Li Y, Liu B, et al. (2015) Realization of controlling eMMC 5.0 device based on FPGA for automatic test system. 2015 IEEE AUTOTESTCON. National Harbor.MD. 251-255.
[10] Tyhach J, Wang B, Sung C, et al. (2005) A 90-nm FPGA I/O buffer design with 1.6-Gb/s data rate for source-synchronous system and 300-MHz clock rate for external memory interface. IEEE journal of solid-state circuits, 40(9): 1829-1838.

# Research on Peak Current Predictive Control Algorithm for Boost Converter with Discrete Disturbance Observer

**Dong Zhang[1*], Yiming Zhang[1], Longfei Cui[1] and Mingxin Wang[1]**

[1] Faculty of Information Technology, Beijing University of Technology, Beijing, 100124, China

[*] Corresponding author's e-mail: 2223617048@qq.com

**Abstract.** In this paper, a peak current predictive control algorithm based on discrete disturbance observer is proposed for BOOST converter. Compared with typical peak current predictive control algorithms, the proposed algorithm does not require additional a sensor to measure the input voltage in real time. The minimum beat tuning method of discrete disturbance observer is presented. It is proved that the inductance current tracks the reference value after two sampling cycles and eliminates input voltage disturbance after five sampling cycles. Simulation and experimental show the effectiveness of the proposed algorithm.

## 1. Introduction

Boost converter has the characteristics of time-varying, nonlinear and non-minimum phase, and the dynamic response of voltage-type control algorithm is slow, so many advanced control strategies are proposed[1]. Peak current type control can realize cycle by cycle overcurrent protection, widely used in analog chip design[2]. With the rise of digital control, more and more researchers use digital system to realize the methods in analog control, combined with the basic idea of model predictive control and peak current type control, put forward many digital predictive control algorithms[3]. The integrated current observer is proposed to replace the current sensor. Although the hardware structure of the system is simplified, the software complexity is increased and the observation accuracy of the voltage sensor is highly dependent[4]. It is proposed to sample the valley value of current and eliminate the error of parameter mismatch by calculating the slope in real time. However, this method is sensitive to noise measurement and requires higher requirements for sensors[5]. A predictive control algorithm for peak current based on real-time inductance identification is proposed. The dynamic response of inductance current is two sampling periods[6]. However, compared with other control algorithms such as PI, this algorithm requires additional a sensor to measure the input voltage, which increases the hardware complexity and cost. In addition, measurement noise is introduced into the control loop, which affects the robustness.

A peak current predictive control algorithm based on discrete disturbance observer is proposed in this paper. The input voltage is regarded as constant disturbance, the disturbance is observed by discrete disturbance observer, and the disturbance is suppressed by feedforward compensation. The effectiveness of the discrete disturbance observer parameter tuning method is verified by simulation and experimental.

## 2. Discrete domain mathematical model of boost converter

### 2.1. Boost converter modelling

A typical structure of boost converter is shown in Figure 1, where $L$ is the filter inductance, $S$ is the switch tube, $D$ is the power diode , $C$ is the output capacitor and $R$ is the load resistance.

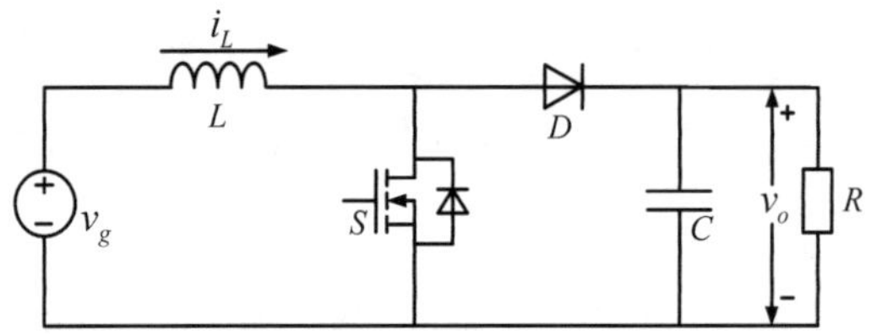


Figure 1. Circuit of boost converter

To investigate the current loop digital control algorithm of boost converter, a discrete domain mathematical model in inductance current continuous mode (CCM) is deduced by inductance current waveform diagram. When the switch is on and off, the slope of inductance current is $S_u$ and $S_d$ respectively.

$$\begin{cases} S_u = \dfrac{v_g}{L} \\ S_d = \dfrac{v_g - v_o}{L} \end{cases} \tag{1}$$

The discrete domain mathematical model is

$$i_L(k+1) = i_L(k) + \frac{T_s}{L}(v_g(k) - v_o(k)\cdot(1-d(k))) \tag{2}$$

Where: $i_L(k)$ is the peak value of inductance current, $v_g(k)$ and $v_o(k)$ denote the values of input voltage and output voltage, $T_s$ is the sampling and control period and $d(k)$ represents the duty cycle of PWM.

### 2.2. Typical peak current predictive control algorithm

As ADC conversion and control algorithm calculation take time, the duty cycle of PWM needs to be updated in the next sampling and control cycle in digital control. The discrete domain mathematical model of inductance current is modified as

$$i_L(k+1) = i_L(k) + \frac{T_s}{L}(v_g(k) - v_o(k)\cdot(1-d(k\text{-}1))) \tag{3}$$

Predictive control is to predict the actual peak value of inductance current one sampling period in advance. The control signal $d(k\text{-}1)$ at the current moment and the actual value of inductance current $i_L(k)$ are used to predict the inductance current $i_L(k+1)$ at the next moment according to Equation (3), and then the predicted value $i_L$(k+1) and reference value $i_c$(k) are substituted into Equation (4)

$$i_c(k) = i_L(k+1) + \frac{T_s}{L}(v_g(k) - v_o(k)\cdot(1-d(k))) \tag{4}$$

Substitute (3) into (4) to obtain control signal $d(k)$

$$d(k) = \frac{L_s}{v_o(k)T_s}\cdot(i_c(k) - i_L(k)) - \frac{2v_{in}(k)}{v_o(k)} - d(k-1) + 2 \tag{5}$$

Equation (5) indicates that compared with other control algorithms such as PI, the peak current predictive control algorithm requires additional a sensor to measure the input voltage, increasing the hardware complexity and cost.

### *2.3. Design of discrete disturbance observer*

When the control algorithm lacks the input voltage term

$$d(k)=\frac{L_s}{v_o(k)T_s}\cdot(i_c(k)-i_L(k))-d(k-1)+2 \tag{6}$$

*Z* transform is applied to Equation (6) and substituted into Equation (3)

$$i_L(z)=z^{-2}i_c(k)+\frac{T_s}{L}(z^{-1}+z^{-2})v_g(k) \tag{7}$$

Let

$$i_L(k)=i_c(k-2)+f(k-2) \tag{8}$$

Equation (8) can be written as the following discrete state space equation

$$\begin{cases} x_1(k+1)=x_2(k) \\ x_2(k+1)=x_3(k)+i_c(k) \\ x_3(k+1)=x_3(k) \\ y(k)=x_1(k) \end{cases} \tag{9}$$

Where

$$\begin{cases} x_1(k)=i_L(k-2) \\ x_2(k)=i_L(k-1) \\ x_3(k)=f(k) \\ y(k)=x_1(k) \end{cases} \tag{10}$$

The error $f(k)$ is estimated by establishing the state observer of the above equation, and the state observer can be written as

$$\begin{cases} \hat{x}_1(k+1)=\hat{x}_2(k)+l_2(y(k)-\hat{y}(k)) \\ \hat{x}_2(k+1)=\hat{x}_3(k)+i_c(k)+l_1(y(k)-\hat{y}(k)) \\ \hat{x}_3(k+1)=\hat{x}_3(k)+l_0(y(k)-\hat{y}(k)) \\ \hat{y}(k)=\hat{x}_1(k) \end{cases} \tag{11}$$

$L=[l_2\ \ l_1\ \ l_0]$ is the observer gain. If the observation error is zero by adjusting the gain matrix of the observer, then let

$$i_c(k)=i_{c0}(k)-\hat{f}(k) \tag{12}$$

Substitute into (8)

$$i_L(k)=i_{c0}(k-2) \tag{13}$$

It shows that the inductance current can accurately track the reference value and eliminate the input voltage disturbance.

### *2.4. Parameter tuning of discrete disturbance observer*

Let the state space equation of the observation error is obtained by combining equations (9) and (11)

$$e(k+1)=A_e e(k) \tag{14}$$

Where

$$A_e=\begin{bmatrix} -l_2 & 1 & 0 \\ -l_1 & 0 & 1 \\ -l_0 & 0 & 1 \end{bmatrix} \tag{15}$$

The characteristic equation of system matrix $A_e$ is

$$\left|zI - A_e\right| = z^3 + (l_2 - 1)z^2 + (l_1 - l_2)z + (l_0 - l_1) \tag{16}$$

In order to make the discrete disturbance observer stable, the absolute value of the characteristic root of the system matrix $A_e$ should be less than 1. In addition, if $l_2 = l_1 = l_0 = 1$, the closed-loop characteristic root can be set at the origin to achieve the minimum beat observation of $f(k)$.

In this case, the control system after adding the discrete disturbance observer is shown in Figure 2.

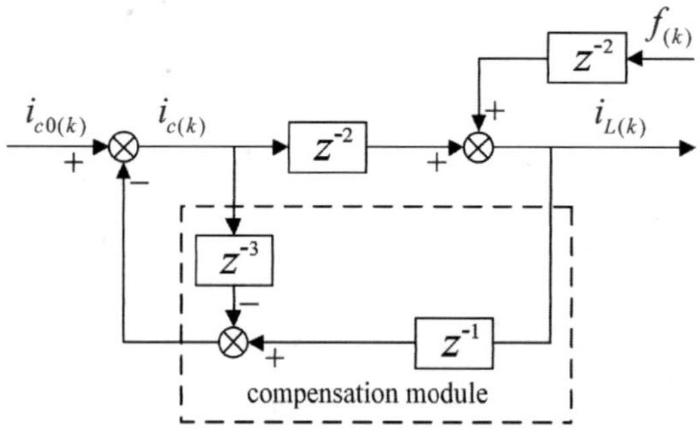


Figure 2. Block diagram of equivalent control system

Set the reference $i_{c0}(k)$ to zero, the closed-loop pulse transfer function of the inductance current to the input voltage disturbance is

$$\frac{i_L(z)}{f(z)} = (1 - z^{-3})z^{-2} \tag{17}$$

Set the disturbance quantity $f(k)$ to zero, the closed-loop pulse transfer function of the inductance current to the reference value is

$$\frac{i_L(z)}{i_{c0}(z)} = z^{-2} \tag{18}$$

Equation (17) shows that after adding the discrete disturbance observer, the control algorithm can eliminate the influence of input voltage disturbance after five sampling cycles. Equation (18) shows that without input voltage disturbance, the inductive current tracks the reference value after two sampling periods, as with typical peak current predictive control.

## 3. Experimental verification

The simulation and experimental parameters of the boost converter are shown in Table 1, and the sampling frequency is 20kHz.

Table 1 Boost converter simulation parameters

| | |
|---|---|
| $v_g$/V | 10 |
| $v_o$/V | 18-22 |
| $L$/$u$H | 300 |
| $C$/$u$F | 660 |
| $R$/Ω | 10 |

Figure 3 and figure 4 show the step response of inductance current after adding a discrete disturbance observer. In accordance with the typical peak current predictive control algorithm, the step response is two sampling periods.

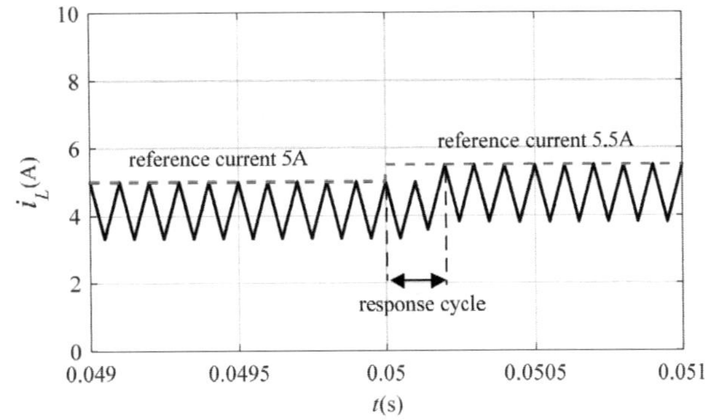


Figure 3. Simulation of step response of inductance current

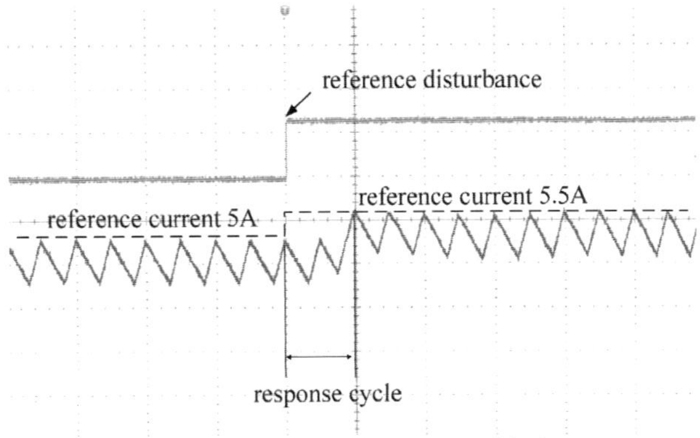


Figure 4. Experimental of step response of inductance current

As can be seen from figure 5 and figure 6, the inductance current will be lower than the reference current after the step disturbance of the input voltage (from 10V to 12V). After five sampling periods, the inductance current track the current reference value.

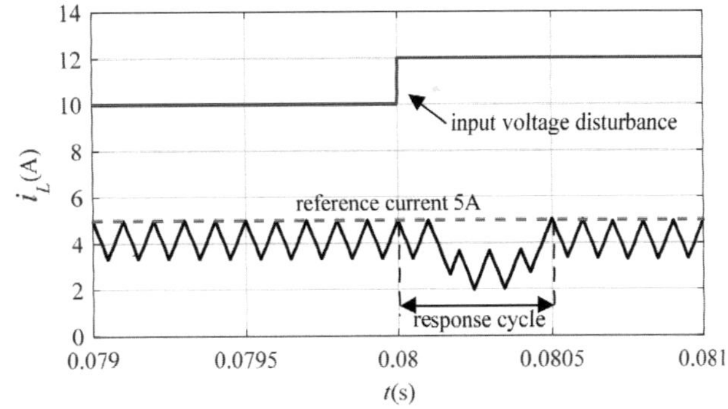


Figure 5. Simulation of input voltage disturbance response

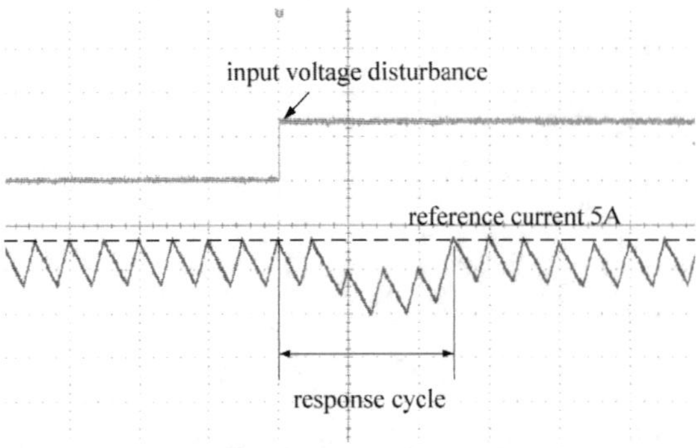


Figure 6. Experimental of input voltage disturbance response

## 4. Conclusion

The peak current predictive control algorithm based on discrete disturbance observer is studied. The input voltage can be measured in a limited sampling period, and the input voltage can be measured in real time without additional sensors, which reduces the cost of the control system. The simulation and experimental show that the control algorithm can eliminate the input voltage disturbance after five sampling periods.

## References

[1] Hejri M , Mokhtari H . Hybrid predictive control of a DC–DC boost converter in both continuous and discontinuous current modes of operation[J]. Optimal Control Applications & Methods, 2011, 32(3):270-284.

[2] Samanes, Javier, Barrios . Control of a Photovoltaic Array Interfacing Current-Mode-Controlled Boost Converter Based on Virtual Impedance Emulation[J]. IEEE Transactions on Industrial Electronics, 2019.

[3] Choi Y J , Kim T J , Kim R Y . An Active Partial Switching Method in Tertiary Loop for a High-Efficiency Predictive Current-Mode Control PFC Converter[J]. IEEE Transactions on Industrial Electronics, 2018:7818-7828.

[4] Min R , Chen C , Tong Q , et al. Sensorless predictive current control with a charge balance current estimator for boost converters based on discrete-time model considering current ripple[J]. Iet Power Electronics, 2018.

[5] Chang Y T , Lai Y S . Parameter Tuning Method for Digital Power Converter With Predictive Current-Mode Control[J]. IEEE Transactions on Power Electronics, 2009, 24(12):2910-2919.

[6] Zhang Y , Zhang Y , X Wang, et al. Adaptive Digital Predictive Peak Current Control Algorithm for Buck Converters[J]. JOURNAL OF POWER ELECTRONICS, 2019, 19.

# AUTHOR INDEX